Antonio Ambrosetti[†], Shair Ahmad
Differential Equations

Also of Interest

Partial Differential Equations
An Unhurried Introduction
Vladimir A. Tolstykh, 2020
ISBN 978-3-11-067724-9, e-ISBN (PDF) 978-3-11-067725-6

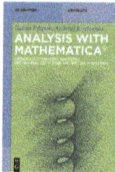

Analysis with Mathematica®
Volume 3 Differential Geometry, Differential Equations, and Special Functions
Galina Filipuk, Andrzej Kozłowski, 2022
ISBN 978-3-11-077454-2, e-ISBN (PDF) 978-3-11-077464-1

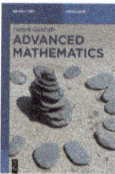

Advanced Mathematics
An Invitation in Preparation for Graduate School
Patrick Guidotti, 2022
ISBN 978-3-11-078085-7, e-ISBN (PDF) 978-3-11-078092-5

in De Gruyter Studies in Mathematics:
ISSN 0179-0986

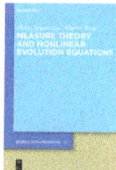

Measure Theory and Nonlinear Evolution Equations
Flavia Smarrazzo, Alberto Tesei, 2022
ISBN 978-3-11-055600-1, e-ISBN (PDF) 978-3-11-055690-2

Complex Delay-Differential Equations
Kai Liu, Ilpo Laine, Lianzhong Yang, 2021
ISBN 978-3-11-056016-9, e-ISBN (PDF) 978-3-11-056056-5

Antonio Ambrosetti[†], Shair Ahmad

Differential Equations

A First Course on ODE and a Brief Introduction to PDE

2nd edition

DE GRUYTER

Mathematics Subject Classification 2020
Primary: 34-01, 35-01; Secondary: 34A12, 34A34, 34C10

Authors
Prof. Dr. Antonio Ambrosetti[†]

Prof. Dr. Shair Ahmad

ISBN 978-3-11-118524-8
e-ISBN (PDF) 978-3-11-118567-5
e-ISBN (EPUB) 978-3-11-118578-1

Library of Congress Control Number: 2023942973

Bibliographic information published by the Deutsche Nationalbibliothek
The Deutsche Nationalbibliothek lists this publication in the Deutsche Nationalbibliografie;
detailed bibliographic data are available on the Internet at http://dnb.dnb.de.

© 2024 Walter de Gruyter GmbH, Berlin/Boston
Cover image: koto_feja / E+ / Getty Images
Typesetting: VTeX UAB, Lithuania
Printing and binding: CPI books GmbH, Leck

www.degruyter.com

Memorial Dedicated to Professor Antonio Ambrosetti

Professor Antonio Ambrosetti was born in Bari, Italy in the year 1944. He died in Venice, Italy on November 20 in the year 2020.

He graduated from the University of Padua, Italy, in 1966. He has established significant groundbreaking results in the general area of nonlinear analysis, particularly related to pde, calculus of variations, Hamiltonian systems, bifurcation theory, and applications of pde to differential geometry.

Professor Ambrosetti was a co-founder of SISSA (the International Center for Advanced Studies), where he taught. SISSA has recently started an award in his honor; it is called the Antonio Ambrosetti Medal. It is offered on a bi-annual basis to exceptionally talented young researchers who have already made significant contributions in the general area of Nonlinear Analysis. He also won the Caccioppoli prize in 1982, and the Amerio Prize by the Istituto Lombardo Accademia di Scienze e Lettere in 2008.

Professor Ambrosetti was not only a great mathematician but also a great educator. He has produced several Ph. D. students who are internationally renowned mathematicians in their own rights. His former students include (chronologically ordered):

1. Coti Zelati, Vittorio
2. Solimini, Sergio
3. Malaguti, Luisa
4. Majer, Pietro
5. Vitillaro, Enzo
6. Cingolani, Silvia
7. Berti, Massimiliano

https://doi.org/10.1515/9783111185675-201

8. Malchiodi, Andrea
9. Biasco, Luca
10. Secchi, Simone
11. Felli, Veronica
12. Baldi, Pietro
13. Calahorrano, Marco
14. Ianni, Isabella
15. Mercuri, Carlo
16. Vaira, Giusi

Preface

The first edition of this book was published in 2021, just prior to the time when Antonio passed away. Perhaps the publication was somewhat rushed due to the surge of the pandemic. Antonio and I agreed that we would revise it as soon as feasible. Unfortunately, Antonio passed away quite suddenly and unexpectedly, at about the same time as the book was being printed.

The revision has mostly addressed correcting some misprints and errors that are fairly common in the first edition of a book. Many new examples and exercises have been added or substituted. Also, a lot of effort has been made to make the presentation, especially of the first half, as lucid as possible. The first half should be quite suitable for a one-semester undergraduate course in differential equations, taught at most universities in the US. It is suitable for a well-motivated student with a minimal knowledge of Algebra and Calculus. Complete solutions to all the even number problems are given in the back of the book.

I wish to acknowledge the efficiency, competence, and courtesy of the DeGruyter personnel; particularly Ute Skambraks and Vilma Vaičeliūnienė. Their help and technical support were indispensable. I also wish to thank Steven Elliot, the publishing editor, for his support and advice.

Finally, I wish to dedicate this edition of the book to my former students from Venezuela. They were a hard-working, fun-loving and lively bunch who were totally dedicated to the cause of advancing education in their country. Working with them was fun and gratifying. All of them went on to play important roles as teachers and administrators, including ranks of Deans, Vice Presidents and Rectors of universities. Unfortunately, I cannot name all of those wonderful people.

My six Ph. D. students from Venezuela include: **Ramon Mogollon**, **Ramon Navarro**, **Francisco Montes de Oca**, **Jorge Salazar** (deceased), **Jose Sarabia**, and **Ennodio Torres**. All of them published in international journals and held professorial ranks at various universities in Venezuela.

A few other friends and former students, who received their terminal degrees under the supervision of other professors, include **Ernesto Asuaje**, **Jose Bethelmy**, **Yetja Cordero**, **Betty** and **Wilfred Duran**, **Ramon Gomez**, **Justina Guierra** (deceased), **Myriam Hernandez**, **Raul Lopez**, **Anna Montes de Oca**, **Rafael Morello**, and **Cruz Daniel Zambrano**.

Shair Ahmad

https://doi.org/10.1515/9783111185675-202

Chapter Contents

Chapter 1 contains a quick review of some topics from multivariable calculus. It includes topics such as stationary points, local extrema, mixed derivatives, gradient, line integral, Implicit Function Theorem, and Fourier series.

Chapter 2 deals with first order linear differential equations. It includes solving such equations by the method of integrating factors. After introducing the notion of initial value problem, simple proofs of existence and uniqueness of solutions are given, using only elementary calculus. Several examples, applications, and some important properties of solutions are discussed.

Chapter 3 deals with theoretical aspects of the general first order equations. Local as well as global existence and uniqueness are discussed with some of the proofs presented in an Appendix. It discusses the importance of learning to analyze the qualitative behavior of solutions, particularly when dealing with problems that cannot be readily solved by known methods. This chapter may be considered optional, some instructors may wish to simply state the existence and uniqueness results and move on to the next chapter, depending on the backgrounds of the particular students in the class.

Chapter 4 deals with separable equations and application to the logistic equation, homogeneous equations, Bernoulli's equation, and Clairaut's equation.

Chapter 5 is dedicated to the study of exact equations and integrating factors. Examples are given to show that one can choose among four options, some of which may be more convenient for a given problem than others.

Chapter 6 is about linear second order equations, homogeneous and nonhomogeneous, and includes the Wronskian, linear independence, general solutions, equations with constant coefficients, methods of variation of parameters and undetermined coefficients.

Chapter 7 is essentially an extension of Chapter 6 to higher order equations.

Chapter 8 is about systems of differential equations, changing linear scalar equations to linear systems in matrix form, eigenvalues and eigenvectors, etc.

Chapter 9 deals with phase space analysis of the trajectories of second order autonomous equations, equilibrium points, periodic solutions, homoclinic and heteroclinic solutions, and limit cycles. Among applications, there is a discussion of the mathematical pendulum, the Kepler problem, the Lienard equation and the Lotka–Volterra system in population dynamics.

Chapter 10 treats the topic of stability, classifying stability of the equilibrium of 2×2 linear systems based on properties of eigenvalues. It includes Lyapunov direct method, a brief discussion of stability of limit cycles, stable and unstable manifolds, and bifurcation of equilibria.

Chapter 11 discusses how to obtain solutions by power series methods. There is also a discussion of singular points, ordinary points, the Frobenius method, and Bessel functions.

Chapter 12 introduces the basic properties of the Laplace transform, its inverse and application to solving initial value problems of linear differential equations, including a brief introduction to the convolution of two functions. This deep subject is presented in a simple and concise manner suitable for readers with minimum background in calculus.

Chapter 13 treats oscillation theory of selfadjoint second order differential equations, the Sturm–Liouville eigenvalue problems. Once again, an effort is made to keep the presentation at an elementary level, but it includes some challenging examples and problems.

Chapters 14 and 15 deal with a short introduction to linear PDEs in two dimensions. The former contains first order equations such as the transport equation, including an Appendix on the inviscid Burgers' equation. The latter deals with the most classical linear second order equations such as the Laplace equation, the heat equation and the vibrating string equation.

Chapter 16 gives an elementary introduction to the subject of Calculus of Variation, starting with an explanation of a functional and the Euler–Lagrange equation. There is also a discussion of the brachistochrone problem, the Fermat problem in optics and the isoperimetric problem.

A final section is devoted to revisiting the Sturm–Liouville problem.

Contents

1 A brief survey of some topics in calculus

Here we recall some results from calculus of functions of two variables. Some additional more specific results that we will need in this book will be stated later. For a textbook on Calculus, see e. g. G. B. Thomas Jr., *Calculus*, 14th ed., Pearson, 2018.

1.1 First partial derivatives

Let $F(x,y)$ denote a function of two variables, defined on an open set $S \subseteq \mathbb{R}^2$, which possesses first partial derivatives F_x, F_y.
- The gradient ∇F is the vector defined by setting

$$\nabla F = (F_x, F_y) \in \mathbb{R}^2.$$

- We say that $F \in C^1(S)$ if F_x, F_y exist and are continuous in S.
- A point $(x_0, y_0) \in S$ is a stationary point of F if $\nabla F(x_0, y_0) = (0,0)$, namely if $F_x(x_0, y_0) = F_y(x_0, y_0) = 0$.
- A point $(x_0, y_0) \in S$ is a (local) maximum, resp. minimum, of F if there exists a neighborhood $U \subset S$ of (x_0, y_0) such that

$$F(x,y) \le F(x_0, y_0), \quad \forall (x,y) \in U, \quad \text{resp.} \quad F(x,y) \ge F(x_0, y_0), \quad \forall (x,y) \in U.$$

- A stationary point of F which is neither a local maximum nor a local minimum is called a saddle.

Theorem 1.1. *Suppose $F(x,y)$ has first partial derivatives F_x, F_y at $(x_0, y_0) \in S$. If (x_0, y_0) is a local maximum or minimum of F, then (x_0, y_0) is a stationary point of F, that is, $\nabla F(x_0, y_0) = (0,0)$, or $F_x(x_0, y_0) = F_y(x_0, y_0) = 0$.*

The following result allows us to solve, locally, the equation $F(x,y) = 0$ for x or y.

Theorem 1.2 (Implicit function theorem). *Let $F(x,y)$ be a continuously differentiable function on $S \subseteq \mathbb{R}^2$ and let $(x_0, y_0) \in S$ be given such that $F(x_0, y_0) = c_0$. If $F_y(x_0, y_0) \ne 0$, resp. $F_x(x_0, y_0) \ne 0$, there exists a neighborhood I of x_0, resp. neighborhood J of y_0, and a unique differentiable function $y = g(x)$ $(x \in I)$, resp. a unique differentiable function $x = h(y)$ $(y \in J)$, such that $g(x_0) = y_0$ and $F(x, g(x)) = c_0$ for all $x \in I$, resp. $h(y_0) = x_0$ and $F(h(y), y) = c_0$ for all $y \in J$.*

https://doi.org/10.1515/9783111185675-001

1.2 Second partial derivatives

Let $F(x,y)$ possess second partial derivatives,

$$F_{xx} = \frac{\partial^2 F}{\partial x^2}, \quad F_{yy} = \frac{\partial^2 F}{\partial y^2}, \quad F_{xy} = \frac{\partial}{\partial y}\left(\frac{\partial F}{\partial x}\right), \quad F_{yx} = \frac{\partial}{\partial x}\left(\frac{\partial F}{\partial y}\right).$$

F_{xy} and F_{yx} are called mixed partial derivatives.

Theorem 1.3. *Suppose that the mixed partial derivatives F_{xy} and F_{xy} are continuous at $(x_0, y_0) \in S$. Then*

$$F_{xy}(x_0, y_0) = F_{yx}(x_0, y_0).$$

- The Hessian of F is the matrix

$$H = \begin{pmatrix} F_{xx} & F_{xy} \\ F_{yx} & F_{yy} \end{pmatrix}.$$

- The determinant of H is given by

$$\det(H) = \begin{vmatrix} F_{xx} & F_{xy} \\ F_{yx} & F_{yy} \end{vmatrix} = F_{xx}F_{yy} - F_{xy}F_{yx}.$$

If Theorem 1.3 applies, then we find simply

$$\det(H) = F_{xx}F_{yy} - F_{xy}^2.$$

- We say that $F \in C^2(S)$ if $F_{xx}, F_{xy}, F_{yx}, F_{yy}$ exist and are continuous in S.

Theorem 1.4. *Let $(x_0, y_0) \in S$ be a stationary point of $F(x,y)$ and suppose $F \in C^2$ in a neighborhood of (x_0, y_0). Then:*
1. *if $\det(H) > 0$ and $F_{xx}(x_0, y_0) > 0$, then (x_0, y_0) is a local minimum;*
2. *if $\det(H) > 0$ and $F_{xx}(x_0, y_0) < 0$, then (x_0, y_0) is a local maximum;*
3. *if $\det(H) < 0$, then (x_0, y_0) is a saddle point.*

Another way to state the previous theorem is to consider the eigenvalues $\lambda_{1,2}$ of H, namely the roots of the second order algebraic equation

$$\det(H - \lambda I) = \begin{vmatrix} F_{xx} - \lambda & F_{xy} \\ F_{yx} & F_{yy} - \lambda \end{vmatrix} = 0,$$

i. e.,

$$(F_{xx} - \lambda)(F_{yy} - \lambda) - F_{xy}^2 = 0, \quad \text{or} \quad \lambda^2 - (F_{xx} + F_{yy})\lambda + F_{xx}F_{yy} - F_{xy}^2 = 0.$$

Then

1. λ_1 and λ_2 are both positive $\Rightarrow (x_0, y_0)$ is a local minimum;
2. λ_1 and λ_2 are both negative $\Rightarrow (x_0, y_0)$ is a local maximum;
3. $\lambda_1 \cdot \lambda_2 < 0 \Rightarrow (x_0, y_0)$ is a saddle point.

1.3 Line integrals

Let y be a planar piecewise $C^1(^1)$ curve with components $x(t), y(t), t \in [a, b]$.
- The length of y is given by

$$\ell(y) = \int_a^b \sqrt{x'^2(t) + y'^2(t)}\, dt.$$

In particular, if y is a Cartesian curve with equation $y = y(x), x \in [a, b]$, we find

$$\ell(y) = \int_a^b \sqrt{1 + y'^2(x)}\, dx.$$

- The curvilinear abscissa of a point $(u, v) = y(\tau)$ is defined as the length of the arc from $y(a, b)$ to (u, v), namely

$$s = \int_a^\tau \sqrt{x'^2(t) + y'^2(t)}\, dt.$$

Introducing the arc differential

$$ds = \sqrt{x'^2(t) + y'^2(t)}\, dt,$$

we can simply write

$$\ell(y) = \int_a^b ds.$$

---

1 Recall that a function $f(t)$ is piecewise continuous on $S \subset \mathbb{R}$ if there is a discrete set of points $D \subset S$ such that:

1. f is continuous on $S \setminus D$;
2. f has a jump discontinuity at any $t_0 \in D$.

We say that f is piecewise C^1 on S if it is continuous on S and there is a discrete set of points $D' \subset S$ such that the restriction of f on $S \setminus D'$ is continuously differentiable.

- Given a continuous $F(x, y)$, $(x, y) \in S$ and the piecewise C^1 curve $\gamma : [a, b] \mapsto \mathbb{R}^2$ such that $\gamma([a, b]) \subset S$, the curvilinear integral F on γ is defined by

$$\int_\gamma F(x, y)\, ds = \int_a^b F(x(t), y(t)) \cdot \sqrt{x'^2(t) + y'^2(t)}\, dt.$$

1.4 The divergence theorem

Given a vector field $\overline{V} : S \ni (x, y) \mapsto (F(x, y), G(x, y)) \in \mathbb{R}^2$ with $F, G \in C^1(S)$, the divergence of $\overline{V}$ is defined as $\operatorname{div} \overline{V} = F_x + G_y$.

Theorem 1.5 (Divergence theorem). *Let $\overline{V}(x, y)$ be a C^1 vector field defined on S and suppose $\Omega \subset S$ is a domain (i. e., a bounded connected open set) such that its boundary $\partial\Omega$ is a piecewise continuous curve with components $(x(t), y(t))$. Then*

$$\int_\Omega \operatorname{div} \overline{V}\, dx\, dy = \int_{\partial\Omega} \overline{V} \cdot \overline{n}\, ds$$

where $\overline{n} = (y', -x')$ denotes the outer unit normal at $\partial\Omega$.

Using the components F, G of $\overline{V}$, we find $\operatorname{div} \overline{V} = F_x + G_y$ and $\overline{V} \cdot \overline{n} = Fy' - Gx'$. Hence the preceding equality can also be written more explicitly as

$$\int_\Omega (F_x + G_y)\, dx\, dy = \int_{\partial\Omega} (Fy' - Gx')\, ds. \tag{1.1}$$

From the divergence theorem we can derive an important formula.
Let $F = gf_x$ and $G = gf_y$. Then $F_x = g_x f_x + gf_{xx}$ and $G_y = g_y f_y + gf_{yy}$ yield

$$F_x + G_y = g_x f_x + gf_{xx} + g_y f_y + gf_{yy}$$

and hence from (1.1)

$$\int_\Omega (g_x f_x + gf_{xx} + g_y f_y + gf_{yy})\, dx\, dy = \int_{\partial\Omega} (gf_x y' - gf_y x')\, ds$$

or, introducing the laplacian operator $\Delta = \partial_{xx}^2 + \partial_{yy}^2 = \operatorname{div} \nabla$,

$$\int_\Omega (g\Delta f + \nabla g \cdot \nabla f)\, dx\, dy = \int_{\partial\Omega} (gf_x y' - gf_y x')\, ds.$$

Assuming that $g = 0$ on $\partial\Omega$ we find

$$\int_{\Omega} (g\Delta f + \nabla g \cdot \nabla f)\, dx\, dy = 0 \implies \int_{\Omega} g\Delta f\, dx\, dy = -\int_{\Omega} \nabla g \cdot \nabla f\, dx\, dy,$$

which can be seen as an integration by parts for functions of two variables. Exchanging the roles of f, g we find

$$\int_{\Omega} f\Delta g\, dx\, dy = -\int_{\Omega} \nabla f \cdot \nabla g\, dx\, dy,$$

from which immediately follows that

$$\int_{\Omega} g\Delta f\, dx\, dy = \int_{\Omega} f\Delta g\, dx\, dy.$$

1.5 Fourier series

Fourier series are series associated to periodic functions. Recall that $f(t)$, $t \in \mathbb{R}$, is T-periodic if $f(t + T) = f(t)$, $\forall t \in \mathbb{R}$.

The Fourier series associated to a 2π-periodic function $f(t)$ is the trigonometric series

$$\sum_0^\infty [a_n \sin nt + b_n \cos nt],$$

where

$$a_n = \frac{1}{\pi} \int_{-\pi}^{\pi} f(t) \sin nt\, dt, \quad b_n = \frac{1}{\pi} \int_{-\pi}^{\pi} f(t) \cos nt\, dt, \quad n \in \mathbb{N},$$

are called the *Fourier coefficients of f*. Here and below it is understood that f is such that the previous integrals make sense.

In particular,

- if f is 2π-periodic and odd, then the associated Fourier series is given by $\sum_1^\infty a_n \sin nt$;
- if f is 2π-periodic and even, then the associated Fourier series is given by $\sum_0^\infty b_n \cos nt$.

If the Fourier series of f is convergent we will write

$$f(t) = \sum_0^\infty [a_n \sin nt + b_n \cos nt].$$

We recall below some simple *convergence criteria for Fourier series*. It is understood that $f(t)$ is 2π-periodic.

Theorem 1.6. *If f is piecewise continuous, then*

(Parseval identity) $$\sum (a_n^2 + b_n^2) = \frac{1}{2\pi} \int_{-\pi}^{\pi} f^2(t)\, dt.$$

Therefore $\lim_{n \to \infty} a_n = \lim_{n \to \infty} b_n = 0$.

Theorem 1.7. *Let f be piecewise continuous and let $t_0 \in [-\pi, \pi]$ be such that*

$$f'(t_0+) \overset{\text{def}}{=} \lim_{h \to 0+} \frac{f(t_0 + h) - f(t_0+)}{h} \quad and \quad f'(t_0-) \overset{\text{def}}{=} \lim_{h \to 0-} \frac{f(t_0 + h) - f(t_0-)}{h}$$

exist and are finite. Then

$$\sum_{0}^{\infty} [a_n \sin nt_0 + b_n \cos nt_0] = \frac{1}{2}[f(t_0+) + f(t_0-)],$$

where

$$f(t_0+) = \lim_{t \to t_0+} f(t) \quad and \quad f(t_0-) = \lim_{t \to t_0-} f(t).$$

Corollary 1.1. *If f is continuous, then the Fourier series of f converges pointwise to $f(t)$ provided f is differentiable at t.*

Theorem 1.8. *If f is of class C^1 then the Fourier series of f is uniformly convergent to f.*

Theorem 1.9. *Suppose that for some integer $k \geq 1$ one has*

$$\sum_{0}^{\infty} n^k (a_n^2 + b_n^2) < +\infty.$$

Then $f(t) = \sum_{0}^{\infty} [a_n \sin nt + b_n \cos nt]$ is of class C^k and

$$\frac{d^k f(t)}{dt^k} = \sum_{1}^{\infty} \left[a_n \frac{d^k \sin nt}{dt^k} + b_n \frac{d^k \cos nt}{dt^k} \right].$$

For example:

(1) If $\sum_{1}^{\infty} n(a_n^2 + b_n^2) < +\infty$ then $f'(t) = \sum_{1}^{\infty} [na_n \cos nt - nb_n \sin nt]$.

(2) If $\sum_{1}^{\infty} n^2(a_n^2 + b_n^2) < +\infty$ then $f''(t) = \sum_{1}^{\infty} [-n^2 a_n \sin nt - n^2 b_n \cos nt]$.

If $f(t)$ is *T*-periodic, all the preceding results can be extended, yielding, under the appropriate assumptions,

$$f(t) = \sum_{0}^{\infty} \left[a_n \sin\left(\frac{2\pi}{T} nt\right) + b_n \cos\left(\frac{2\pi}{T} nt\right) \right],$$

where

$$a_n = \frac{2}{T} \int_{-\frac{T}{2}}^{\frac{T}{2}} f(t) \sin\left(\frac{2\pi}{T} nt\right) dt, \quad b_n = \frac{2}{T} \int_{-\frac{T}{2}}^{\frac{T}{2}} f(t) \cos\left(\frac{2\pi}{T} nt\right) dt, \quad n \in \mathbb{N}.$$

For a broader discussion on Fourier series we refer, e. g., to the book by S. Suslov, *An Introduction to Basic Fourier Series*, Springer US, 2003.

2 First order linear differential equations

2.1 Introduction

A *differential equation* is any equation that expresses a relationship between a function and its derivatives.

In this introductory chapter we focus on studying first order linear differential equations of the form

$$x' + p(t)x = q(t) \tag{2.1}$$

where p, q are continuous functions defined on some interval I. The main features of equation (2.1) are:

1. it is an *ordinary* differential equation because there is only one independent variable t and hence the derivative is ordinary;
2. it is *first order*, because the highest derivative of the function $x(t)$ is the first derivative;
3. it is *linear*, because the dependence on x and x' is linear.

A *solution* to (2.1) is a differentiable function $x(t)$, defined on the interval I, such that

$$x'(t) + p(t)x(t) = q(t), \quad \forall t \in I.$$

For example, $x(t) = e^{2t}$ is a solution of the differential equation $x' + 2x = 4e^{2t}$ since replacing $x(t)$ by e^{2t} results in $(e^{2t})' + 2e^{2t} = 2e^{2t} + 2e^{2t} = 4e^{2t}$ for all real numbers t; hence it satisfies the equation $x' + 2x = 4e^{2t}$.

Let us point out explicitly that the fact that solutions are defined on the whole interval I, where the coefficient functions $p(t)$ and $q(t)$ are continuous, is a specific feature of linear equations.

If $q = 0$, the resulting equation

$$x' + p(t)x = 0 \tag{2.2}$$

is called *homogeneous*, to distinguish it from the more general equation (2.1), which is a *nonhomogeneous* equation.

The rest of the chapter is organized as follows. In section 2.2 we deal with homogeneous equations. In section 2.3 we address nonhomogeneous equations. Finally, section 2.4 deals with applications to electric circuits.

2.2 Linear homogeneous equations

Let us consider the homogeneous equation (2.2). First of all, we note that the zero function $x(t) \equiv 0$ is a solution, regardless of what kind of a function $p(t)$ might be. The zero

https://doi.org/10.1515/9783111185675-002

solution is also called the *trivial solution*. Of course, we are interested in finding non-trivial solutions of (2.2).

To outline the idea of the procedure for solving such equations, we first consider the simple equation

$$x' - 2x = 0.$$

We note that if we multiply both sides of the equation by the function $\mu(t) = e^{-2t}$, we obtain

$$e^{-2t}x' - 2e^{-2t}x = 0,$$

which is equivalent to the equation

$$\left(e^{-2t}x\right)' = 0.$$

Therefore, integrating both sides yields $e^{-2t}x = c$ and $x(t) = ce^{2t}$, $c \in \mathbb{R}$.

As we just saw, the function $\mu(t)$ played an important role in solving the above equation for $x(t)$. We now search for a differentiable function $\mu(t)$ that will help us to solve such linear equations in general.

Definition 2.1. A differentiable function $\mu(t)$ is called an *integrating factor* of

$$x' + p(t)x = 0$$

if

$$\mu(t)x' + \mu(t)p(t)x = \left(\mu(t)x\right)'.$$

In order for $\mu(t)$ to be an integrating factor, we must have

$$\mu(t)x' + p(t)\mu(t)x = \left(\mu(t)x\right)',$$

which is equivalent to

$$\mu(t)x' + p(t)\mu(t)x = \mu'(t)x + \mu(t)x'.$$

Simplifying, we obtain

$$p(t)\mu(t)x(t) = \mu'(t)x(t).$$

Let us further impose the condition that $\mu(t) \neq 0$, $x(t) \neq 0$. Then we can write $\frac{\mu'(t)}{\mu(t)} = p(t)$, or $\frac{d\mu(t)}{\mu(t)} = p(t)\,dt$. Integrating both sides and suppressing the constant of integrations, since we need only one function, we obtain $\ln|\mu(t)| = \int p(t)\,dt$, or

$$\mu(t) = e^{\int p(t)\,dt}.$$

Notice that $\mu(t) > 0$ so that the preceding calculations make sense. The function $\mu(t)$ is an integrating factor since it can easily be verified, using the fundamental theorem of calculus, that

$$e^{\int p(t)\,dt}x' + p(t)e^{\int p(t)\,dt}x = \left(e^{\int p(t)\,dt}x\right)'.$$

Remark 2.1.

1. Let us agree by convention that $\int f(t)\,dt$ stands for a fixed antiderivative of $f(t)$ and not the whole family of antiderivatives. When we wish to discuss the family of antiderivatives of a function $f(t)$, we will write $\int f(t)\,dt + c$, where c is an arbitrary constant.
2. We often divide both sides of an equation by a function that may be 0 somewhere. To be correct, we should always mention that we are assuming that it is not 0. For example, if we divide both sides of $tx' + t^2 = t$ by t, it would not be valid for $t = 0$. But it gets to be rather cumbersome to keep saying that we are assuming $\cdots \neq 0$. So, let us implicitly agree that division is valid only if the divisor is not zero without explicitly stating so each time.

Let $x(t)$ be any nontrivial solution satisfying (2.2). If we multiply both sides of the equation by $\mu(t) = e^{\int p(t)\,dt}$, it will automatically make the left side of the equation to be the exact derivative of the product $e^{\int p(t)\,dt}x(t)$, yielding the equation

$$\left(e^{\int p(t)\,dt}x(t)\right)' = 0.$$

Integrating both sides, we obtain $e^{\int p(t)\,dt}x(t) = c$, or $x(t) = ce^{-\int p(t)\,dt}$, where c is an arbitrary constant. Notice that if $c = 0$, then $x(t)$ is the zero solution. For $c \neq 0$, $x(t)$ never vanishes; it is either always positive or always negative.

We have shown that any nontrivial solution $x(t)$ has to be of the form $x(t) = ce^{-\int p(t)\,dt}$. On the other hand, it is easy to see that, for any constant c, $x(t) = ce^{-\int p(t)\,dt}$ is a solution, simply by substituting $ce^{-\int p(t)\,dt}$ for $x(t)$ in (2.2) and verifying that it satisfies the equation. Hence the family of solutions

$$x(t) = ce^{-\int p(t)\,dt}, \tag{2.3}$$

where c is an arbitrary constant, includes all the solutions and is called the *general solution* of (2.2).

Remark 2.2. In determining the integrating factor, instead of using the indefinite integral, sometimes it is convenient to use the definite integral, obtaining $\mu(t) = e^{\int_{t_0}^{t} p(t)\,dt}$ and the general solution $x(t) = ce^{-\int_{t_0}^{t} p(t)\,dt}$, where t_0 is some number in I.

2.2.1 Cauchy problem

Both, in application and in theoretical studies, one often needs to find a particular solution $x(t)$ of a differential equation that has a certain value x_0 at a fixed value t_0 of the independent variable t. The problem of finding such a solution is referred to as an *initial value problem* or a *Cauchy problem*. The initial value problem for a homogeneous first order equation can be expressed as

$$x' + p(t)x = 0, \quad x(t_0) = x_0. \tag{2.4}$$

Example 2.1. (a) Solve the Cauchy problem

$$x' - 2x = 0, \quad x(0) = 3.$$

We have shown above that the general solution to this problem is given by $x(t) = ce^{2t}$. So, in order to find the solution satisfying $x(0) = 3$, all we have to do is substitute the initial data into the equation and solve for the constant c. Thus we have $3 = ce^0 = c$ and the desired solution is $x(t) = 3e^{2t}$. The solution can be verified by noting that $(3e^{2t})' - 2.3e^{2t} = 6e^{2t} - 6e^{2t} = 0$ and $x(0) = 3$.

(b) Find k such that the solution of the Cauchy problem

$$x' = kx, \quad x(0) = 1$$

satisfies $x(1) = 2$.

The general solution of $x' = kx$ is given by $x(t) = ce^{kt}$. The initial condition yields $1 = ce^0 = c$. Thus the solution of the Cauchy problem is $x(t) = e^{kt}$. Since $x(1) = 2$ we find $2 = x(1) = e^k$ whereby $k = \ln 2$. □

Lemma 2.1. *A nontrivial solution of the homogeneous equation (2.4) cannot vanish anywhere.*

Proof. Since $x(t)$ is nontrivial, there exists a number $\bar{t}$ in I such that $x(\bar{t}) = \bar{x} \neq 0$. Recall that the general solution is given by $x(t) = ce^{-\int p(t)\,dt}$. Therefore, there exists a constant $\bar{c}$ such that $x(t) = \bar{c}e^{-\int p(t)\,dt}$. Since $x(t)$ is nontrivial, we must have $\bar{c} \neq 0$, otherwise we would have $x(t) \equiv 0$ because $e^{-\int p(t)\,dt} \neq 0$. This shows that $x(t)$ cannot vanish anywhere. □

Theorem 2.1 (Existence–uniqueness, homogeneous equations). *If $p(t)$ is continuous in an interval I, then there is a unique solution to the initial value problem*

$$x' + p(t)x = 0, \quad x(t_0) = x_0. \tag{2.5}$$

Furthermore, the solution is defined for all t in I.

Proof. Existence. Recall that the general solution is given by $x(t) = ce^{-\int p(t)\,dt}$. We claim that $x(t) = x_0 e^{-\int_{t_0}^t p(t)\,dt}$ is a particular solution that satisfies the initial condition $x(t_0) = x_0$. This can easily be verified by direct substitution, since using the fundamental theorem of calculus, $(x_0 e^{-\int_{t_0}^t p(t)\,dt})' = -p(t)x_0 e^{-\int_{t_0}^t p(t)\,dt}$, and hence

$$\left(x_0 e^{-\int_{t_0}^t p(t)\, dt}\right)' + p(t)\left(x_0 e^{-\int_{t_0}^t p(t)\, dt}\right) = -p(t)x_0 e^{-\int_{t_0}^t p(t)\, dt} + p(t)x_0 e^{-\int_{t_0}^t p(t)\, dt} = 0.$$

Finally, it is clear from $x(t) = x_0 e^{-\int_{t_0}^t p(t)\, dt}$ that $x(t_0) = x_0 e^0 = x_0$ and it is defined on the interval where $p(t)$ is continuous.

An alternate proof is: using the integrating factor $\mu(t) = e^{\int_{t_0}^t p(t)\, dt}$, the general solution is given by $x(t) = ce^{-\int_{t_0}^t p(t)\, dt}$. Hence $x(t_0) = x_0$ implies that $x_0 = ce^0 = c$ and therefore a solution satisfying the required initial condition is given by $x(t) = x_0 e^{-\int_{t_0}^t p(t)\, dt}$, which can be checked by direct substitution.

Furthermore, $x(t) = x_0 e^{-\int_{t_0}^t p(t)\, dt}$ implies that $x(t_0) = x_0$.

Uniqueness. Suppose that $x(t)$ and $y(t)$ are both solutions. Then we must have $x' + p(t)x = 0$, $x(t_0) = x_0$, and $y' + p(t)y = 0$, $y(t_0) = x_0$. Subtracting the second equation from the first, we obtain $x' - y' + p(t)(x - y) = 0$. Let $z(t) = x(t) - y(t)$. Then $z(t)$ satisfies the equation $z' + p(t)z = 0$ and $z(t_0) = 0$. It follows from Lemma 2.1 that $z(t) \equiv 0$ and hence $x(t) - y(t) \equiv 0$, or $x(t) \equiv y(t)$. □

Remark 2.3. Let $\bar{x}_0$ be another initial value and let $\bar{x}(t)$ denote the corresponding solution of the ivp (2.5). Evaluating $|x(t) - \bar{x}(t)|$ we find

$$\left|x(t) - \bar{x}(t)\right| = \left|x_0 e^{-\int_{t_0}^t p(t)\, dt} - \bar{x}_0 e^{-\int_{t_0}^t p(t)\, dt}\right| = e^{-\int_{t_0}^t p(t)\, dt} \cdot |x_0 - \bar{x}_0|.$$

For any bounded interval $T \subseteq I$, setting $\max_{t \in T}[e^{-\int_{t_0}^t p(t)\, dt}] = C$, a constant depending on T, we infer

$$\max_{t \in T}|x(t) - \bar{x}(t)| \le C \cdot |x_0 - \bar{x}_0|.$$

We will refer to this result by saying that *the solutions of* (2.5) *depend continuously on the initial values.*

The homogeneous equation $x' + p(t)x = 0$ has some important properties that do not hold in the nonhomogeneous case. We list them in a corollary.

Corollary 2.1.
(a) *If $x(t_0) = 0$ for some number t_0 in I, then $x(t) \equiv 0$ for all t in I.*
(b) *Every nontrivial solution is either always positive or always negative.*
(c) *If $x_1(t)$ and $x_2(t)$ are any solutions of the homogeneous differential equation $x' + p(t)x = 0$, then $x(t) = c_1 x_1(t) + c_2 x_2(t)$ is also a solution, where c_1 and c_2 are any constants.*

Proof. (a) This immediately follows from Lemma 2.1; in fact, it is a contrapositive of the statement there.

(b) Suppose that $x(t)$ is a nontrivial solution such that it is neither always positive nor always negative. Then there exist two numbers t_1, t_2 such that $x(t_1) > 0$ and $x(t_2) < 0$. This, however, would imply the existence of a number $\bar{t}$, between t_1 and t_2, such that $x(\bar{t}) = 0$, by the intermediate value theorem, contradicting Lemma 2.1.

(c) This can be verified just by substitution and regrouping (see problem 22). $\qquad \square$

None of the three properties listed above are valid for nonhomogeneous equations (see exercises at the end of the chapter).

2.3 Linear nonhomogeneous equations

There are some significant differences between the homogeneous and nonhomogeneous equations. For example, simple substitution shows that contrary to the case for homogeneous equations, the zero function is not a solution of the nonhomogeneous equation (2.1). However, the method for solving the nonhomogeneous equation (2.1) is similar to that for solving the homogeneous equation (2.2), although the integration can be more complicated. First, some examples are in order.

Example 2.2. (a) Solve

$$x' + kx = h, \quad h, k \in \mathbb{R}.$$

In order to find the general solution, we multiply both sides of the equation by the integrating factor $\mu(t) = e^{\int k\, dt} = e^{kt}$, obtaining

$$x' e^{kt} + kx e^{kt} = h e^{kt},$$

which makes the left side the exact derivative of the product $x e^{kt}$. So we have $(x e^{kt})' = h e^{kt}$. Integrating both sides, we obtain $x e^{kt} = h \int e^{kt}\, dt = \frac{h}{k} e^{kt} + c$. Therefore, the general solution is given by

$$x(t) = c e^{-kt} + \frac{h}{k}.$$

If we let $c = 0$ in the general solution, we obtain the constant solution $x(t) = \frac{h}{k}$. In the tx-plane, this solution traces out a straight horizontal line such that all other solution curves either lie above it or below it. This is because of the uniqueness of the solution to initial value problem, *no two curves representing different solutions can intersect*. This will follow from the general property that we will see more precisely later on: solution to initial value problem (2.6) is unique and thus *no two curves representing different solutions can intersect*. In this case, another, more direct way to prove the claim is to notice that $e^{-kt} > 0$ and hence $x(t) > \frac{h}{k}$ if $c > 0$ whereas $x(t) < \frac{h}{k}$ if $c < 0$, see Fig. 2.1.

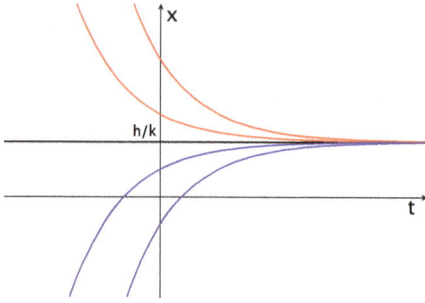

Figure 2.1: Plot of the general solution $x(t) = ce^{-kt} + \frac{h}{k}$. In red the integrals for $c > 0$, in blue the integrals for $c < 0$, in black the line $x = \frac{h}{k}$.

(b) Solve

$$x' - 2tx = t.$$

An integrating factor is now $\mu(t) = e^{\int -2t\,dt} = e^{-t^2}$. Repeating the preceding calculation we find $x'e^{-t^2} - 2te^{-t^2}x = te^{-t^2}$ whereby $(e^{-t^2}x)' = te^{-t^2}$. Integrating both sides, we obtain $e^{-t^2}x(t) = -\frac{1}{2}e^{-t^2} + c$. Therefore, the general solution is given by

$$x(t) = -\frac{1}{2} + ce^{t^2}.$$

Once again, the constant solution $x(t) = -\frac{1}{2}$ (corresponding to $c = 0$) divides the tx-plane into two disjoint regions that contain all the non-constant solutions; see Fig. 2.2.

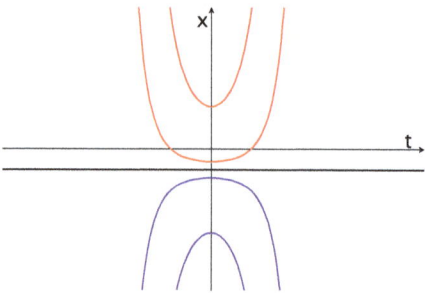

Figure 2.2: Plot of the general solution $x(t) = -\frac{1}{2} + ce^{t^2}$. In red are the integrals for $c > 0$, in blue are the integrals for $c < 0$, in black is the line $x = -\frac{1}{2}$.

(c) In order to solve

$$x' + \sin t \cdot x = \sin t,$$

we multiply both sides of the equation by the integrating factor $\mu(t) = e^{\int \sin t \, dt} = e^{-\cos t}$, obtaining

$$x'e^{-\cos t} + e^{-\cos t} \sin t \cdot x = e^{-\cos t} \sin t$$

which is equivalent to

$$\left(xe^{-\cos t}\right)' = e^{-\cos t} \sin t.$$

Integrating both sides yields

$$xe^{-\cos t} = e^{-\cos t} + c$$

and hence

$$x(t) = 1 + ce^{\cos t}$$

is the general solution.

If we let $c = 0$ in the general solution, we obtain the constant solution $x(t) = 1$, which again divides all the solutions into two groups: those lying above the horizontal line $x = 1$ and those lying below it. We note that, in this case, *all solutions are bounded*, since $-1 \le \cos t \le 1$. □

Similar to the case for homogeneous equations, we now consider the ivp for non-homogeneous equations, that is,

$$x' + p(t)x = q(t), \quad x(t_0) = x_0. \tag{2.6}$$

The initial condition $x(t_0) = x_0$ corresponds to a unique constant c.

Examples 2.2 revisited. (a) Solve

$$x' + kx = h, \quad x(0) = 0.$$

We know that the general solution is $x(t) = ce^{-kt} + \frac{h}{k}$. For $t = 0$ we obtain $x(0) = ce^0 + \frac{h}{k}$. Then $x(0) = 0$ yields $0 = c + \frac{h}{k}$. Solving for c, we get $c = -\frac{h}{k}$ and thus the solution of the ivp is given by

$$x(t) = -\frac{h}{k}e^{-kt} + \frac{h}{k} = \frac{h}{k} \cdot (1 - e^{-kt}).$$

(b) Solve the ivp

$$x' - 2tx = t, \quad x(0) = 1.$$

If we put $t = 0$ and $x(0) = 1$ in the general solution $x(t) = -\frac{1}{2} + ce^{t^2}$ we obtain the algebraic equation $1 = -\frac{1}{2} + c$, which we can solve for the constant c, obtaining $c = \frac{3}{2}$. Therefore, the desired solution is given by $x(t) = -\frac{1}{2} + \frac{3}{2}e^{t^2}$.

(c) Solve the ivp

$$x' + \sin t \cdot x = \sin t, \quad x\left(\frac{\pi}{2}\right) = -2.$$

Multiplying both sides of the equation by $e^{-\cos t}$ and integrating, we find the general solution $x(t) = 1 + ce^{\cos t}$. Then we find the required solution by substituting $t = \frac{\pi}{2}$ and $x = -2$ in the general solution and solving for c, which yields $x(t) = 1 - 3e^{\cos t}$. □

Theorem 2.2 (Existence–uniqueness, nonhomogeneous). *If $p(t)$ and $q(t)$ are continuous on an interval I, then there exists a unique solution $x(t)$ satisfying the initial value problem (2.6). Furthermore, this solution is defined for all t in I.*

Proof. Existence. Multiplying both sides of the equation by the integrating factor $\mu(t) = e^{\int_{t_0}^{t} p(t)\,dt}$, we have

$$\left(x(t)\mu(t)\right)' = \mu(t)q(t).$$

Integrating both sides from t_0 to t, we have

$$x(t)\mu(t) - x(t_0)\mu(t_0) = \int_{t_0}^{t} \mu(t)q(t)\,dt.$$

Recall that $\mu(t) = e^{\int_{t_0}^{t} p(t)\,dt}$ implies that $\mu(t_0) = e^0 = 1$ and hence

$$x(t) = \frac{1}{\mu(t)}\left(\int_{t_0}^{t} \mu(t)q(t)\,dt + x_0\right) = e^{-\int_{t_0}^{t} p(t)\,dt}\left(\int_{t_0}^{t} e^{\int_{t_0}^{t} p(t)\,dt}q(t)\,dt + x_0\right)$$

is the desired solution, which can be verified by substituting it in the equation.

Uniqueness. One can argue that the uniqueness of the solution can be derived from the above discussion or by finding the general solution and then showing that there is only one value of the constant that will satisfy the prescribed initial condition. But the simple proof below shows that it is essentially a corollary of its own special case (the homogeneous equation), which is an interesting phenomenon in its own right.

Suppose that $x(t)$ and $y(t)$ are any solutions of (2.6). We will show that $x(t) \equiv y(t)$. We have $x' + p(t)x = q(t)$, $x(t_0) = x_0$, and $y' + p(t)y = q(t)$, $y(t_0) = x_0$. Subtracting the second equation from the first, we obtain $x' - y' + p(t)(x - y) = 0$. Let $z(t) = x(t) - y(t)$. Then $z(t)$ satisfies the homogeneous equation $z' + p(t)z = 0$ and $z(t_0) = 0$, which implies that $z(t) \equiv 0$ by the uniqueness part of Theorem 2.1 (or from Lemma 2.1), and hence $x(t) - y(t) \equiv 0$, or $x(t) \equiv y(t)$. ☐

Remark 2.4. Similar to homogeneous equations, solutions of (2.6) also depend continuously on the initial values. To see this, we can simply repeat the calculation carried out in Remark 2.3. Another way is to set $z = x - \overline{x}$ and notice that, as before, $z' + pz = 0$. Moreover, $z(t_0) = x(t_0) - \overline{x}(t_0) = x_0 - \overline{x}_0$. Thus, according to (2.3),

$$z(t) = z(t_0) \cdot e^{-\int_{t_0}^t p(t)\, dt} = (x_0 - \overline{x}_0) \cdot e^{-\int_{t_0}^t p(t)\, dt}.$$

Then, for any bounded interval $T \subseteq I$, one has

$$\max_{t \in T} |x(t) - \overline{x}(t)| = \max_{t \in T} |z(t)| = \max_{t \in T}\big[e^{-\int_{t_0}^t p(t)\, dt}\big] \cdot |x_0 - \overline{x}_0|,$$

and the conclusion follows as in Remark 2.3.

In solving problems, it is important to first express them in convenient and familiar forms that we know how to handle. For example, $t + x = x' - 1$ can be easily written in the form $x' - x = t + 1$, which is a particular case of $x' + p(t)x = q(t)$ for which we have discussed a method of solving it.

Example 2.3.

$$x'' - x' = 1, \quad x(0) = 1.$$

At first sight this seems to be a second order equation and outside the scope of this chapter. But on closer look, we notice that it can be easily written as a first order linear equation which we can handle. To this end, let $x' = z$. Then the problem can be expressed as $z' - z = 1$. Multiplying both sides by the integrating factor $e^{\int e^{-t}\, dt}$ we obtain the equation

$$(e^{-t}z)' = e^{-t}.$$

Integrating both sides, we have $e^{-t}z = -e^{-t} + c$ yielding $z = -1 + ce^t$. Recalling that $z = x'$, we have $x' = -1 + ce^t$. Integrating once again, we obtain the general solution $x = -t + ce^t + k$. Substituting the initial values $t = 0$ and $x = 1$, we have $k = 1 - c$. Therefore we have

$$x(t) = -t + ce^t + 1 - c = -t + c(e^t - 1) + 1.$$

We note that since c is an arbitrary constant, in this example there are infinitely many solutions to the given initial value problem and thus the uniqueness property does not hold. The reason for this is that the given equation is actually a second order equation which required two integrations, each producing an arbitrary constant.

2.4 Applications to RC electric circuits

An RC circuit, is an electric circuit composed of a Resistor and a Capacitor.

As a first application we consider an RC circuit with capacitor voltage $V(t)$ depending on time $t \geq 0$, constant resistance R and constant capacity C and with no external current or voltage source; see Fig. 2.3.

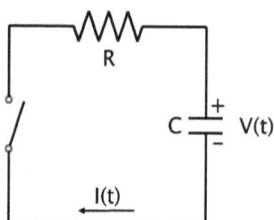

Figure 2.3: An RC circuit.

Let
- $I(t)$ denote the current circulating in the circuit;
- $V_0 = V(0)$ denote the voltage at the initial rime $t = 0$.

The Kirchhoff law and the constitutive law of capacitor yield, respectively,

$$R \cdot I(t) + V(t) = 0, \quad I(t) = C \cdot \frac{dV(t)}{dt}.$$

Putting together these two relationships, we obtain the first order differential equation

$$RC \cdot V'(t) + V(t) = 0.$$

Taking into account the initial condition $V(0) = V_0$ we are led to the initial value problem (ivp)

$$\begin{cases} V'(t) + \frac{1}{RC} \cdot V(t) = 0, \\ V(0) = V_0. \end{cases} \tag{2.7}$$

Notice that the differential equation in (2.7) is of the form $x' + kx = 0$, with $x = V$ and $k = 1/RC$.

According to the previous results, the solution of the ivp (2.7) is given by

$$V(t) = V_0 \cdot e^{-t/RC}.$$

Introducing the so called RC *time constant* $\tau = RC$, we can write the solution in the form

$$V(t) = V_0 \cdot e^{-t/\tau}.$$

We infer that, based on experience, the capacitor voltage $V(t)$ is a decreasing function and tends exponentially to 0 as $t \to +\infty$, with a decay rate that depends on τ: the bigger is τ the slower is the decay. Notice also that for $t = \tau$ one has $V(\tau) = V_0 e^{-1}$: hence τ is the time after which the voltage $V(t)$ decays to V_0/e, see Fig. 2.4.

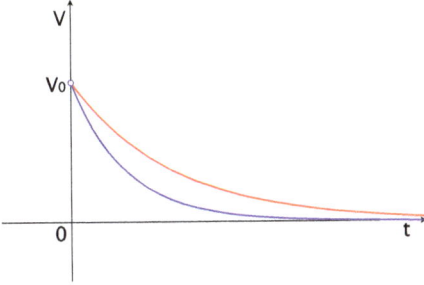

Figure 2.4: Plot of the solutions of (2.7) with different values of $\tau = RC$: the red curve corresponds to a bigger τ, the blue curve to a smaller τ.

Since $I = -\frac{V}{R}$ we get

$$I(t) = -\frac{V_0}{R} e^{-t/\tau},$$

which implies that the current intensity $I(t) \nearrow 0$ as $t \to +\infty$; see Fig. 2.5.

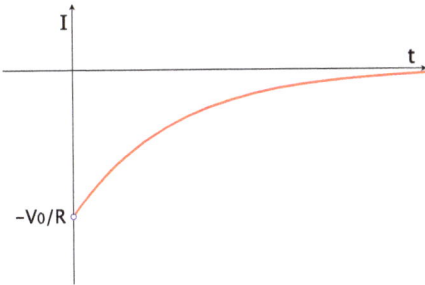

Figure 2.5: Plot of the current intensity $I(t)$.

Our next application deals with an RC circuit in which a generator of constant voltage $\overline{V}$ is inserted; see Fig. 2.6.
Repeating the preceding arguments we can see that $V(t)$ satisfies the ivp

$$\begin{cases} V'(t) + \frac{1}{RC} \cdot V(t) = \frac{1}{RC} \overline{V} \\ V(0) = V_0 \end{cases} \tag{2.8}$$

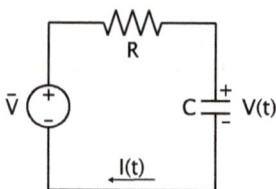

Figure 2.6: RC circuit with a generator of voltage.

The equation above is of the type $x' + kx = h$, with $x = V$, $k = 1/RC$ and $h = \frac{\overline{V}}{RC}$. We know, see (2.2)-(a), that the general solution of $x' + kx = h$ is given by

$$x(t) = ce^{-kt} + \frac{h}{k}.$$

With the current notation, we get

$$V(t) = ce^{-t/RC} + \overline{V}.$$

Setting $t = 0$ in the preceding equation, we get

$$V(0) = c + \overline{V} \quad \Longrightarrow \quad c = V_0 - \overline{V}.$$

Hence the unique solution of (2.8) is

$$V(t) = (V_0 - \overline{V})e^{\frac{-t}{\tau}} + \overline{V}. \tag{2.9}$$

This equation shows that the presence of the constant voltage $\overline{V}$ changes the behavior of $V(t)$. First of all, $V(t)$ does not decay to 0 but tends, as $t \to +\infty$, to the constant voltage $\overline{V}$:

$$\lim_{t \to +\infty} V(t) = \overline{V}.$$

Furthermore, $V(t)$ is increasing or decreasing according to the sign of $V_0 - \overline{V}$. Precisely:
- if $V_0 - \overline{V} < 0$, i. e. $V_0 < \overline{V}$, then $(V_0 - \overline{V})e^{\frac{-t}{\tau}}$ is negative and increasing. Thus, $V(t) < \overline{V}$ for all $t \geq 0$ and, in particular, $V(t) \to \overline{V}$ from below, as $t \to +\infty$, see Fig. 2.7;

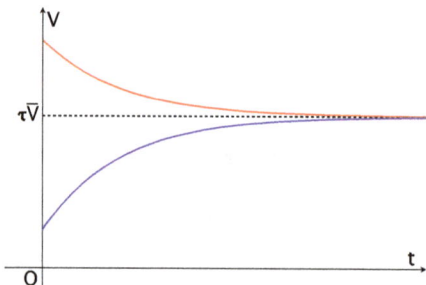

Figure 2.7: Plot of solutions: $V_0 < \tau\overline{V}$ in blue, $V_0 > \tau\overline{V}$ in red.

- if $V_0 - \overline{V} > 0$, i. e. $V_0 > \overline{V}$, then $(V_0 - \overline{V})e^{-t/\tau}$ is positive and decreasing. Thus, $V(t)\overline{V}$ for all $t \geq 0$ and, in particular, $V(t) \to \overline{V}$ from above, as $t \to +\infty$, see Fig. 2.7;
- if $V_0 - \overline{V} = 0$, i. e. $V_0 = \overline{V}$, then $V(t) = \overline{V}$ for all $t \geq 0$.

Other equations arising in the electric circuit theory will be discussed later.

2.5 Exercises

1. Solve $x' - 4x = 8$.
2. Solve $x' + (\ln 2)x = 2$.
3. Solve $x' + 2tx = 6t$.
4. Solve $x' - 4t = 2x, x(0) = 2$.
5. Solve $x' + 2x = 4t, x(0) = 2$.
6. Solve $x' - t = x + 1, x(0) = 1$
7. Solve the initial value problem $x' + 2tx = 4t, x(0) = 1$.
8. Solve $x' + x = \sin t + \cos t$.
9. Show that for any positive number c, all solutions of $x' + cx = 0$ tend to 0 as t tends to infinity.
10. Solve $tx'' + x' = t + 2$.
11. $x' + (\sin t)x = 0, x(\frac{\pi}{2}) = 1$.
12. Show that if $k < 0$ then there is no solution of $x' + kx = h$ such that $\lim_{t\to+\infty} x(t)$ is finite, but the constant one.
13. Solve $tx' + x = 4t + e^t$.
14. Show that for any $x_0 \in \mathbb{R}$ the solution of $x' - x = h, x(0) = x_0$ is such that $\lim_{t\to-\infty} x(t) = -h$.
15. Find x_0 such that the solution of $x' = \frac{1}{1+t} x, x(0) = x_0$ satisfies $\lim_{t\to\pm\infty} x(t) = 0$.
16. * Explain why the boundary value problem $x' + kx = 0, x(1) = 2, x(3) = -1$ has no solution. This shows that, unlike the initial value problems, boundary value problems for first order homogeneous equations don't always have solutions.
17. Solve $x' = t^2 - 1 - x, x(0) = -1$.
18. Solve $x' + (\cot t)x = \cos t, t \neq n\pi$.
19. Solve $(\sin t)x' + (\cos t)x = \cos 2t, x(\frac{\pi}{2}) = \frac{1}{2}$.
20. Show that the boundary value problem $x' + p(t)x = q(t), x(t_1) = x_1, x(t_2) = x_2, p(t)$ and $q(t)$ continuous, cannot have more than one solution.
21. Solve $\cos t \cdot x' + \sin t \cdot x = \sin 2t$.
22. Find the largest interval in which the solution to the following Cauchy Problems exist.
    1. $x' + \frac{1}{t^2-4}x(t) = 0, x(-1) = 10$.
    2. $x' + \frac{1}{t^2-4}x(t) = 0, x(3) = 1$.
    3. $x' + \frac{1}{t^2-4}x(t) = 0, x(-10) = 1$.

23. Show that if x_1 and x_2 are solutions of the nonhomogeneous equation $x' + p(t)x = q(t) \neq 0$, then $x(t) = x_1 + x_2$ is not a solution.

24. (a) Show that if $x_1(t)$ is a solution of $x' + p(t)x = f(t)$ and $x_2(t)$ is a solution of $x' + p(t)x = g(t)$, then $x_1 + x_2$ is a solution of $x' + p(t)x = f(t) + g(t)$.

    (b) Use the method in part (a) to solve $x' + x = 2e^t - 3e^{-t}$.

25. Show that the ivp $tx' + x = 0$, $x(0) = a \neq 0$ has no solution. Explain why this does not contradict the existence and uniqueness Theorem 2.1.

26. * Find a number k such that $x' + kx = 0$, $x(1) = 2$, $x(2) = 1$ has a solution.

27. (a) Show that if $f(t)$ is a differentiable function, then all solutions of $x' + x = f(t) + f'(t)$ are asymptotic to $f(t)$, that is, they approach $f(t)$ as $t \to \infty$.

    (b) Find a differential equation such that all of its solutions are asymptotic to $t^2 - 1$.

28. * Let $x_1(t)$ be any solution of (2.1) $x' + p(t)x = q(t)$. Recall that $x(t) = ce^{-\int_{t_0}^t p(t)\, dt}$ is the general solution of (2.2) $x' + p(t)x = 0$. Show that $x(t) = x_1(t) + ce^{-\int_{t_0}^t p(t)\, dt}$ is the general solution of the nonhomogeneous equation (2.1).

29. Find the general solution of $x' - x = \cos t - \sin t$.

30. Explain why the function $x(t) = e^t - 1$ cannot be a solution of any homogeneous equation $x' + p(t)x = 0$, where $p(t)$ is some continuous function.

31. Solve $x' + x = t + \sin t$, by using the method of problem 24.

32. Explain why $x(t) = t^2 - 1$ cannot be a solution of a homogeneous equation $x' + p(t)x = 0$, $p(t)$ continuous.

3 Analytical study of first order differential equations

3.1 General first order differential equations

Before discussing methods of solving differential equations in general, in this chapter we study and analyze the theoretical aspects of such equations. In contrast to Chapter 2, where we mainly learned methods of solving differential equations, here we stress a more rigorous treatment of the rationale for drawing conclusions concerning the qualitative behavior of solutions. This is particularly important since there are no known methods to solve such equations in general. Of course, these general equations include the linear equations covered in Chapter 2.

Let $A \subseteq \mathbb{R}^2$ be a set, and let $f = f(t, x)$ be a real valued function defined on A.

Definition 3.1. A first order ordinary differential equation is an equation of the form

$$x' = f(t, x), \quad \text{where } x' = \frac{dx}{dt}. \tag{3.1}$$

We note that the linear equation $x' + p(t)x = q(t)$, discussed in the previous chapter, is included in (3.1), with $f(t, x) = -p(t)x + q(t)$ and $A = I \times \mathbb{R}$, I being the interval where p and q are defined.

Definition 3.2. A solution (or integral) of (3.1) is a differentiable real valued function $x(t)$ defined on an interval $I \subseteq \mathbb{R}$ such that
(i) $(t, x(t)) \in A$ for all $t \in I$;
(ii) $x'(t) = f(t, x(t))$ for all $t \in I$.

More precisely, to see that a differentiable function $x(t)$ solves $x' = f(t, x)$ one evaluates $x'(t)$ and checks to see if $x' = f(t, x)$ holds true for all t in I.

Let $x(t), t \in I$, be a solution to (3.1). Let us recall that the tangent line to $x(t)$ at a point $t^* \in I$ has equation $x = x'(t^*)(t - t^*) + x(t^*)$. Since $x'(t^*) = f(t^*, x(t^*))$ we find

$$x = f(t^*, x(t^*))(t - t^*) + x(t^*).$$

So, from a geometrical point of view, a solution to (3.1) is a differentiable curve $x = x(t)$ such that the slope of the tangent line at each point $t^* \in I$ equals $f(t^*, x(t^*))$.

For example, in Fig. 3.1 red arrows are plotted to indicate the slope of each tangent line to the family of solutions $x = ce^t$.

Remark 3.1. An important difference between differential equations and algebraic equations is that the former ones have functions as solutions, whereas the solutions of the latter ones are real or complex numbers.

https://doi.org/10.1515/9783111185675-003

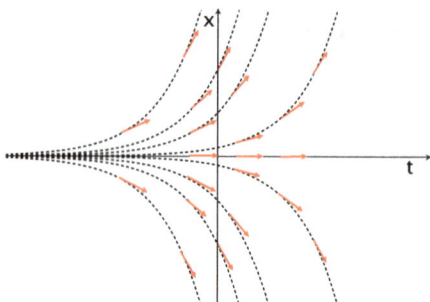

Figure 3.1: Slope of tangents at a grid of points in the (t, x) plane; the dotted curves are the graphs of $x = ce^t$, the solutions to $x' = x$.

Furthermore, it is worth pointing out that in (3.1) there are two unknowns: the function $x(t)$ satisfying $x' = f(t, x)$ and the interval I where the solution is defined. It may happen that I not only depends upon the given function f but also on other data. This is the case in Example 3.1 below, where $x(t)$ is defined on an open half-line $(-\infty, t_0)$, where t_0 depends on the value $x(0)$.

We also notice that in (3.1) we used the notation t and x to indicate, respectively, the independent and dependent variable. But the reader should be aware that we might have chosen different notations: for example, y might be the dependent variable and x the independent variable. In such a case, (3.1) would be written as

$$\frac{dy}{dx} = f(x, y).$$

In particular, this notation will be used in Chapters 5 and 16.

More generally, let $F(t, x, \xi)$ is a real valued function of three variables, defined for $(t, x, \xi) \in S \subseteq \mathbb{R}^3$. Then a first order ordinary differential equation can be written as

$$F(t, x, x') = 0,$$

where Equation (3.1) corresponds to the case in which $S = A \times \mathbb{R}$ and $F(t, x, x') = x' - f(t, x)$.

We will be mainly dealing with equation (3.1), which will be referred to as an *equation in normal form*.

Moreover, we use the following terminology:
1. we say that (3.1) is *linear* if $f(t, x) = a(t)x + b(t)$, otherwise it is *nonlinear*;
2. we say that (3.1) is *autonomous* if $f = f(x)$ does not depend upon t, otherwise it is *non-autonomous*.

For example:
1. $x' = x$ is linear autonomous, $x' = x(1 - x)$ is nonlinear autonomous;
2. $x' = x + t$ is linear non-autonomous, $x' = \sin x + t$ is nonlinear non-autonomous;

3.1.1 Initial value problem

Similar to the case for linear equations, we can add to a general first order equation an *initial condition*. Focusing on the equation in normal form, $x' = f(t, x)$, the resulting *initial value problem* (ivp for short) or *Cauchy problem* is given by

$$\begin{cases} x' = f(t, x), \\ x(t_0) = x_0. \end{cases}$$

In the next section, we will see that, under appropriate assumptions on f, the initial condition allows us to single out one integral among a family of solutions of the equation.

3.2 Existence and uniqueness results

The main purpose of this section is to discuss some fundamental results yielding the existence and uniqueness of the Cauchy problem

$$\begin{cases} x' = f(t, x), \\ x(t_0) = x_0. \end{cases} \tag{3.2}$$

We suggest that the reader examines these theoretical results in depth and becomes familiar with them, since they provide the groundwork for the theory of ordinary differential equations as well as a basic tool used throughout the book.

We start with the following lemma which shows the equivalence of (3.2) in the form of a convenient integral equation.

Lemma 3.1.

(1) *Let $x = x(t)$ be a solution of (3.2) in an interval I, with $t_0 \in I$. Then $x(t)$ satisfies the integral equation*

$$x(t) = x_0 + \int_{t_0}^{t} f(s, x(s))\, ds, \quad t \in I. \tag{3.3}$$

(2) *Conversely, if (3.3) holds, then $x(t)$ solves the ivp (3.2).*

Proof. (1) Let $x(t)$ satisfy

$$x'(t) = f(t, x(t)), \quad \forall t \in I, \quad x(t_0) = x_0.$$

Integrating from t_0 to t and denoting the variable of integration by s, we find

$$\int_{t_0}^{t} x'(s)\, ds = \int_{t_0}^{t} f(s, x(s))\, ds, \quad t \in I.$$

Since $\int_{t_0}^{t} x'(s)\,ds = x(t) - x(t_0)$, it follows that

$$x(t) - x(t_0) = \int_{t_0}^{t} f(s, x(s))\,ds.$$

Recalling that $x(t_0) = x_0$ we get

$$x(t) - x_0 = \int_{t_0}^{t} f(s, x(s))\,ds \quad \Longleftrightarrow \quad x(t) = x_0 + \int_{t_0}^{t} f(s, x(s))\,ds, \quad \forall t \in I.$$

(2) Conversely, let $x(t)$ satisfy (3.3). Then $x(t)$ is differentiable on I. Moreover, by taking the derivatives of both sides and using the fundamental theorem of calculus, one finds

$$x'(t) = \frac{d}{dt}\left[\int_{t_0}^{t} f(s, x(s))\,ds\right] = f(t, x(t)), \quad \forall t \in I,$$

as well as $x(t_0) = x_0 + \int_{t_0}^{t_0} f(s, x(s))\,ds = x_0$. □

Theorem 3.1 (Local existence and uniqueness). *Suppose that f is continuous in $A \subseteq \mathbb{R}^2$ and has continuous partial derivative f_x with respect to x. Let (t_0, x_0) be a given point in the interior of A. Then there exists a closed interval I containing t_0 in its interior such that the Cauchy problem (3.2) has one and only solution, defined in I.*

The complete proof as well as a more general result will be given in the appendix at the end of this chapter.

Here we are going to outline the argument of the existence part in the specific case where $f(t, x) = x$ and $t_0 = 0$. The method was introduced by Picard and is based on an approximation scheme, interesting in itself.

According to Lemma 3.1, the ivp $x' = x$, $x(0) = x_0$, is equivalent to the integral equation

$$x(t) = x_0 + \int_{0}^{t} x(s)\,ds.$$

We now indicate a method for finding a solution of this integral equation. Define a sequence of approximate solutions for this equation by setting

$$x_0(t) = x_0,$$

$$x_1(t) = x_0 + \int_{0}^{t} x_0(s)\,ds = x_0 + \int_{0}^{t} x_0\,ds$$

$$= x_0 + x_0 \cdot t = x_0 \cdot [1 + t],$$

$$x_2(t) = x_0 + \int_0^t x_1(s)\, ds = x_0 + \int_0^t x_0 \cdot (1 + s)\, ds$$

$$= x_0 \left[1 + \int_0^t (1 + s)\, ds \right] = x_0 \cdot \left[1 + \frac{1}{2}(1 + t)^2 - \frac{1}{2} \right]$$

$$= x_0 \cdot \left[1 + t + \frac{1}{2} t^2 \right],$$

$$x_3(t) = x_0 \cdot \left[1 + t + \frac{1}{2} t^2 + \frac{1}{3!} t^3 \right],$$

$$\cdots \quad \cdots$$

$$x_k(t) = x_0 \cdot \left[1 + t + \frac{1}{2} t^2 + \cdots + \frac{1}{k!} t^k \right].$$

The approximated solutions x_1, x_2, x_3 are plotted in Fig. 3.2.

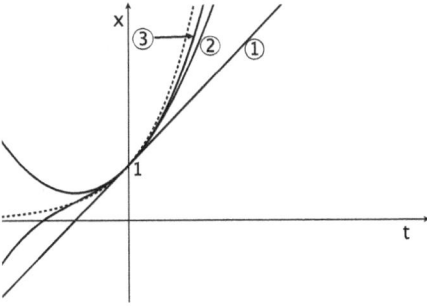

Figure 3.2: Plot of $x_1(t), x_2(t), x_3(t)$, the first three approximated solutions of $x' = x$, $x(0) = 1$. The dotted curve is the actual solution $x(t) = e^t$.

The reader should recognize that the polynomials in brackets are nothing but the Taylor polynomials of the exponential function e^t. Thus the sequence $x_k(t)$ converges uniformly on any bounded interval to

$$x(t) = x_0 \cdot \sum_0^{+\infty} \frac{t^k}{k!} = x_0 \cdot e^t,$$

which is the solution of $x' = x$, $x(0) = x_0$.

For a general function f the approximating sequence is given by

$$x_0(t) = x_0, \quad x_{k+1}(t) = x_0 + \int_{t_0}^t f(s, x_k(s))\, ds, \quad k = 0, 1, \ldots$$

One can show that, under the given assumptions, there exists $\delta > 0$ such that $f(s, x_k(s))$ makes sense and $x_k(t)$ converges to some function $x(t)$, uniformly in $[t_0 - \delta, t_0 + \delta]$. This allows us to find

$$\lim_{k \to +\infty} \int_{t_0}^{t} f(s, x_k(s)) \, ds = \int_{t_0}^{t} \left[\lim_{k \to +\infty} f(s, x_k(s)) \right] ds = \int_{t_0}^{t} f(s, x(s)) \, ds,$$

for all $t \in [t_0 - \delta, t_0 + \delta]$. Therefore, passing to the limit, it turns out that x satisfies

$$x(t) = x_0 + \int_{t_0}^{t} f(s, x(s)) \, ds, \quad \forall t \in [t_0 - \delta, t_0 + \delta].$$

According to Lemma 3.1, $x(t)$ solves the ivp (3.2) in the interval $[t_0 - \delta, t_0 + \delta]$.

It is worth noticing that Theorem 3.1 provides a (unique) solution defined on an interval $[t_0 - \delta, t_0 + \delta]$, for *some* $\delta > 0$. Completing Remark 3.1, we will show in the following example that δ depends not only on f but also on the initial condition $x(t_0) = x_0$. This will make it clear that we could not find a solution of (3.2) defined on a *given* interval I.

Example 3.1. Consider the ivp

$$\begin{cases} x' = x^2, \\ x(0) = 1. \end{cases} \tag{3.4}$$

We will solve (3.4) by using some intuition.

First of all, let us try to find some functions $x(t)$ such that $x' = x^2$. It is obvious that $x(t) \equiv 0$ is one such function, but it does not satisfy the initial condition $x(0) = 1$. Recalling that the derivative of $x(t) = \frac{1}{t}$ is $x' = -\frac{1}{t^2}$, we can verify that $x(t) = -\frac{1}{t}$ is another such function; in fact, for any constant c, $x(t) = \frac{1}{c-t}$ satisfies the equation $x' = x^2$. It is now clear that if we choose $c = 1$, then the function $x(t) = \frac{1}{1-t}$ solves the given initial value problem. The solution is not defined at $t = 1$. Moreover, we have to take into account that the solution of a differential equation is differentiable and hence continuous. Thus we cannot take both branches of the hyperbola $\frac{1}{1-t}$, but only the upper branch. In other words, the domain of definition of the solution of the ivp (3.4) is the open half-line $(-\infty, 1)$. See Fig. 3.3. In general, we may see that the domain of definition of the solution depends upon the value of $x(t_0)$, see exercise n. 6 later. □

This simple example shows that a differential equation $x' = f(t, x)$, where $f(t, x)$ is a perfectly well behaved function, with continuous derivatives everywhere, can have solutions that are not defined for some unpredictable values of t. Although it is important to learn methods for solving differential equations, we also need to learn the theory for a complete comprehension of the concepts.

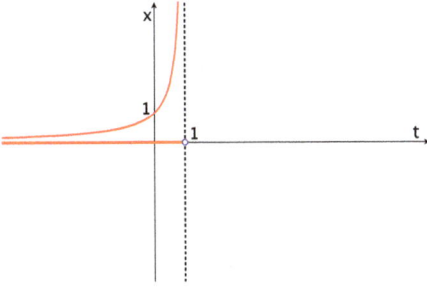

Figure 3.3: Solution of $x' = x^2$, $x(0) = 1$.

Remark 3.2. If the equation is given as $F(t, x, x') = 0$, we first try to put it in normal form, $x' = f(t, x)$, and then apply the existence and uniqueness results stated above.

3.2.1 On uniqueness

Dealing with linear equations in the previous chapter, we noticed that two solutions cannot have points in common. Theorem 3.1 allows us to show that nonlinear equations (3.1) possesses the same feature. Roughly, if two distinct solutions of (3.1) cross at a point, then they are both solutions of the same Cauchy problem, contrary to the uniqueness stated in Theorem 3.1.

In the sequel it is understood that the assumptions of Theorem 3.1 are satisfied.

Lemma 3.2. *Given a pair of solutions of* (3.1), $x_1(t), x_2(t)$ *defined on an interval I as in Theorem 3.1, let us suppose $\exists\,\tau \in I$ such that $x_1(\tau) = x_2(\tau)$. Then $x_1(t) = x_2(t)$, $\forall\,t \in I$.*

Proof. Consider the set $S = \{s \in I : x_1(t) = x_2(t), \ \forall\,t \in [\tau, s]\}$ which is not empty since $\tau \in S$. Letting $\sigma = \sup S$ we notice that, by continuity, one has $x_1(\sigma) = x_2(\sigma)$. If $\sigma = \delta$, we have done. Otherwise, we set $\xi = x_1(\sigma) = x_2(\sigma)$ and consider the Cauchy problem

$$\begin{cases} x' = f(t, x), \\ x(0) = \xi. \end{cases}$$

Applying Theorem 3.1 we infer that $\exists\,\delta' > 0$ such that this ivp has a unique solution defined on an interval $[\sigma, \sigma + \delta']$. Thus $x_1(t) = x_2(t)$ on $[\tau, \sigma + \delta']$, a contradiction for $\sigma = \sup S$. This proves that $x_1(t) = x_2(t)$, for any $t \in I$, $t \geq \tau$. A similar argument yields that $x_1(t) = x_2(t)$, for any $t \in I$, $t \leq \tau$.

Let us set $\xi = x_1(\tau) = x_2(\tau)$. Applying Theorem 3.1 to the Cauchy problem

$$\begin{cases} x' = f(t, x), \\ x(0) = \xi, \end{cases}$$

we infer that $\exists \delta > 0$ such that this ivp has a *unique solution* defined on an interval $[\xi - \delta, \xi + \delta]$. In particular, this implies that $x_1(\xi \pm \delta) = x_2(\xi \pm \delta)$. Repeating the same argument for the ivps

$$\begin{cases} x' = f(t,x), \\ x(0) = \xi \pm \delta, \end{cases}$$

we easily conclude that x_1 and x_2 coincide on all the intervals I where x_1, x_2 are defined. ☐

Example 3.2. Show that the solutions $x(t)$ of $x' = \sin(tx)$ are even.

Let $z(t) = x(-t)$. Then $z'(t) = -x'(-t) = -\sin(-tx(-t)) = \sin(tx(-t)) = \sin(tz)$. Furthermore, $z(0) = x(-0) = x(0)$. By uniqueness, $z(t) \equiv x(t)$. In other words, $x(t) = x(-t)$ which means that $x(t)$ is even. ☐

In Theorem 3.1 we assumed f is differentiable w. r. t. x. In the sequel we will improve this result, requiring a weaker hypothesis which, roughly, includes functions $f(x)$ such that f_x has jump discontinuities. On the other hand, we are going to see an example that shows that when this milder assumption does not hold, the solution may not be unique.

Example 3.3. Let us show that the Cauchy problem

$$\begin{cases} x' = 2\sqrt{x}, \\ x(0) = 0, \end{cases} \tag{3.5}$$

has, in addition to the trivial solution $x(t) \equiv 0$, also infinitely many nontrivial solutions. By direct inspection, we see that one of these solutions is given by

$$x(t) = t^2, \quad t \geq 0,$$

because $\frac{d}{dt}(t^2) = 2t = 2\sqrt{t^2} = 2|t| = 2t$ for $t \geq 0$. Note that the decreasing branch of the parabola $x = t^2$ is not a solution since for $t < 0$ one has $2\sqrt{t^2} = 2|t| = -2t$. This may also be deduced from the fact that the solutions cannot be decreasing, for $x' = 2\sqrt{x} \geq 0$.
Finally, $\forall a > 0$ the functions

$$x_a(t) = \begin{cases} 0, & \text{for } 0 \leq t \leq a, \\ (t-a)^2, & \text{for } t \geq a, \end{cases}$$

are solutions; see Fig. 3.4. First of all, they satisfy the initial condition $x_a(0) = 0$. Moreover,

$$\frac{d}{dt}x_a(t) = \begin{cases} 0, & \text{for } 0 \leq t \leq a \\ 2(x-a), & \text{for } t \geq a \end{cases} = 2\sqrt{x_a(t)}.$$

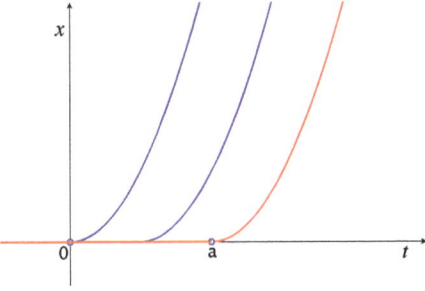

Figure 3.4: Plot of some solutions to $x' = \sqrt{x}$, $x(0) = 0$; in red the solution $x_a(t)$.

This drastic lack of uniqueness is due to the fact that the function $f(x) = 2\sqrt{|x|}$ is not differentiable at $x = 0$ (f_x has a jump discontinuity); hence the uniqueness theorem does not apply.

On the other hand, since $f(x) = 2\sqrt{x}$ is differentiable in the half open plane $x > 0$, the uniqueness proved in Theorem 3.1 applies therein and the ivp $x' = 2\sqrt{x}$, $x(0) = x_0 > 0$, has a unique solution. See Exercise n. 8. □

3.2.2 Local vs. global existence

As mentioned before, Theorem 3.1 is local. The following *global* result holds, provided the set A where f is defined is a strip, namely a set as $I \times \mathbb{R}$ where $I \subseteq \mathbb{R}$ is an interval, and f_x is bounded w. r. t. x.

Theorem 3.2 (Global existence and uniqueness). *Let f satisfy the assumptions of Theorem 3.1. In addition, we assume:*
(i) *A is a strip: $A = I \times \mathbb{R}$, $I \subseteq \mathbb{R}$ an interval, and*
(ii) *for every bounded closed interval $[a, b] \subset I$, there exists $C > 0$ such that*

$$|f_x(t, x)| \le C, \quad \forall (t, x) \in [a, b] \times \mathbb{R}. \tag{3.6}$$

Then the Cauchy problem

$$\begin{cases} x' = f(t, x), \\ x(t_0) = x_0, \end{cases}$$

has one and only one solution defined on all of I.

The proof will be given later in the appendix.

Compared with the local existence and uniqueness Theorem 3.1, the new feature of Theorem 3.2 is that now the solution $x(t)$ to $x' = f(t, x)$ is defined on the whole interval I.

Remark 3.3. In Example 3.1 one has $A = \mathbb{R} \times \mathbb{R}$ and $f = x^2$. Then $f_x = 2x$ and assumption (3.6) is not satisfied no matter what $[a, b] \subset \mathbb{R}$ is. This shows that the condition (3.6) cannot be removed.

Example 3.4. If $f(t, x) = -p(t)x + q(t)$, with p, q continuous on an interval $I \subseteq \mathbb{R}$, the equation $x' = f(t, x)$ becomes the linear equation

$$x' = -p(t)x + q(t),$$

discussed in the previous chapter. Taking any bounded closed interval $[a, b] \subset I$, one has

$$\max_{(t,x) \in [a,b] \times \mathbb{R}} |f_x(t, x)| = \max_{t \in [a,b]} |p(t)|.$$

Since p is continuous, $\max_{t \in [a,b]} |p(t)|$ is finite, showing that (3.6) holds. Then Theorem 3.2 applies and yields a solution $x(t)$ defined on all of I. In particular, this argument provides another proof of Theorem 2.2 of Chapter 2. □

3.3 Qualitative properties of solutions

In this section we will use Theorem 3.1 and Theorem 3.2 to study the behavior of the solutions, without solving the equation $x' = f(t, x)$ explicitly. The reader should be aware that quite often it might be rather complicated, and in some cases impossible, to express the solutions of a nonlinear differential equation by means of known functions like those usually studied in Calculus. With the exception of some specific equations discussed in the next chapter, studying the qualitative properties of solutions can be the only way to understand certain qualitative properties of solutions, such as monotonicity, convexity, periodicity, oscillation etc.

To avoid technicalities, we will understand that the assumptions of Theorems 3.1 or 3.2 are fulfilled and that f is defined on $A = \mathbb{R} \times \mathbb{R}$, unless specified otherwise.

Rather than carrying out a general theoretical discussion we prefer to discuss some specific cases from which one can infer the general idea that lies behind.

Consider the equation

$$x' = h(t)g(x). \tag{3.7}$$

Lemma 3.3. If $h(t)$ is not identically zero, then $x(t) = k$ is a constant solution (or equilibrium solution) if and only if $g(k) = 0$.

Proof. If $g(k) = 0$, then $x(t) \equiv k$ is a constant solution of (3.7). Conversely, if $x(t) = k$ is an equilibrium solution, then we have $h(t)g(k) = 0, \forall t \in \mathbb{R}$. Since $h(t) \not\equiv 0$, there exists t_0 such that $h(t_0) \neq 0$. Then $h(t_0)g(k) = 0$ yields $g(k) = 0$. □

By uniqueness, we infer the following.

Corollary 3.1. *The non-constant solutions of* (3.7) *cannot cross the equilibrium solutions.*

For example, if (3.7) has precisely two constant solutions $x = k_1, x = k_2$, with $k_1 < k_2$, the non-constant solutions of (3.7) are trapped within the three horizontal open strips

$$S = \{x < k_1\}, \quad S' = \{k_1 < x < k_2\}, \quad S'' = \{x > k_2\}.$$

We note that Lemma 3.3 and Corollary 3.1 are extensions of results proved for linear equations in Chapter 2 to nonlinear equations.

Next, let $h(t) \equiv 1$, $(t_0, x_0) \in \mathbb{R}^2$, and let $x(t)$ denote the solution of the ivp

$$\begin{cases} x' = g(x), \\ x(t_0) = x_0. \end{cases} \tag{3.8}$$

Lemma 3.4. *Solutions of* (3.8) *do not have maxima or minima.*
(i)   *if* $g(x_0) = 0$ *then* $x(t) = x_0$ *is an equilibrium,*
(ii) *if* $g(x_0) > 0$ *then* $x(t)$ *is increasing,*
(iii) *if* $g(x_0) < 0$ *then* $x(t)$ *is decreasing.*

Proof. (i) follows directly from Lemma 3.3.
(ii) If $g(x_0) > 0$ then $x'(t_0) = g(x(t_0)) = g(x_0) > 0$. Moreover, x' does not change sign. To prove this claim, we argue by contradiction: if t_1 is such that $x'(t_1) = 0$ then $g(x(t_1)) = 0$. Setting $x_1 = x(t_1)$ one has $g(x_1) = 0$, namely x_1 is a constant solution. Thus $x(t)$ crosses an equilibrium, contradicting Corollary 3.1.
(iii) The argument is quite similar to (ii). ☐

If, as in Lemma 3.4, $x(t)$ is increasing or decreasing, we know from Calculus that $x(t)$ has limits $L_\pm$ (finite or infinite) as $t \to \pm\infty$.

Lemma 3.5. *The finite limits are zeros of* g.

Proof. If $L_+ < +\infty$ then it is easy to see that $x'(t) \to 0$ as $t \to +\infty$. Passing to the limit in the identity $x'(t) = g(x(t))$ we find

$$0 = \lim_{t \to +\infty} g(x(t)) = \lim_{x \to L_+} g(x) = g(L_+).$$

A similar argument shows that $g(L_-) = 0$ provided $L_- > -\infty$. ☐

Example 3.5. Let us study the behavior of the solutions to $x' = 1 - x^3$.
The only constant solution is $x = 1$, the unique zero of $g(x) = 1 - x^3$. All the other solutions do not cross this line. Since $g(x) = 1 - x^3 > 0$ if and only if $x < 1$, it follows that the solutions such that $x(t) < 1$ are increasing, whereas the solutions such that $x(t) > 1$ are decreasing. Moreover, noticing that $\lim_{t \to +\infty} x(t)$ is finite, we can use Lemma 3.5 to infer that $\lim_{t \to +\infty} x(t) = 1$. ☐

We can also study the convexity of solutions. Indeed, if g is differentiable, then $x(t)$ is twice differentiable and, taking into account that $x' = g(x)$, one infers

$$x''(t) = \frac{d}{dt}x'(t) = \frac{d}{dt}g(x(t)) = g'(x(t))x'(t) = g'(x(t))g(x(t)).$$

Hence the convexity of x depends on the sign of $g'(x)g(x)$. For example, in the case of the previous equation $x' = 1 - x^3$ we have $g'(x)g(x) = -3x^2 \cdot (1 - x^3)$ which implies $g'(x)g(x) > 0$ if $x > 1$ and $g'(x)g(x) < 0$ if $x < 1$. Thus the solutions greater than 1 are convex, whereas the solutions smaller than 1 are concave, as shown in Fig. 3.5.

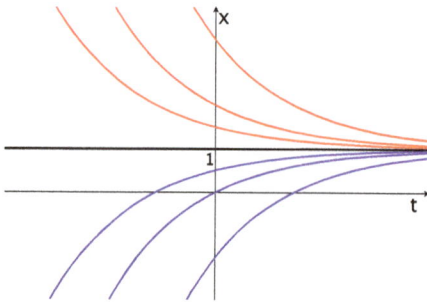

Figure 3.5: Behavior of solutions to $x' = 1 - x^3$.

Example 3.6. Let $g : \mathbb{R} \to \mathbb{R}$ be the smooth function plotted in Fig. 3.6 and let $x(t)$ denote the solution of the ivp

$$\begin{cases} x' = g(x), \\ x(t_0) = x_0, \end{cases}$$

where $g(a) = g(b) = 0$ and $a < x_0 < b$. Notice that $x = a, x = b$ are constant solutions. According to the preceding discussion $a < x(t) < b$, $x(t)$ is increasing, and

$$\lim_{t \to -\infty} x(t) = a, \qquad \lim_{t \to +\infty} x(t) = b.$$

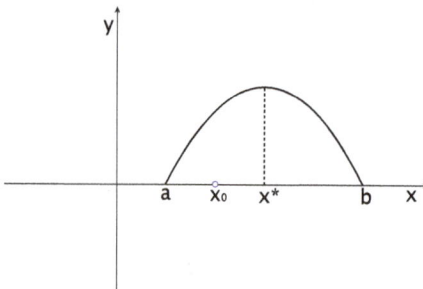

Figure 3.6: Plot of $y = g(x)$, $a \leq x \leq b$.

Furthermore, from $x'' = g'(x)x' = g'(x)g(x)$ and the fact that $g(x) > 0$ for $a < x < b$, we infer that $x'' > 0$ if and only if $g' > 0$. Thus, letting t^* be such that $x(t^*) = x^*$ (notice that t^* is unique because x is increasing), we find that $x''(t) > 0$ (hence x is convex) for $t < t^*$ and $x''(t) < 0$ (hence x is concave) for $t > t^*$. Notice that the slope of x at $t = t^*$ depends on $g(x^*)$ because $x'(t^*) = g(x(t^*)) = g(x^*)$. The approximated graph of $x(t)$ is plotted in Fig. 3.7. □

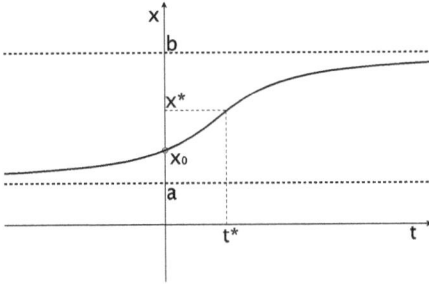

Figure 3.7: Solution of $x' = g(x), x(0) = x_0$.

In the case of autonomous equations we have seen that non-constant solutions are monotonic and thus cannot have maxima or minima. On the contrary, solutions of equations such as

$$x' = f(t, x) \tag{3.9}$$

where f depends explicitly on t, might oscillate.

For brevity, let us focus only on possible extrema of a solution $x(t)$.

If t_0 is a minimum or maximum of $x(t)$, then $x'(t_0) = 0$ and hence $f(t_0, x(t_0)) = 0$. Thus the maxima or minima belong to the curve defined implicitly by $f(t, x) = 0$. Moreover,

$$x''(t) = \frac{d}{dt} f(t, (x(t))) = f_t(t, (x(t))) + f_x(t, (x(t))) \cdot x'(t).$$

In particular, at t_0 one has

$$x''(t_0) = f_t(t_0, (x(t_0))) + f_x(t, (x(t_0))) \cdot x'(t_0) = f_t(t_0, (x(t_0))),$$

because $x'(t_0) = 0$. Therefore from the sign of $f_t(t_0, (x(t_0)))$ we may deduce whether $x(t)$ has a minimum or a maximum at $t = t_0$.

Example 3.7. Let $x(t)$ be any solution of the equation $x' = x^2 - t^2$. The locus of maxima/minima is $x^2 - t^2 = 0$, namely $x = \pm t$. Since $f_t(t, x) = -2t$ we infer that the points

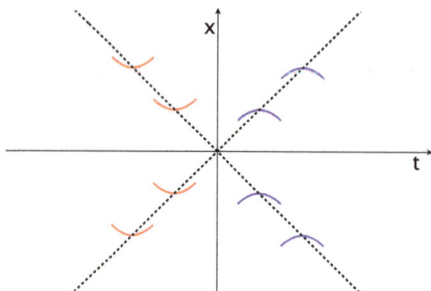

Figure 3.8: Plot of solutions to $x' = x^2 - t^2$ near the points of maxima and minima.

(t, x) such that $x = \pm t$, are points where $x(t)$ has minima if $t < 0$, while they are points where $x(t)$ has maxima if $t > 0$; see Fig. 3.8.

Moreover, to see where solutions are increasing or decreasing we observe that $x' = x^2 - t^2 > 0$ if and only if $x(t) \in \{x > |t|\} \cup \{x < |t|\}$. It follows that the solutions have a behavior similar to that indicated in Fig. 3.9. An exception is the solution through $(0, 0)$ which is always decreasing and has horizontal tangent at $t = 0$. $\square$

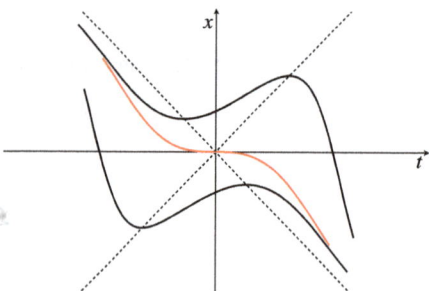

Figure 3.9: Plot of solutions to $x' = x^2 - t^2$.

3.4 Appendix

This appendix contains the proofs of the main existence and uniqueness theorems, stated earlier, and some additional topics related to the general first order equations.

A1: Proof of Theorem 3.1. Let us restate it as follows: *suppose that f is continuous in $A \subseteq \mathbb{R}^2$ and has continuous partial derivative f_x with respect to x. Given (t_0, x_0) in the interior of A, there exists $\delta > 0$ such that the Cauchy problem*

$$\begin{cases} x' = f(t, x), \\ x(t_0) = x_0. \end{cases} \tag{3.2}$$

has one and only one solution defined in $[t_0 - \delta, t_0 + \delta]$.

Existence. In order to use Lemma 3.1, we will show that the sequence

$$x_{k+1}(t) = x_0 + \int_{t_0}^{t} f(s, x_k(s))\, ds$$

converges uniformly in a suitable neighborhood of t_0.

Choose $r > 0$ such that the fixed closed square $K = K_r = [t_0 - r, t_0 + r] \times [x_0 - r, x_0 + r]$ is contained in A. We let:

- M denote the maximum of $|f(t, x)|$ for $(t, x) \in K$;
- C denote the maximum of $|f_x(t, x)|$ for $(t, x) \in K$. Notice that $C < \infty$ because f_x is continuous by assumption. Moreover, for any $(t, x), (t, y) \in K$ the mean value theorem implies that $f(t.x) - f(t, y) = f_x(t, \tilde{x})(x - y)$, for some $\tilde{x} \in (x, y)$, yielding

$$|f(t.x) - f(t, y)| \le \max_{K} f_x(t.x) \cdot |x - y| \le C \cdot |x - y|. \tag{A1}$$

We choose δ to be a number such that

$$0 < \delta < \min\left[1, \frac{1}{C}, \frac{r}{M}\right]$$

which implies the inequalities

$$\delta < 1, \quad C\delta < 1, \quad M\delta < r.$$

Set $T = T_\delta = [x_0 - \delta, x_0 + \delta]$.

Step 1. We wish to show that $\max_T |x_k(t) - x_0| < r$ for all integers $k \ge 1$. This will imply that $(t, x_k(t)) \in K \subset A$ for each $k \ge 1$, so that $f(t, x_k(t))$ is well defined.

One has

$$|x_1(t) - x_0| \le \left| \int_{t_0}^{t} |f(s, x_0)|\, ds \right| \le M\delta \Rightarrow \max_{T} |x_1(t) - x_0| \le M\delta < r, \tag{A2}$$

since $M\delta < r$. Similarly,

$$|x_2(t) - x_0| \le \left| \int_{t_0}^{t} |f(s, x_1(s))|\, ds \right|.$$

Since $(s, x_1(s)) \in K$ by (A2), we have $|f(s, x_1(s))| \le M$ and we deduce

$$|x_2(t) - x_0| \le M\delta < r \Rightarrow \max_{T} |x_2(t) - x_0| \le M\delta < r.$$

It is now clear that the general statement follows from mathematical induction.

Step 2. Next, we show that

$$\max_T |x_k(t) - x_{k-1}(t)| \le \frac{M}{C} \cdot (C\delta)^k, \quad \forall k \ge 1.$$

We note that

$$|x_2(t) - x_1(t)| = \left| \int_{t_0}^t [f(s, x_1(s)) - f(s, x_0)]\, ds \right| \le \left| \int_{t_0}^t |f(s, x_1(s)) - f(s, x_0)|\, ds \right|.$$

Using (A1) and (A2) we infer

$$|x_2(t) - x_1(t)| \le C \left| \int_{t_0}^t |x_1(s) - x_0|\, ds \right| \le C\delta \cdot M\delta = MC\delta^2.$$

Since the right hand side is independent of t, we deduce

$$\max_T |x_2(t) - x_1(t)| \le MC\delta^2 \le \frac{M}{C} \cdot (C\delta)^2.$$

Similarly,

$$|x_3(t) - x_2(t)| \le C \left| \int_{t_0}^t |x_2(s) - x_1(s)|\, ds \right| \le C\delta \cdot MC\delta^2 \Rightarrow \max_T |x_3(t) - x_2(t)| \le MC^2\delta^3 = \frac{M}{C} \cdot (C\delta)^3.$$

In general, one can show (for example, by induction) that

$$\max_T |x_k(t) - x_{k-1}(t)| \le \frac{M}{C} \cdot (C\delta)^k, \quad \forall k \ge 1.$$

Since $C\delta < 1$, we have $(C\delta)^k \to 0$ as $k \to +\infty$. It follows that $\max_T |x_k(t) - x_{k-1}(t)| \to 0$ as $k \to +\infty$, and this proves that $x_k(t)$ converges uniformly on T to some continuous function $x(t)$. Taking the limit in

$$x_{k+1}(t) = x_0 + \int_{t_0}^t f(s, x_k(s))\, ds$$

we obtain $x(t) = x_0 + \int_{t_0}^t f(s, x(s))\, ds$. According to Lemma 3.1, this is equivalent to saying that $x(t)$ solves the ivp $x' = f(t,x)$, $x(t_0) = x_0$ on T, completing the proof of existence.

Uniqueness. Let $y(t), z(t), t \in T$, be two solutions satisfying $x' = f(t,x)$, $x(t_0) = x_0$. One has

$$|y(t) - z(t)| = \left|\int_{t_0}^{t} |f(s,y(s)) - f(s,z(s))| \, ds\right|.$$

Taking $\delta > 0$ sufficiently small, we can assume that $(t, y(t)) \in K$ and $(t, z(t)) \in K$. Then once again we can use (A1) to infer

$$\left|\int_{t_0}^{t} |f(s,y(s)) - f(s,z(s))| \, ds\right| \le C \left|\int_{t_0}^{t} |y(s) - z(s)| \, ds\right| \le C\delta \max_T |y(t) - z(t)|,$$

yielding

$$|y(t) - z(t)| \le C\delta \max_T |y(t) - z(t)|.$$

Since the right hand side in independent of t, we can take the maximum of the left hand side yielding

$$\max_T |y(t) - z(t)| \le C\delta \max_T |y(t) - z(t)|.$$

Since $C\delta < 1$ we deduce that $\max_T |y(t) - z(t)| = 0$, namely $y(t) = z(t)$ for all $t \in T$. ☐

Remark A1. The preceding proof makes it clear that both the existence and the uniqueness assertions would follow if, instead of requiring $f(t, x)$ to have a continuous partial derivative, we assume that f is continuous on A and satisfies following condition: there exist a neighborhood U of (t_0, x_0) and $C > 0$ such that

$$|f(t, x) - f(t, y)| \le C \cdot |x - y|, \quad \forall (t, x), (t, y) \in U.$$

Such functions are called (locally) *Lipschitzian*. Notice that the mean value theorem guarantees that every function with continuous partial derivative f_x is locally Lipschitzian.

On the other hand, there are Lipschitzian functions that are not differentiable: for example, $f(x) = |x|$ is obviously Lipschitzian, since $|f(x) \ f(y)| = ||x| \ |y|| \le |x \ y|$, but it is not differentiable at $x = 0$.

A2: Gluing solutions. According to Theorem 3.1 a solution of the ivp

$$x' = f(t, x), x(t_0) = x_0$$

is defined in a neighborhood $[t_0 - \delta, t_0 + \delta]$ of t_0, for some $\delta > 0$. If the point $(t_0 + \delta, x(t_0 + \delta))$ belongs to A (the set where f is defined), we can apply Theorem 3.1 again and find a solution $x_1(t)$, defined on $[t_0 + \delta, t_0 + \delta_1]$, for some $\delta_1 > \delta$. The two solutions x and x_1 glue together and give rise to the function defined on $[t_0, t_0 + \delta_1]$ by setting

$$\tilde{x}(t) = \begin{cases} x(t) & t \in [t_0, t_0 + \delta], \\ x_1(t) & t \in [t_0 + \delta, t_0 + \delta_1]. \end{cases}$$

Since $x(t_0 + \delta) = x_1(t_0 + \delta)$, $\tilde{x}$ is continuous. Moreover, $\tilde{x}$ is differentiable. This is obvious for $t \neq t_0 + \delta$. At $t = t_0 + \delta$, the left derivative of $\tilde{x}$ is given by

$$\tilde{x}'^- = \lim_{t \uparrow t_0 + \delta} x'(t) = \lim_{t \uparrow t_0 + \delta} f(x(t)) = f(x(t_0 + \delta))$$

and the right derivative is given by

$$\tilde{x}'^+ = \lim_{t \downarrow t_0 + \delta} x_1'(t) = \lim_{t \downarrow t_0 + \delta} f(x_1(t)) = f(x_1(t_0 + \delta)).$$

Since $x(t_0 + \delta) = x_1(t_0 + \delta)$ it follows that $\tilde{x}'^- = \tilde{x}'^+$.

Iterating the procedure, after k steps we find $\delta_k > 0$ such that the solution $\tilde{x}_k$ is defined on $[t_0, t_0 + \delta_k]$. If $\delta_k \to +\infty$ we will find a solution defined on $[t_0, +\infty)$. On the other hand, the numerical sequence δ_k might converge to a finite δ^*, either because $\tilde{x}_k(t_0 + \delta_k)$ approaches the boundary ∂A, or because $\tilde{x}_k(t_0 + \delta_k) \to +\infty$: the latter is indeed the case arising in Example 3.1. In a quite similar way we can possibly extend the solution $x(t)$ for $t < t_0$.

The largest interval I (containing t_0) where the solution $x(t)$ of an ivp can be defined is called the *maximal interval of definition* of $x(t)$.

We claim that the maximal interval cannot be closed. For, let α, resp. β, be the lower bound, resp. the upper bound, of the maximal interval of definition I. We claim that *if $\alpha > -\infty$, resp. $\beta < +\infty$, then $\alpha \notin I$, resp. $\beta \notin I$.* Otherwise, let, e. g., $\beta < +\infty$ belong to I and consider the ivp

$$y' = f(t, y), \quad y(\beta) = x(\beta),$$

which has a solution $y(t)$ defined on $[\beta, \beta + \epsilon]$, for some $\epsilon > 0$. Gluing x and y we find a solution to the preceding ivp which is defined on $[t_0, \beta + \epsilon]$, a contradiction, for I is maximal, proving the claim.

Lemma A1. *Let $f(t, x)$ be continuous and have a continuous partial derivative f_x on $A = \mathbb{R}^2$. If the solution $x(t)$ to the ivp*

$$x' = f(t, y), \quad x(t_0) = x_0$$

is bounded and monotone, then its maximal interval of definition I is all of $\mathbb{R}$.

Proof. Otherwise, let I have a finite upper bound β (the case of finite lower bound is quite similar). Since x is bounded and monotone it has finite limit $x(\beta) = \lim_{t \uparrow \beta} x(t)$. Then $I = [t_0, \beta]$; hence I is closed, contradicting the preceding claim. $\square$

A3: Proof of Theorem 3.2. Let us recall that f is defined on a strip $A = I \times \mathbb{R}, I \subseteq \mathbb{R}$ an interval, and satisfies the assumptions of Theorem 3.1. Furthermore,

$$\exists C > 0 : |f_x(t,x)| \le C, \quad \forall (t,x) \in [a,b] \times \mathbb{R},$$

for any interval $[a,b] \subset I$. Then we wish to show that the ivp (3.2) has one and only one solution, defined on all of I.

We can repeat the arguments carried out in the proof of Theorem 3.1 yielding a solution $x(t)$ defined on the interval $T = [t_0 - \delta, t_0 + \delta] \subseteq [a,b]$. Notice that here Step 1 is unnecessary since f is defined on the strip $A = I \times \mathbb{R}$ and therefore $\delta > 0$ can be chosen by requiring merely that $\delta < 1$ and $C\delta < 1$. So, if T is a proper subset of $[a,b]$, we repeat the above arguments for the ivp

$$y' = f(t,y), \quad y(t \pm \delta) = x(t_0 \pm \delta).$$

We can iterate the procedure, gluing the solutions as in A2. After a finite number of steps we get a solution defined on all of $[a,b]$. In the case when I is an unbounded interval, we repeat the procedure on any $[a,b] \subset I$. Since any point in I can be contained in an interval $[a,b] \subset I$, and we have shown the existence on arbitrary intervals $[a,b]$, it follows that our existence result holds on all of I. $\qquad\square$

A4: A general existence theorem. In order to complete the theoretical study, we state, without proof, a general result which provides the existence (but not the uniqueness) to a Cauchy problem assuming merely the continuity of f.

Theorem A1 (Peano–Cauchy). *Suppose that f is continuous in $A \subseteq \mathbb{R}^2$. Given (t_0, x_0) in the interior of A, there exists $\delta > 0$ such that the Cauchy problem (3.2) has at least one solution defined in $[t_0 - \delta, t_0 + \delta]$.*

A5: A comparison theorem. In some cases it might be useful to compare the solutions of two different equations. Let x, y be the solutions of the Cauchy problems

$$\begin{cases} x' = f(t,x), \\ x(t_0) = x_0, \end{cases}$$

$$\begin{cases} y' = g(t,y), \\ y(t_0) = y_0. \end{cases}$$

For the sake of simplicity we will assume that $f, g \in C^1(\mathbb{R}^2)$ and that $x(t), y(t)$ are defined on $\mathbb{R}$.

Theorem A2. *If $f(t,x) < g(t,x)$ and $x_0 < y_0$, then $x(t) < y(t)$ for all $t \ge t_0$.*

Proof. We shall show that the set $T = \{t \ge t_0 : x(t) \ge y(t)\}$ is the empty set. Otherwise, let ℓ be its greatest lower bound. Since $x(t_0) = x_0 < y_0 = y(t_0)$, the sign permanence theorem implies that $\ell > t_0$; see Fig. 3.10.

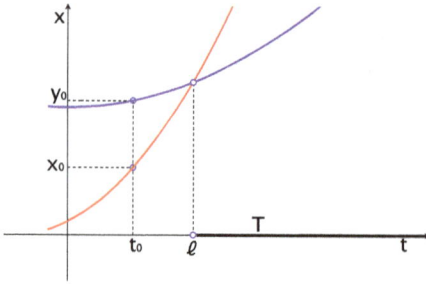

Figure 3.10: Plot of $x(t)$ (in red), $y(t)$ (in blue) and the set T.

Moreover, it is easy to see that $x(\ell) = y(\ell)$. This and the assumption $f(t,x) < g(t,x)$ yield

$$x'(\ell) = f(\ell, x(\ell)) < g(\ell, x(\ell)) = g(\ell, y(\ell)) = y'(\ell).$$

Let $z(t) = x(t) - y(t)$. Then $z(\ell) = 0$ and $z'(\ell) < 0$. It follows that for $h > 0$ small enough one has $z(\ell - h) > 0$, namely $x(\ell - h) > y(\ell - h)$. Thus $\ell - h \in T$ for some $h > 0$, a contradiction, for ℓ is the greatest lower bound of T. □

Theorem A2 can be employed to establish the behavior of solutions of a differential equation. For example, let $y(t)$ be given such that: (i) $y' = f(t,y)$, (ii) it is defined on $[0, +\infty)$, (iii) $y(t) > 0$ for all $t \geq 0$ and (iv) $\lim_{t \to +\infty} y(t) = 0$. Consider the solution $x(t)$ to the ivp

$$x' = g(t,x), \quad x(0) = x_0 > 0$$

and suppose that it is also positive and defined on $[0, +\infty)$. If $g(t,x) < f(t,x)$ and $0 < x_0 < y(0)$, then Theorem A2 implies that $0 < x(t) < y(t)$ for all $t \geq 0$. Then, using the squeeze theorem, we infer that $x(t)$ decays to 0 as $t \to +\infty$; see Fig. 3.11.

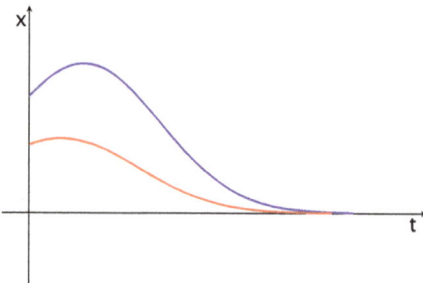

Figure 3.11: Plot of $x(t)$ (red) and $y(t)$ (blue).

3.5 Exercises

1. Verify that the local existence and uniqueness theorem applies to the ivp $x' = \sqrt{t} + e^x$, $x(0) = 0$.

2. Verify that for any $(t_0, x_0) \in \mathbb{R}^2$ the local existence and uniqueness theorem applies to the ivp $x' = \ln(1 + x^4)$, $x(t_0) = x_0$.

3. Find the first 3 terms of the Picard approximation scheme for $x' = tx$, $x(0) = 1$.

4. Find the domain of definition of the solution to the ivp $x' = x^2$, $x(0) = a$.

5. Verify that $x(t) = (t + \sqrt{a})^2$ for $t \geq -\sqrt{a}$ and $x(t) = 0$ for $t < -\sqrt{a}$ is the unique solution to the ivp $x' = 2\sqrt{x}$, $x(0) = a > 0$.

6. Show that the Cauchy problem $x' = x^{1/3}$, $x(0) = 0$, does not have a unique solution.

7. Show that the solutions of $x' = \cos^2 x$ are defined for all $t \in \mathbb{R}$.

8. Show that the solutions of $x' = \sqrt{1 + x^2}$ are defined on all $t \in \mathbb{R}$.

9. Show that the solution $x(t)$ of the Cauchy problem $x' = 1 + x^2$, $x(0) = 0$, cannot vanish for $t > 0$.

10. Show that the solution of the Cauchy problem $x' = \sin x$, $x(0) = 1$, is such that $0 < x(t) < \pi$ and is increasing.

11. Prove that if $f(x)$ is odd and $x(t)$ is a solution of $x' = f(x)$, then $-x(t)$ is also a solution.

12. Let $f(x)$ be an even $C^1(\mathbb{R})$ function and let $x_0(t)$ be a solution of $x' = f(x)$ such that $x(0) = 0$. Prove that $x_0(t)$ is an odd function.

13. If f is an odd function, show that the solutions of $x' = f(tx)$ are even.

14. Solve $x' = |x| - 1$, $x(0) = 0$.

15. Show that for any $x_0 \in \mathbb{R}$ the solutions to $x' = x^3 - 1$, $x(0) = x_0$, tend to 1 as $t \to -\infty$.

16. Find the $\lim_{t \to +\infty} x(t)$, where $x(t)$ is the solution of the Cauchy problem $x' = 1 - e^x$, $x(0) = 1$.

17. Find $\lim_{t \to +\infty} x(t)$, where $x(t)$ is the solution of the Cauchy problem $x' = 1 - \arctan x$, $x(0) = 0$.

18. Let $x(t)$ be the solution of the Cauchy problem $x' = \ln(1 + x^2)$, $x(0) = 0$. Show that $\lim_{t \to +\infty} x(t) = +\infty$ and $\lim_{t \to -\infty} x(t) = -\infty$.

19. Study the qualitative behavior of the solution to $x' = 3x - x^2$, $x(0) = 1$.

20. Study the convexity of the solutions to $x' = x^3 - 1$.

21. Study the convexity of the solutions to $x' = x(2 - x)$.

22. Study the convexity of the solutions to $x' = x(1 + x)$.

23. Show that the solution to $x' = x^2 - t^3$, $x(1) = 1$, has a strict maximum at $t = 1$.

24. Find the locus of maxima of $x' = x^3 - t$.

25. Show that the solutions to $x' = e^x - t - t^5$ cannot have minima.

26. Show that the solution of $x' = x^4$, $x(0) = x_0 \neq 0$, is strictly convex if $x_0 > 0$ and strictly concave if $x_0 < 0$. Extend the result to $x' = x^p$, $x(0) = x_0 \neq 0$, $p \geq 1$.

27. Show that for any x_0 the solution of the ivp $x' = 1 + 2t + \sin^2 x$, $x(0) = x_0$, is such that $\lim_{t \to +\infty} x(t) = +\infty$.

28. Let $x(t)$ be the solution to $x' = f(t, x)$, $x(0) = x_0 > 0$, where f is smooth for $(t, x) \in \mathbb{R}^2$. If $x(t)$ is defined on $[0, +\infty)$ and $f(t, x) > x$, show that $\lim_{t \to +\infty} x(t) = +\infty$.

29. Let $x(t)$ be the solution of $x' = -1 - t + h(x)$, $x(0) = a$, with $0 < a < 2$, where h is a smooth function such that $h(x) \leq -x$. If $x(t)$ is defined for all $t \geq 0$, show that the equation $x(t) = 0$ has at least one solution in $(0, 1)$.

30. Find a function $f(x)$ such that

$$x(t) = \begin{cases} e^t & \text{if } t \geq 0, \\ t + 1 & \text{if } t \leq 0, \end{cases}$$

is the solution of the Cauchy problem $x' = f(x)$, $x(0) = 1$.

4 Solving and analyzing some nonlinear first order equations

In this chapter we discuss methods of solving several classes of some of the basic first order nonlinear differential equations.

4.1 Separable equations

In theory, these are among the simplest nonlinear equations to solve. *Separable equations* are equations that can be put in the form

$$x' = h(t)g(x) \tag{4.1}$$

where, for the sake of simplicity of notations, we assume that $h \in C(\mathbb{R})$ (the case $h \in C([a,b])$ requires obvious modifications), and that $g \in C^1(\mathbb{R})$. We also suppose that $h(t)$ is not identically zero, otherwise the equation becomes trivial.

Given any $(t_0, x_0) \in \mathbb{R}^2$, the Cauchy problem associated with (4.1) is given by

$$\begin{cases} x' = h(t)g(x), \\ x(t_0) = x_0. \end{cases} \tag{4.2}$$

Under the previous assumptions on h and g, the local existence and uniqueness Theorem 3.1 in Chapter 3 applies and yields, locally, a unique solution to (4.2).

Essentially, to solve (4.1) one simply separates the variables and then integrates. Of course, the integration may be quite challenging. Also, solving for $x(t)$ explicitly may be difficult, or even impossible; in such cases, we may let it be defined implicitly by an algebraic equation.

Recall that if $g(x_0) = 0$ then $x(t) \equiv x_0$ is a constant solution of (4.2). For example, in the equation $x'(t) = e^t \sin x$, $x(t) \equiv \pi$ is a constant solution. Similarly, since $\sin 2\pi = 0$, $x(t) \equiv 2\pi$ is also a solution. Thus it has infinitely many constant solutions.

Consider a non-constant solution $x(t)$. According to Corollary 3.1 of Chapter 3, such solutions never cross the constant ones and this implies that $g(x(t))$ never vanishes. So we can divide the identity $x'(t) = h(t)g(x(t))$ by $g(x(t)) \neq 0$ to obtain

$$\frac{x'(t)}{g(x(t))} = h(t).$$

Integrating both sides with respect to t, one has

$$\int \frac{x'(t)}{g(x(t))}\, dt = \int h(t)\, dt + c.$$

Changing variable on the left hand side and noting that $dx = x'dt$, we get

https://doi.org/10.1515/9783111185675-004

$$\int \frac{dx}{g(x)} = \int h(t)\, dt + c. \tag{4.3}$$

Thus, if G, H denote antiderivatives of $1/g$ and h, respectively, (4.3) implies

$$G(x) = H(t) + c,$$

which provides the solutions of (4.1) in an implicit form.

Example 4.1. Solve

$$x' = \frac{dx}{dt} = \frac{8t^3 - 2t + 1}{6x^2 + 1}.$$

There are no constant solutions here. Separating the variables and integrating, we have $\int (6x^2 + 1)\, dx = \int (8t^3 - 2t + 1)\, dt + c$ and hence

$$2x^3 + x = 2t^4 - t^2 + t + c,$$

which defines the solutions implicitly. If we wish to verify the answer, we take the derivative on both sides, obtaining $(6x^2 + 1)x' = 8t^3 - 2t + 1$, and then solve for x'. Alternatively, we can take the differentials on both sides of $2x^3 + x = 2t^4 - t^2 + t + c$, obtaining $(6x^2 + 1)\, dx = (8t^3 - 2t + 1)\, dt$ and

$$\frac{dx}{dt} = \frac{8t^3 - 2t + 1}{6x^2 + 1}.$$

As a final remark, we can observe that the function $G(x) = 2x^3 + x$ is globally invertible, for $G'(x) = 6x^2 + 1 > 0$, and this implies that the domain of definition of the solution $x(t)$ is all of $\mathbb{R}$. The same conclusion could have been deduced by the general Theorem 3.2 of Chapter 3. □

To find the solution to (4.2) we simply substitute the initial value $x(t_0) = x_0$ into (4.3) and solve for c.

We could also use the definite integral and integrate the identity $\frac{x'(t)}{g(x(t))} = h(t)$ from t_0 to t, which would yield

$$\int_{t_0}^{t} \frac{x'(t)}{g(x(t))}\, dt = \int_{t_0}^{t} h(t)\, dt \quad \Longrightarrow \quad \int_{x_0}^{x} \frac{dx}{g(x)} = \int_{t_0}^{t} h(t)\, dt.$$

Example 4.2. Solve the ivp $x' = tx^3$, $x(0) = 1$. There are no constant solutions other than $x \equiv 0$ and the remaining solutions do not change sign. Thus the solution we are looking for is positive because $x(0) = 1$. Separating the variables and integrating, we have

$$\int\limits_1^x \frac{dx}{x^3} = \int\limits_0^t t\,dt,$$

which yields

$$-\frac{1}{2x^2} + \frac{1}{2} = \frac{1}{2}t^2.$$

Solving for x we get $x^2 = \frac{1}{1-t^2}$. Since $x > 0$ we finally find $x = \frac{1}{\sqrt{1-t^2}}$. Notice that in the present case the maximal interval of definition is given by $(-1, 1)$. □

Consider the separable differential equation

$$x' = \frac{M(t)}{N(x)}, \quad N(x) \neq 0.$$

It is sometimes convenient to write separable equations using differentials and replacing t and x by x and y, respectively. This can be accomplished by first replacing t and x by x and y, respectively, and rewriting the equation as $y' = \frac{M(x)}{N(y)}$. Then recalling that the differential of a function $y(x)$ is given by $y'(x)\,dx$, we have

$$dy = y'(x)\,dx = \frac{M(x)}{N(y)}\,dx$$

whereby

$$N(y)\,dy = M(x)\,dx.$$

An advantage of dealing with $N\,dy = M\,dx$ is that we do not need to take care of the case when M or N vanishes.

The equation $N(y)\,dy = M(x)\,dx$ is easily solved by simply integrating both sides to obtain $\int N(y)\,dy = \int M(x)\,dx + c$, c a constant.

For example, integrating the equation $y\,dy + x\,dx = 0$ we find $\int y\,dy + \int x\,dx = c$ yielding $\frac{1}{2}y^2 + \frac{1}{2}x^2 = c$, which for $c > 0$ is a family of circles centered at $(0, 0)$ with radius $\sqrt{2c}$.

Example 4.3. Consider the ivp

$$x' = t\left(\frac{x^3 + 1}{x^2}\right), \quad x(0) = 1.$$

First we note that $x = -1$ is a constant solution since $g(x) = \frac{x^3+1}{x^2} = 0$ at $x = -1$. For $x \neq -1$, we write it in the form

$$\left(\frac{x^2}{x^3 + 1}\right) dx = t\,dt.$$

Integrating, we obtain

$$\frac{1}{3}\ln|x^3 + 1| = \frac{1}{2}t^2 + C.$$

Substituting the initial values $t = 0$, $x = 1$ and solving for C, we obtain $C = \frac{1}{3}\ln 2$, which yields the desired solution

$$\frac{1}{3}\ln|x^3 + 1| = \frac{1}{2}t^2 + \frac{1}{3}\ln 2$$

in implicit form, which can be further simplified as

$$\frac{x^3 + 1}{2} = e^{\frac{3}{2}t^2}.$$

4.2 An application: the logistic equation in population dynamics

As a remarkable example of a separable equation we consider a problem arising in population dynamics.

The malthusian model: the exponential growth

Let us consider a species, whose size is denoted by x, living in a bounded environment. If the species breeds at a rate $\beta > 0$ and dies off at a rate $\delta > 0$, then the growth rate of the species over time t is given by

$$\frac{\Delta x}{\Delta t} = \lambda x, \quad \text{where } \lambda = \beta - \delta,$$

leading to the differential equation

$$x' = \lambda x. \tag{4.4}$$

This model goes back to the Belgian scientist T. R. Malthus.

As is well known, the general solution of (4.4) is given by $x = ce^{\lambda t}$ and the solution such that $x(0) = x_0 > 0$ is

$$x(t) = x_0 e^{\lambda t}.$$

So we find (see Fig. 4.1):
1. if $\beta = \delta$, i. e., if the birth rate equals the death rate, then $\lambda = 0$ and $x(t) \equiv x_0$;
2. if $\beta > \delta$, i. e., if the birth rate is greater than the death rate, then $\lambda > 0$ and $x(t)$ grows exponentially and approaches $+\infty$ as $t \to +\infty$;
3. if $\beta < \delta$, i. e., if the birth rate is smaller than the death rate, then $\lambda < 0$ and $x(t)$ decays exponentially to 0 as $t \to +\infty$.

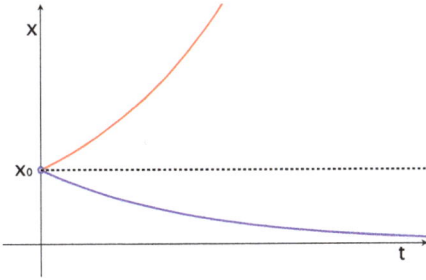

Figure 4.1: Plot of $x(t) = x_0 e^{\lambda t}$ with $\lambda > 0$ in red and $\lambda < 0$ in blue; the dotted line corresponds to $\lambda = 0$.

The Verhulst model: the logistic equation

In a more realistic model, proposed by Verhulst, one assumes that the growth rate λ is not constant but equals the function $a - bx$, $a, b > 0$. In other words, the growth rate λ changes sign depending on x, which is more reasonable than having a constant growth rate. Actually, when the population size x is small the birth rate β will be greater that the death rate δ and hence $\lambda > 0$. On the contrary, if x becomes large, the resources to survive do not suffice any more and this will lead to a negative growth rate $\lambda < 0$. The function $\lambda = a - bx$ is the simplest function with this feature.

The corresponding equation,

$$x' = x(a - bx), \tag{4.5}$$

is called the *logistic equation*. Equation (4.5) is separable, because it is in the form (4.1) with $h(t) = 1$ and $g(x) = x(a - bx)$.

The equilibrium solutions are given by $x(t) \equiv 0$ and $x(t) \equiv a/b > 0$. The other solutions cannot cross these two lines, namely are such that $x(t) \neq 0$ and $x(t) \neq a/b$ for all t. As a consequence, $x = 0$ and $x = \frac{a}{b}$ divide the trajectories into two regions: those that lie above the line $x = \frac{a}{b}$ and those that lie between the lines $x = 0$ and $x = \frac{a}{b}$. More precisely, the solutions above $x = \frac{a}{b}$ are decreasing, because $x(a - bx) < 0$ for $x > \frac{a}{b}$; on the other hand, the solutions between $x = 0$ and $x = \frac{a}{b}$ are increasing, because $x(a - bx) > 0$ for $0 < x < \frac{a}{b}$.

Let us find the general integral of (4.5). Separating the variables and integrating we obtain

$$\int \frac{dx}{x(a - bx)} = \int dt = t + c. \tag{4.6}$$

To evaluate the integral in the left hand side, we use the partial fractions method: we search for constants A, B such that

$$\frac{1}{x(a - bx)} = \frac{A}{x} + \frac{B}{a - bx}.$$

We find

$$A = \frac{1}{a}, \quad B = \frac{b}{a},$$

and hence

$$\int \frac{dx}{x(a - bx)} = \frac{1}{a} \int \frac{dx}{x} + \frac{b}{a} \int \frac{dx}{a - bx} = \frac{1}{a} \ln |x| - \frac{1}{a} \ln |a - bx| = \frac{1}{a} \ln \left| \frac{x}{a - bx} \right|.$$

Substituting into (4.6) we find

$$\frac{1}{a} \ln \left| \frac{x}{a - bx} \right| = t + c \Rightarrow \ln \left| \frac{x}{a - bx} \right| = at + ac \Rightarrow \left| \frac{x}{a - bx} \right| = e^{at + ac} = e^{ac} \cdot e^{at}.$$

Relabeling $C = \pm e^{ac} \in \mathbb{R}$ we obtain

$$\frac{x}{a - bx} = Ce^{at},$$

which is the general integral in implicit form.

To find the integral as $x = x(t)$, we solve for x, yielding $x = Ce^{at} \cdot (a - bx) = aCe^{at} - bCxe^{at}$. With straightforward algebraic calculations we get $x + bCxe^{at} = aCe^{at}$ and finally

$$x = \frac{aCe^{at}}{1 + bCe^{at}}. \tag{4.7}$$

Next, we impose the initial condition $x(0) = x_0 > 0$. Of course, we can take $x_0 \neq \frac{a}{b}$; otherwise $x \equiv \frac{a}{b}$. Inserting $t = 0$ and $x = x_0$ in (4.7) we find

$$x_0 = \frac{aC}{1 + bC},$$

which yields $x_0 + x_0 bC = aC \Rightarrow (a - x_0 b)C = x_0 \Rightarrow C := C_0 = \frac{x_0}{a - x_0 b}$. In conclusion, the solution of the ivp

$$\begin{cases} x' = x(a - bx), \\ x(0) = x_0, \end{cases}$$

is given by

$$x = \frac{aC_0 e^{at}}{1 + bC_0 e^{at}}.$$

The graph for $t \geq 0$ of positive solutions is reported in Fig. 4.2.

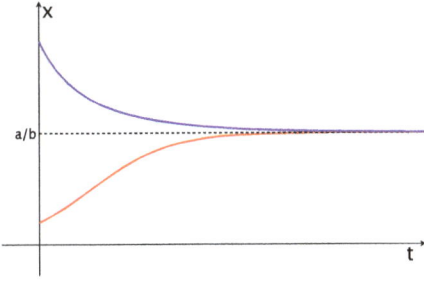

Figure 4.2: Solutions of the logistic equation for $x_0 > \frac{a}{b}$ (blue curve) and $x_0 < \frac{a}{b}$ (red curve).

The previous graph shows that in the Verhulst model there is neither a population explosion nor a population extinction, as in the malthusian model. On the contrary, here one has

$$\lim_{t \to +\infty} x(t) = \frac{a}{b}.$$

Precisely, the population size asymptotically decreases, resp. increases, to the positive equilibrium $x = \frac{a}{b}$, provided $x_0 > \frac{a}{b}$, resp. $0 < x_0 < \frac{a}{b}$.

Example 4.4. (i) Find the solution of the logistic equation $x' = x(3-2x)$ such that $x(0) = 1$. We first separate the variables and write the equation as

$$\frac{dx}{x(3 - 2x)} = dt.$$

Then using partial fractions, we write

$$\frac{\frac{1}{3}}{x} + \frac{\frac{2}{3}}{3 - 2x} = dt.$$

Integrating and simplifying, we find the general solution implicitly given by

$$\frac{x}{3 - 2x} = Ce^{3t}.$$

To find the particular solution, we substitute $t = 0$, $x = 1$ and obtain $C = 1$. Thus

$$\frac{x}{3 - 2x} = e^{3t}$$

whereby $x = (3 - 2x)e^{3t}$. Rearranging, we get $x + 2xe^{3t} = 3e^{3t}$ and finally

$$x = \frac{3e^{3t}}{1 + 2e^{3t}}.$$

Notice that, for $t > 0$, $x(t)$ is increasing with $1 < x(t) < \frac{3}{2}$ and $\lim_{t \to +\infty} x(t) = \frac{3}{2}$.

(ii) Solve the i. v. p. $2x' = x(2 - x)$, $x(0) = 3$.
Separating the variable, we have

$$\frac{2dx}{x(2-x)} = dt$$

which, using partial fractions, can be written as

$$\left[\frac{1}{x} + \frac{1}{2-x}\right] dx = dt.$$

Integrating and utilizing the initial value, we obtain the solution

$$x = \frac{-6e^t}{1 - 3e^t}.$$

Notice that, for $t > 0$, $x(t)$ is decreasing with $2 < x(t) < 3$ and $\lim_{t\to+\infty} x(t) = 2$. □

4.3 Exercises on separable variables

1. Solve $x' = 4t^3 x^5$, $x(0) = 1$.
2. Solve $x' = e^{-2x} e^t$.
3. Solve $x' = \frac{t^2+1}{5x^6}$, $x(0) = 2$.
4. Solve $x' = \frac{t(x+1)}{t+1}$.
5. Solve $(x + 1) dt + (t - 1) dx = 0$.
6. Solve $x' = x(x + 1)$.
7. Solve the logistic equation $x' = x(1 - x)$
   (a) by using the general solution developed in the text,
   (b) by using the separation of variables method directly.
8. Solve $x' = \frac{2t^p+1}{3x^q+1}$, p, q positive numbers.
9. Solve $xx' = \sqrt{x^2 - 1}$.
10. Solve $e^{t+2x} x' = e^{2t-x}$.
11. Solve $2\sin t \sin x + (1 + \cot t)x' = 0$.
12. Solve $\sqrt{1 - t^2}x' = x$.
13. Solve $x' = (1 + 2t - t^2) \cdot \frac{x+x^2}{1+2x}$.
14. Solve $x' = \frac{x+1}{1+t^2}$.
15. Solve $x' = \frac{t-1}{t} \cdot \frac{x^3-x^2+1}{9x^2-6x}$.
16. Solve $x' = 4t\sqrt{x}$, $x > 0$, $x(0) = 1$.
17. Solve $x' = \frac{x+1}{2x-1}$, $x(0) = -2$.
18. Solve $x' = 4t^3\sqrt{x}$, $x \geq 0$, $x(0) = 1$.
19. Find the limits as $t \to \pm\infty$ of the solution of $x' = t^{-2}x^2$, $x(1) = -1$.

20. Solve $x' = x(2 - x)$, first with the initial value $x(0) = -1$ and then with the initial value $x(0) = 1$.
21. Show that if $x_1(t)$ is a solution of the homogeneous equation $x' + p(t)x = 0$ and $x_2(t)$ is a solution of the nonhomogeneous equation $x' + p(t)x = h(t)$ then $x_1 + x_2$ is a solution of the nonhomogeneous equation $x' + p(t)x = h(t)$.
22. Solve $e^{2x-t} dx = 2e^{x+t} dt$.

4.4 Homogeneous equations

The equation

$$x' = f(t, x)$$

is called *homogeneous* if there exists a function g of the variable $\frac{x}{t}, t \neq 0$, such that

$$f(t, x) = g\left(\frac{x}{t}\right), \quad t \neq 0.$$

For example,

$$x' = \frac{x^2 + t^2}{tx}, \quad t \neq 0,$$

is homogeneous because dividing numerator and denominator by t^2 we obtain

$$x' = \frac{\left(\frac{x}{t}\right)^2 + 1}{\frac{x}{t}}, \quad t \neq 0,$$

and the right hand side is a function of $\frac{x}{t}$.
    On the other hand

$$x' = xe^t$$

is not homogeneous because it is impossible to express it as a function of $\frac{x}{t}$.
    So homogeneous equations can be written in the form

$$x' = g\left(\frac{x}{t}\right). \tag{4.8}$$

Equation (4.8) can be transformed into a separable equation by introducing a new dependent variable z such that $x(t) = tz(t)$. Then $x'(t) = z(t) + tz'(t)$ and hence (4.8) becomes

$$z + tz' = g\left(\frac{tz}{t}\right) = g(z) \implies tz' = g(z) - z,$$

which is separable.

Example 4.5. Solve the homogeneous equation

$$x' = \frac{t^2 + tx + x^2}{t^2}.$$

Since

$$\frac{t^2 + tx + x^2}{t^2} = 1 + \frac{x}{t} + \left(\frac{x}{t}\right)^2$$

the given equation can be written as

$$x' = 1 + \frac{x}{t} + \left(\frac{x}{t}\right)^2.$$

Setting $x = tz$ we find

$$tz' + z = 1 + z + z^2 \Rightarrow tz' = 1 + z^2 \Rightarrow \frac{dz}{z^2 + 1} = \frac{dt}{t}.$$

Integrating

$$\int \frac{dz}{z^2 + 1} = \int \frac{dt}{t} = \ln|t| + c.$$

Since

$$\int \frac{dz}{z^2 + 1} = \arctan z$$

we find $\arctan z = \ln|t| + c$. Solving for z we get $z = \tan(\ln|t| + c)$ and thus

$$x = tz = t \cdot \tan(\ln|t| + c), \quad t \neq 0. \qquad \square$$

Let us now consider the more general equation

$$x' = \frac{M(t, x)}{N(t, x)}, \tag{4.9}$$

where M, N are homogeneous functions of the same order k. Let us recall that $M = M(t, x)$ is k-homogeneous function if

$$M(\lambda t, \lambda x) = \lambda^k M(t, x), \tag{4.10}$$

for all λ such that $M(\lambda t, \lambda x)$ and λ^k make sense.
If both M and N are k-homogeneous, then (4.10) yields

$$\frac{M(t, x)}{N(t, x)} = \frac{M(t \cdot 1, t \cdot x/t)}{N(t \cdot 1, t \cdot x/t)} = \frac{t^k M(1, x/t)}{t^k N(1, x/t)} = \frac{M(1, x/t)}{N(1, x/t)}.$$

If we define

$$g\left(\frac{x}{t}\right) := \frac{M(1, x/t)}{N(1, x/t)}$$

we infer that

$$x' = \frac{M(t, x)}{N(t, x)} = g\left(\frac{x}{t}\right)$$

which shows that (4.9) is homogeneous.

It follows from the above discussion that if $M(t, x)$ and $N(t, x)$ are polynomials of the variables t and x, then a quick way to check for homogeneity is to determine if all the terms of the two polynomials have the same total degree (the sum of the exponents of x and t) or not. If they have the same total degree, say k, then M and N are both k-homogeneous and substituting $x = tz$, one has $M(t, x) = t^k M(1, z)$ and $N(t, x) = t^k N(1, z)$. For example, to check whether

$$x' = \frac{t^5 x^2 - t^4 x^3}{t x^6 + t^3 x^4}$$

is homogeneous or not, we note that the total degree of each term is 7 and it is therefore homogeneous. Dividing by t^7 one obtains

$$\frac{(\frac{x}{t})^2 - (\frac{x}{t})^3}{(\frac{x}{t})^6 + (\frac{x}{t})^4}$$

which is a function of $\frac{x}{t}$.

On the other hand,

$$x' = \frac{t^5 x^2 - t^4 x^3}{t x^5 + t^3 x^4}$$

is not homogeneous because the term $t x^5$ has a total degree of 6 while the other terms have total degree of 7.

Example 4.6. (i) Solve the homogeneous equation

$$x' = \frac{x}{x + t}.$$

Setting $x = tz$ we obtain

$$tz' + z = \frac{tz}{tz + t} = \frac{z}{z + 1},$$

which yields

$$tz' = \frac{z}{z+1} - z = \frac{z - z(z+1)}{z+1} = -\frac{z^2}{z+1}.$$

Separating the variables and integrating we find

$$\int \frac{z+1}{z^2} \, dz = -\int \frac{dt}{t} \Rightarrow \ln|z| - \frac{1}{z} = -\ln|t| + c \Rightarrow \ln|tz| - \frac{1}{z} = c.$$

Recalling that $tz = x$ we obtain the solution in an implicit form

$$\ln|x| - \frac{t}{x} = c.$$

(ii) Consider the equation

$$x' = \frac{x^3 + t^3}{tx^2}, \quad tx \neq 0.$$

Both the numerator and the denominator are 3-homogeneous and thus the equation is homogeneous. Setting $x = tz$ we obtain

$$tz' + z = \frac{t^3 z^3 + t^3}{t^3 z^2} = \frac{z^3 + 1}{z^2},$$

namely

$$tz' = \frac{z^3 + 1}{z^2} - z = \frac{z^3 + 1 - z^3}{z^2} = \frac{1}{z^2}.$$

Separating the variables and integrating we get

$$\int z^2 \, dz = \int \frac{dt}{t} \Rightarrow \frac{1}{3} z^3 = \ln|t| + c \Rightarrow z^3 = 3(\ln|t| + c),$$

whereby

$$z(t) = 3^{\frac{1}{3}} \cdot (\ln|t| + c)^{\frac{1}{3}}.$$

Since $x = tz$ we find the solution

$$x(t) = 3^{\frac{1}{3}} t \cdot (\ln|t| + c)^{\frac{1}{3}},$$

which is defined for $t \neq 0$. □

4.5 Exercises on homogeneous equations

1. Identify the homogeneous equations and explain why they are homogeneous.

(a) $x' = \cos(\frac{t+2x}{2t-x})$.

(b) $x' = \frac{e^{2x-t}}{x+t}$.

(c) $x' = \ln x - \ln t - x$.

(d) $x' = \ln t - \ln x - \frac{x+t}{x-t}$.

(e) $x' = \frac{x^5+t^3t^2+tx^4}{x^5+t^5}$.

2. Solve $x' = \frac{x}{t} - (\frac{x}{t})^2$.

3. Solve $x' = \frac{x}{t} + 2(\frac{x}{t})^3$.

4. Solve $x' = \frac{x}{t} + \tan(\frac{x}{t})$.

5. Solve $x' = \frac{x^2+tx}{t^2}$.

6. Solve $x' = \frac{x^2+tx+t^2}{t^2}$.

7. Solve $x' = \frac{x^3-t^3}{tx^2}$.

8. Solve $x' = \frac{x^2+2t^2}{tx}$.

9. Solve $tx.x' - t^2 - x^2 = 0$.

10. Solve $x' = e^{\frac{x}{t}} + \frac{x}{t}$.

4.6 Bernoulli's equation

An equation of the form

$$x' = a(t)x + b(t)x^n \qquad (4.11)$$

where $a(t)$ and $b(t)$ are continuous and n is any number, $n \neq 0$, is called a *Bernoulli equation*. Notice that n is not necessarily an integer. Furthermore, for $n = 1$, the equation is linear. So we restrict our study to the case $n \neq 0, 1$.

Before learning how to solve Bernoulli's equation, let us note the following:

(1) If $n \geq 0$, $x(t) \equiv 0$ is a trivial solution of the equation.

(2) If $n \geq 1$ we can apply the general existence and uniqueness Theorems discussed in Chapter 3. As a consequence, by uniqueness, all the nontrivial solutions are either always positive or always negative.

Notice that if $0 < n < 1$ uniqueness might fail (see e. g. Example 4.7-(ii) below).

Equation (4.11) can be transformed into a linear equation by making the substitution $y = x^{1-n}$ (of course, if $n > 1$ we understand that $x \geq 0$) as follows: since $y' = (1-n)x^{-n}x'$, we have $x' = \frac{1}{1-n}x^n y'$. Also, $y = x^{1-n}$ implies that $x = x^n y$. Substituting these values of x and x' into the equation, we obtain the linear equation $\frac{1}{1-n}x^n y' = a(t)x^n y + b(t)x^n$ or

$$y' = (1-n)a(t)y + (1-n)b(t). \qquad (4.12)$$

Students are advised to go through all the steps in solving such equations and not just use the above formula.

Example 4.7. (i) Solve $x' = x + x^2$.

Setting $y = x^{1-2} = x^{-1}$ and substituting, as suggested above, we obtain the linear equation $y' = -y - 1$, or $y' + y = -1$. Solving this linear equation, we find $y = ce^{-t} - 1$. Finally, since $y = x^{-1}$, we find that

$$x = \frac{1}{y} = \frac{1}{ce^{-t} - 1}.$$

(ii) Solve $x' + x = tx^{\frac{1}{2}}$, $x(2) = 0$.

One solution is $x \equiv 0$. To find the nontrivial solutions we set $y = x^{\frac{1}{2}}$, obtaining the equation $2y' + y = t$. Solving this linear equation for y, we have

$$y = t - 2 + ce^{-\frac{1}{2}t}$$

and $x = y^2 = (t - 2 + ce^{-\frac{1}{2}t})^2$, which is the general solution. To find the particular required solution, we substituting $t = 2$ and $x = 0$, which implies $c = 0$ and $x(t) = (t - 2)^2$. Let us remark explicitly that uniqueness fails. □

Next, let us consider an equation in the form

$$x' = h(t) + a(t)x + b(t)x^2, \tag{4.13}$$

referred to as the *Riccati equation*, and let us suppose that we know one solution $z(t)$ of (4.13). If x is another solution, we set $u = x - z$ and evaluate

$$u' = x' - z' = h + ax + bx^2 - (h + az + bz^2) = ax + bx^2 - az - bz^2.$$

Rearranging, we get

$$u' = a(x - z) + b(x^2 - z^2) = au + b(x - z)(x + z) = au + bu(x + z).$$

Since $x = u + z$ then $x + z = u + 2z$ and we find

$$u' = au + bu(u + 2z) = (a + 2bz)u + bu^2,$$

which is a Bernoulli equation. Thus, *the general solution of* (4.13) is given by $x = z + u_c$, where u_c is the general solution of the corresponding Bernoulli equation and z is a particular solution of the Riccati equation.

Example 4.8. Solve $x' = x^2 - 2x + 1$, $x(1) = 2$.

Note that $z = 1$ satisfies the equation but not the initial conditions. However, we can use it to find the general solution and then the particular one.

As shown above, $u' = (a + 2bz)u + bu^2 = u^2$. Then $u' = u^2 \implies u = \frac{-1}{t+c} = x - 1$. So

$$x = \frac{-1}{t+c} + 1 = \frac{t+c-1}{t+c}$$

is the general solution. In order to find the particular solution, we substitute the initial data and find $c = -2$. Hence

$$x = \frac{t-3}{t-2}.$$ □

4.7 Exercises on Bernoulli equations

1. Solve $x' - x = tx^{\frac{1}{2}}$.
2. Solve the Bernoulli equation $x' = \frac{x}{t} + 3x^3$.
3. Solve the Bernoulli equation $x' = 4x - 2t \cdot \sqrt{x}, x \geq 0$.
4. Solve $x' + x = x^2$.
5. Solve the Bernoulli equation $x' = x + 2tx^{-1}, x(0) = 1$.
6. Solve the Bernoulli equation $x' = -\frac{x}{t} + x^{-2}, x(1) = 1$.
7. Solve $x' + x = x^{\frac{1}{2}}$.
8. Solve $x' - tx^2 = x$.
9. Solve the Riccati equation $x' = -3x + x^2 + 2$.
10. Solve $x' - x = x^{\frac{1}{3}}$.

4.8 Clairaut's equation

In this section we consider a typical equation which is not in normal form, namely

$$x = tx' + h(x'),$$ (4.14)

where h is a smooth function, defined on a set $S \subseteq \mathbb{R}$. This equation is called Clairaut's equation.

A direct inspection shows that the family of lines $x = ct + h(c), c \in S$, are solutions of (4.14). Conversely, if $x(t)$ is a smooth solution of (4.14), differentiating $x = tx' + h(x')$, we find

$$x' = x' + tx'' + h'(x')x'' \implies tx'' + h'(x')x'' = 0 \implies x''(t + h'(x')) = 0 \implies x'' = 0 \quad \text{or}$$
$$t = -h'(x').$$

The former yields $x'' = 0$, namely $x' = c$, constant, which gives rise to $x = ct + h(c)$. Taking $x' = p$ as parameter, the latter gives rise to the system

$$\begin{cases} x = x(p) = -ph'(p) + h(p), \\ t = t(p) = -h'(p). \end{cases} \tag{4.15}$$

If $h'(p)$ is not constant this system defines a parametric curve in the plane (t, x), otherwise it reduces to a point; see Remark 4.1(2). This planar curve solves (4.14), in the sense that

$$x(p) = t(p) \cdot \frac{dx}{dt} + h\left(\frac{dx}{dt}\right), \quad \forall p \in S,$$

since $\frac{dx}{dt} = p$.

We will call (4.15) the *singular solution*[1] to the Clairaut equation (4.14).

We have shown that, apart from the singular solution, the family $x = ct + h(c)$ includes all the solutions of (4.14). For this reason we can say that $x = ct + h(c)$ is the *general solution (or integral) of the Clairaut equation.*

Summing up, the Clairaut equation has two different types of solutions: the general solution $x = ct + h(c)$ and the singular solution which is the parametric curve defined by (4.15).

Notice that the singular solution cannot be obtained from the general solution for any particular value of the real constant c.

Concerning the singular solution, some further remark is in order.

Remark 4.1. (1) The singular solution turns out to be the *envelope* of the general solution $x = ct + h(c)$. To see this, let us recall that an *envelope* to a family of planar curves $\mathcal{G}$ is a curve which is tangent to any curve of $\mathcal{G}$. If the family $\mathcal{G}$ is given by $x = g(t, c), c \in \mathbb{R}$, with g depending smoothly on (t, c), it is known that, if $\mathcal{G}$ has an envelope, then it solves the system

$$\begin{cases} x = g(t, c), \\ \frac{\partial}{\partial c} g(t, c) = 0, \end{cases} \tag{4.16}$$

and, conversely, if (4.16) has a solution, then this is an envelope of $\mathcal{G}$.

In the case of Clairaut equations, one has $g(t, c) = ct + h(c)$ and thus the system (4.16) becomes

$$\begin{cases} x = ct + h(c), \\ t + h'(c) = 0, \end{cases} \quad \Longleftrightarrow \quad \begin{cases} x = ct + h(c), \\ t = -h'(c). \end{cases}$$

---

1 The reader should be aware that the term "singular solution" might have different meanings: some authors define a singular solution to a first order equation as a solution that cannot be obtained from the general integral for any values of c, other authors define a singular solution as a solution for which the initial value problem does not have a unique solution, and the list could continue.

This is exactly the system (4.15) (with $c = p$) which defines the singular solution. This proves that, as claimed, the singular solution to the Clairaut equation is the envelope of its general solution.

(2) We have found the singular solution in the case when h' is not constant. If $h' = k$ (constant), then $h(p) = kp + k_1$ and we obtain

$$\begin{cases} x = -pk + h(p) = -pk + kp + k_1 = k_1, \\ t = -k, \end{cases}$$

which reduces to the single point $(-k, k_1)$. Notice that, in such a case, equation (4.14) becomes $x = tx' + kx' + k_1 = (t+k)x' + k_1$. Solving this equation we find $x = c(t+k) + k_1$, a bundle of straight lines centered at $(-k, k_1)$, which has no envelope.

Example 4.9. (i) Solve $x = tx' - \frac{1}{2}(x')^2$.

The general solution is $x = ct - \frac{1}{2}c^2$. The system (4.15) becomes

$$\begin{cases} x = ct - \frac{1}{2}c^2, \\ t = c, \end{cases}$$

whereby $x = t^2 - \frac{1}{2}t^2 = \frac{1}{2}t^2$, which is the singular solution; see Fig. 4.3.

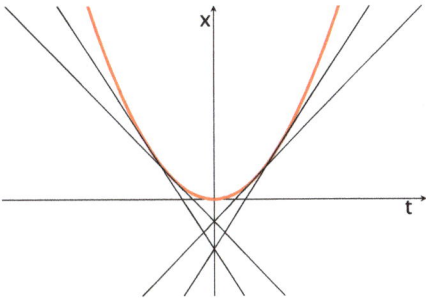

Figure 4.3: Solutions of $x = tx' - \frac{1}{2}(x')^2$ and the singular solution $x = \frac{1}{2}t^2$ (in red).

(ii) Solve $x = tx' - 2\sqrt{x'}$.

In this case $h(x') = \sqrt{x'}$ is defined for $x' \geq 0$. Thus the general solution is $x = tc - 2\sqrt{c}$, $c \geq 0$. As for the singular solution we find

$$\begin{cases} x = tc - 2\sqrt{c}, \\ t = \frac{1}{\sqrt{c}}. \end{cases}$$

Solving for c the last equation we get $c = \frac{1}{t^2}$, $t > 0$, whereby $x = \frac{1}{t} - \frac{2}{t} = -\frac{1}{t}$, $t > 0$; see Fig. 4.4. □

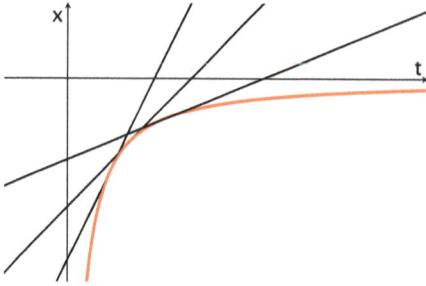

Figure 4.4: Solutions of $x = tx' - 2\sqrt{x'}$ and the singular solution $x = -\frac{1}{t}, t > 0$ (in red).

Remark 4.2. In the preceding examples we have seen two features of equations which are not in normal form, such as the Clairaut equation: dealing with the Cauchy problem

$$x = tx' + h(x'), \quad x(a) = b, \tag{4.17}$$

(1) there are points $(a, b) \in \mathbb{R}^2$ where there is a lack of uniqueness since (4.17) has two solutions—in particular, through any point of the singular solution passes the singular solution itself and the lines $x = ct + h(c)$ tangent to the singular solution at that point;

(2) there are points $(a, b) \in \mathbb{R}^2$ such that (4.17) has no solution.

For example, let us solve the ivp $x = tx' - \frac{1}{2}(x')^2$, $x(0) = b$ (see Example 4.9(i)). Inserting the initial condition into the general solution yields $b = -\frac{1}{2}c^2$ namely $c^2 = -2b$. Thus for $b < 0$ there are two solutions $x = \sqrt{(-2b)} \cdot t + b$ and $x = -\sqrt{(-2b)} \cdot t + b$; for $b = 0$ we find $x = 0$, to which we have to add the singular solution $x = \frac{1}{2}t^2$. Finally, for $b > 0$ there are neither solutions given by the general integral, nor the singular solution. $\square$

4.9 Exercises on Clairaut equations

1. Solve the Clairaut equation $x = (t - 1)x' - (x')^2$ and find the singular solution.
2. Solve the Clairaut equation $x = tx' - e^{x'}$ and find the singular solution.
3. Solve the Clairaut equation $x = tx' - \frac{1}{4}(x')^4$ and find the singular solution.
4. Solve the Clairaut equation $x = tx' + \frac{1}{3(x')^3}$ and find the singular solution.
5. Solve the Cauchy problem $x = tx' + (x')^2$, $x(0) = b$.
6. Solve the Cauchy problem $x = tx' + \frac{1}{x'}$, $x(1) = b$.

5 Exact differential equations

In this chapter we study another important class of first order differential equations that can be solved.

5.1 Exact equations

Let $M(x,y), N(x,y)$ be continuous functions, with continuous partial derivatives, defined on a set $S \subseteq \mathbb{R}^2$ and consider the equation of the form

$$M(x,y)\, dx + N(x,y)\, dy = 0. \tag{5.1}$$

We note that this equation can be written as a first order differential equation in either of the two forms below:

$$N(x,y)\, \frac{dy}{dx} + M(x,y) = 0, \tag{5.2}$$

$$M(x,y)\, \frac{dx}{dy} + N(x,y) = 0. \tag{5.3}$$

Equation (5.1) is said to be *exact* if there exists a function $F(x,y)$ such that

$$dF = M(x,y)\, dx + N(x,y)\, dy, \quad (x,y) \in S$$

where dF denotes the differential of the function $F(x,y)$. Recall that

$$dF = F_x\, dx + F_y\, dy.$$

Hence, one can equivalently say that (5.1) is exact if there exists a function $F(x,y)$ such that

$$F_x(x,y) = M(x,y), \quad F_y(x,y) = N(x,y).$$

We note that $dF = 0$ implies that $F(x,y) = c$, where c is a constant, and may implicitly define either y as a function of x or x as a function of y.

On the other hand, taking the differential of $F(x,y) = c$, c real constant, we find $F_x\, dx + F_y\, dy = 0$ and hence

$$M(x,y)\, dx + N(x,y)\, dy = 0.$$

In other words, $F(x,y) = c$ may define $y = y(x)$ as a solution of the differential equation (5.2) or $x = x(y)$ as a solution of the differential equation (5.3). $F(x,y) = c$ is called the general solution of (5.1).

https://doi.org/10.1515/9783111185675-005

Remark 5.1. An alternative way of looking at solutions is to say that the general solution of an exact equation is given by the level curves $\{(x,y) \in S : F(x,y) = c\}$ of the function F. Precisely, if $F(x,y) = c$ defines a C^1 curve $(x(s),y(s))$, differentiating the identity $F(x(s),y(s)) = c$ we infer that

$$F_x(x(s),y(s))x'(s) + F_y(x(s),y(s))y'(s) = 0.$$

Since $F_x = M$ and $F_y = N$ the preceding equation yields

$$M(x(s),y(s))x'(s) + N(x(s),y(s))y'(s) = 0.$$

In this sense we can say that the curve $(x(s),y(s))$ solves the exact equation $M dx + N dy = 0$.

In particular, it might happen that, for certain values of c, the curve defined by $F(x(s),y(s)) = c$ reduces to an isolated point $P_0 = (x(s_0),y(s_0))$. This point plays the role of the constant solutions for the equations discussed in chapters 2–4 and it is rather natural to consider such a degenerate curve as a solution of $M dx + N dy = 0$, at least in a generalized sense. Notice that it can be shown that at such a point P_0, one has $M(P_0) = N(P_0) = 0$.

5.2 Solving exact equations

Example 5.1. By way of motivation, let us start with looking at one of the simplest types of exact equations, namely separable equations.

Consider, for example, the differential equation

$$e^{-y} dx + \frac{1}{x} dy = 0,$$

which is of the form (5.1). It is separable and can be written as $x \, dx + e^y \, dy = 0$. Its general solution then is given by

$$\frac{1}{2}x^2 + e^y = c.$$

It is now clear that the given differential equation is also exact because if we let

$$F(x,y) = \frac{1}{2}x^2 + e^y$$

then $dF(x,y) = x \, dx + e^y \, dy$.

In general, solving a separable equation $M(x) \, dx + N(y) \, dy = 0$ leads to

$$F(x,y) = \int\limits_a^x M(x) + \int\limits_b^y N(y) = c.$$

It is then clear that $F_x = M(x)$ and $F_y = N(y)$; which shows that the equation $M(x)\,dx + N(y)\,dy = 0$ is exact. □

Remark 5.2. As shown in the example above, separable equations are also exact equations. But one can imagine how challenging it might be to determine whether or not an equation is exact, in general. Even with a simple equation like $dx + (x + y)\,dy = 0$, it is difficult to determine whether or not it is exact by using the definition of exactness, which requires the existence of a function $F(x,y)$ such that $dF = dx + (x + y)\,dy$. Fortunately, Theorem 5.1 in Section 5.3 below gives a simple condition that is both necessary and sufficient for exactness, and also methods for solving exact equations.

As we will see later, sometimes the difficulty in solving exact equations can be in-tegration. In some cases one may leave the answer in integral form or use numerical methods, if needed.

Example 5.2. In order to solve $a^2 x\,dx + b^2 y\,dy = 0$, $a, b \neq 0$, we find the general solution $a^2 x^2 + b^2 y^2 = c \geq 0$. For $c > 0$ this is a family of ellipses (or circles if $a = b$), if $c = 0$ the ellipse reduces to the point $(0,0)$, which can be considered as the constant solution of $a^2 x\,dx + b^2 y\,dy = 0$, see Remark 5.1. □

5.2.1 The initial value problem for exact equations

Given a point $P_0 = (x_0, y_0) \in S$, we may seek a solution of (5.3) that passes through P_0. This will be called the *initial value problem for* (5.3) *at* P_0.

Clearly, if a solution $y(x)$, resp. $x(y)$, of (5.3) is such that $y(x_0) = y_0$, resp. $x(y_0) = x_0$, then it solves the ivp at P_0. In particular, from $F(x,y) = c$ it follows that $F(x_0, y_0) = c$. In other words, the initial condition allows us to single out $c = F(x_0, y_0)$.

For example, since the general solution of $x\,dx + y\,dy = 0$ is given by $x^2 + y^2 = c^2$, then in order to find the solution passing through $(1,2)$, we solve the equation $1^2 + 2^2 = c^2$ for c, obtaining the solution passing through $(1,2)$ to be implicitly given by $x^2 + y^2 = 5$.

Let us assume that $N(x_0, y_0) \neq 0$. Since N is assumed to be continuous on $S \subseteq \mathbb{R}^2$, by continuity, there exists a neighborhood $U \subset S$ of $P_0 = (x_0, y_0)$ such that $N(x,y) \neq 0$ for all $(x,y) \in U$. Dividing by N, equation (5.2) becomes

$$\frac{dy}{dx} = -\frac{M(x,y)}{N(x,y)}, \quad (x,y) \in U.$$

Therefore, the initial value problem for (5.3) at P_0 is

$$\begin{cases} \frac{dy}{dx} = -\frac{M(x,y)}{N(x,y)}, & (x,y) \in U, \\ y(x_0) = y_0. \end{cases} \tag{5.4}$$

Similarly, if $M(x_0, y_0) \neq 0$, there exists a neighborhood $U' \subset S$ of $P_0 = (x_0, y_0)$ such that $M(x, y) \neq 0$ for all $(x, y) \in U'$ and the initial value problem for (5.3) at P_0 becomes

$$\begin{cases} \frac{dx}{dy} = -\frac{N(x,y)}{M(x,y)}, & (x,y) \in U', \\ x(y_0) = x_0. \end{cases} \tag{5.5}$$

If we can apply the local existence and uniqueness theorem, stated in Chapter 3, to (5.4) or to (5.5), we can be sure that there is a unique solution $y = y(x)$ defined in a neighborhood of x_0, or a unique solution $x = x(y)$ defined in a neighborhood of y_0.

Summing up, if either $N(x_0, y_0) \neq 0$ or $M(x_0, y_0) \neq 0$ the local existence and uniqueness theorem yields *a unique solution to the initial value problem for (5.3) at P_0*.

Remark 5.3. Let us briefly elaborate on the preceding uniqueness result. The curve $F(x, y) = F(x_0, y_0)$ might consist of several branches. But if M and N do not vanish simultaneously at the point $P_0 = (x_0, y_0)$, there is only one of these possible branches that passes through the point P_0, giving rise to the unique solution of the initial value problem for (5.3) at P_0. On the other hand, uniqueness might fail when M and N vanish simultaneously.

Example 5.3. (i) Find the solutions of $x\, dx + 2y\, dy = 0$ passing through $P = (1, 2)$.

As in Example 5.2, the general solution is $x^2 + 2y^2 = c$. The initial conditions $x = 1$, $y = 2$ yield $c = 1^2 + 2 \times 2^2 = 9$. Thus we find $x^2 + 2y^2 = 9$, which is the ellipse plotted in Fig. 5.1.

(ii) Find the solutions of $x\, dx - y\, dy = 0$ passing through (a, b).

The general solution is

$$\int_0^x x\, dx - \int_0^y y\, dy = c \Rightarrow \frac{1}{2}x^2 - \frac{1}{2}y^2 = c \Rightarrow x^2 - y^2 = 2c.$$

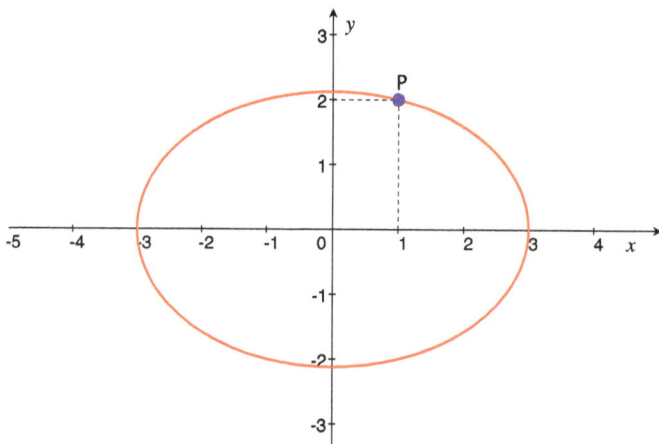

Figure 5.1: Plot of $x^2 + 2y^2 = 9$.

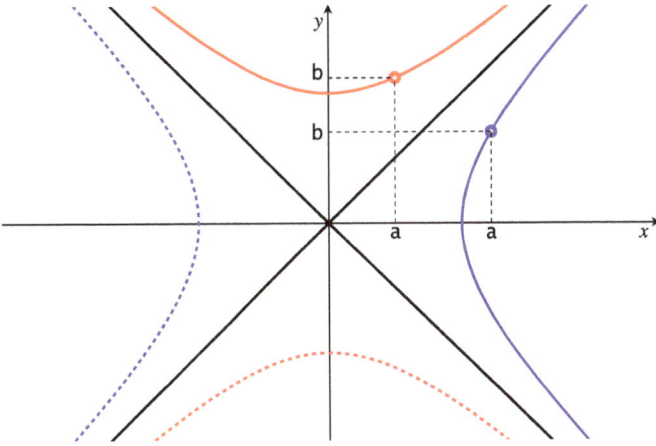

Figure 5.2: Plot of $x^2 - y^2 = 2c$; for some $c > 0$ in red, for some $c < 0$ in blue, for $c = 0$ in black. The continuous curves are the solutions through (a, b); the dotted curves are the branches of the hyperbola which do not pass through (a, b).

If $x = a$, $y = b$ we find $2c = a^2 - b^2$ and hence $x^2 - y^2 = a^2 - b^2$, which is a hyperbole if $a \neq b$, otherwise we get $x^2 - y^2 = 0$, namely $y = \pm x$. We have to distinguish between two cases:

(1) if $(a, b) \neq (0, 0)$ we find a unique solution: the branch of the hyperbola passing through (a, b) if $a^2 \neq b^2$, otherwise the line $y = x$ or $y = -x$. See Fig. 5.2.

(2) if $a = b = 0$, then both of the lines $y = \pm x$ satisfy the equation and the initial condition and hence uniqueness fails. Notice that both $M = x$ and $N = -y$ vanish at $(0, 0)$ so that the existence and uniqueness results stated before do not apply. □

Remark 5.4. Continuing Remark 5.1, let $F(x, y)$ be such that $dF = M dx + N dy$ and consider the equation (*) $F(x, y) = F(x_0, y_0)$. Using the generalized notion of a solution given in Remark 5.1, we can say that the exact equation $M dx + N dy = 0$ has at least one solution (possibly in generalized sense) passing through $P_0 \in S$ provided (*) defines a planar curve or if this curve reduces to the isolated point (x_0, y_0).

Further theoretical results on the existence and uniqueness for exact equations will be discussed in some more detail in the appendix at the end of the chapter.

5.3 General solutions of exact equations

Now we address two important issues concerning exact equations. First of all, how do we identify an exact equation? Secondly, once we have identified it, how do we solve it? The theorem below answers both of these questions.

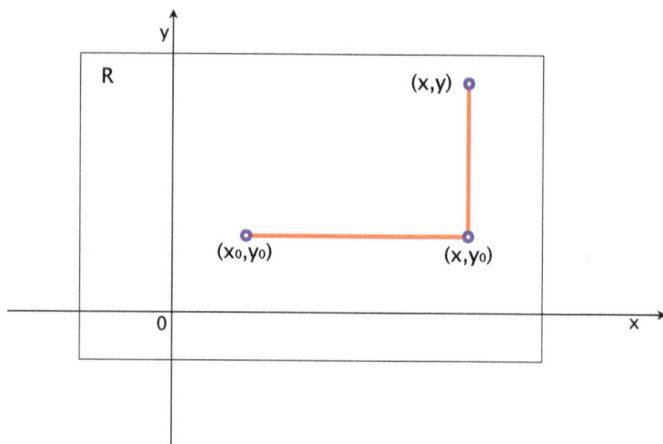

Figure 5.3: The path Γ.

Theorem 5.1. *Suppose that $M(x,y)$ and $N(x,y)$ are continuous and have continuous partial derivatives with respect to x and y, on $R = (a_1, a_2) \times (b_1, b_2)$. Then $M(x,y)\, dx + N(x,y)\, dy = 0$ is exact if and only if $M_y(x,y) = N_x(x,y)$.*

Proof of Theorem 5.1. First, in order to prove the sufficiency, we assume that (5.3) is exact and show that this implies $M_y(x,y) = N_x(x,y)$. Since (5.3) is exact, by definition there exists a function $F(x,y)$ such that $dF = M(x,y)\, dx + N(x,y)\, dy$, which means that $F_x(x,y) = M(x,y)$ and $F_y(x,y) = N(x,y)$. Consequently, we have $F_{xy}(x,y) = M_y(x,y)$ and $F_{yx}(x,y) = N_x(x,y)$. Recalling from calculus that the mixed derivatives of $F(x,y)$ are equal, i. e., $F_{xy} = F_{yx}$, we reach the conclusion that $M_y(x,y) = N_x(x,y)$.

In order to prove the converse, we assume that $M_y = N_x$ and show that there exists a function $F(x,y)$ such that $F_x = M$ and $F_y = N$, i. e., (5.3) is exact. We give two proofs of this part, which will not only prove the existence of the functions $F(x,y)$ but also indicate methods for finding such functions, i. e. solving exact equations.

First method. Let (x_0, y_0), (x,y) be two arbitrary points in the rectangle R shown below. Consider the polygonal path $\Gamma = ([x_0, x] \times \{y_0\}) \cup (\{x\} \times [y_0, y]) \subset R$, see Fig. 5.3, and define $F(x,y)$ by

$$F(x,y) = \int_{x_0}^{x} M(x, y_0)\, dx + \int_{y_0}^{y} N(x, y)\, dy. \tag{5.6}$$

Then by the Fundamental Theorem of Calculus it follows that

$$F_x(x,y) = M(x, y_0) + \frac{\partial}{\partial x} \int_{y_0}^{y} N(x, y)\, dy.$$

Since

$$\frac{\partial}{\partial x} \int_{y_0}^{y} N(x,y)\,dy = \int_{y_0}^{y} \frac{\partial}{\partial x} N(x,y)\,dy,$$

and by assumption $\frac{\partial}{\partial x} N = \frac{\partial}{\partial y} M$, we have

$$F_x(x,y) = M(x,y_0) + \int_{y_0}^{y} \frac{\partial}{\partial y} M(x,y)\,dy.$$

It follows that

$$\int_{y_0}^{y} \frac{\partial}{\partial y} M(x,y)\,dy = \frac{\partial}{\partial y} \int_{y_0}^{y} M(x,y)\,dy = M(x,y) - M(x,y_0)$$

whereby

$$F_x(x,y) = M(x,y_0) + M(x,y) - M(x,y_0) = M(x,y).$$

Moreover, one has

$$\frac{\partial}{\partial y} \int_{x_0}^{x} M(x,y_0)\,dx = 0,$$

because the integral is independent of y, so it is the same as taking the derivative of a constant. Therefore we have

$$F_y = \frac{\partial}{\partial y} \int_{y_0}^{y} N(x,y)\,dy = N(x,y).$$

In conclusion, we have shown that $F_x = M$; $F_y = N$, as claimed.

In the above discussion, we could have also taken the path $\Gamma_1 = (\{x_0\} \times [y_0,y]) \cup ([x_0,x] \times \{y\})$ yielding

$$F(x,y) = \int_{x_0}^{x} M(x,y)\,dx + \int_{y_0}^{y} N(x_0,y)\,dy. \tag{5.7}$$

Second method. Integrating $F_x(x,y) = M(x,y)$ with respect to x we obtain

$$F(x,y) = \int M(x,y)\,dx + H(y),$$

where the constant of integration is expressed as $H(y)$, a function of y. Obviously, $F_x = M$. We now want to choose $H(y)$ in such a way that $F_y = N$. This yields (notice that $H(y) = F(x,y) - \int M(x,y) \, dx$ is differentiable)

$$F_y(x,y) = \frac{\partial}{\partial y} \int M(x,y) \, dx + H'(y) = N(x,y)$$

whereby

$$H'(y) = N(x,y) - \frac{\partial}{\partial y} \int M(x,y) \, dx. \qquad (5.8)$$

If the right hand side depended on both variables, x and y, that would make it impossible to find H as a function of y only. However, as we will see below, in spite of its appearance, it is a function of y only. In order to verify this, we will show that its partial derivative with respect to x is zero. This follows since

$$\frac{\partial}{\partial x}\left[N(x,y) - \frac{\partial}{\partial y} \int M(x,y) \, dx \right] = N_x - \frac{\partial}{\partial x} \int M_y(x,y) \, dx = N_x - M_y = 0.$$

Since the right hand side of (5.8) is independent of x, an integration yields $H(y)$, giving rise to the general solution $F(x,y) = \int M(x,y) \, dx + H(y)$. □

In the above proof, we could just as easily have chosen

$$F(x,y) = \int N(x,y) \, dy + K(x),$$

and determine $K(x)$ as we obtained $H(y)$ above.

Summing up, there are essentially four methods that we can use to solve exact equations:

1. letting $F(x,y) = \int_{x_0}^{x} M(x,y_0) \, dx + \int_{y_0}^{y} N(x,y) \, dy$;
2. letting $F(x,y) = \int_{x_0}^{x} M(x,y) \, dx + \int_{y_0}^{y} N(x_0,y) \, dy$;
3. letting $F(x,y) = \int M(x,y) \, dx + H(y)$ and solving for $H(y)$;
4. letting $F(x,y) = \int N(x,y) \, dy + K(x)$ and solving for $K(x)$.

Remark 5.5. In Theorem 5.1 we can replace the rectangle R by any, possibly unbounded, *simply connected* domain, namely a path-connected domain in which any simple closed curve can be continuously shrunk to a point, while remaining inside the domain. Roughly, a simply connected domain has no hole in it. For example, any convex domain is simply connected, whereas $\mathbb{R}^2 \setminus \{0\}$ is not.

Moreover, instead of defining F as in (5.6) we can just as well integrate $M dx + N dy$ on any smooth curve contained in R, joining the points $P_0 = (x_0, y_0)$ and $P = (x,y)$. For example, if $P_0 = (0,0) \in R$ and the segment $[0, P] \subset R$, we would simply find (notice that $dx = x\, ds$ and $dy = y\, ds$)

$$F(x,y) = \int_0^1 M(sx, sy)x\, ds + \int_0^1 N(sx, sy)y\, ds. \qquad (5.9)$$

Example 5.4. Solve $(x^4 + 2y)\, dx + (2x - 3y^5)\, dy = 0$.

Since $M_y = 2 = N_x$, it is exact. We illustrate four ways to obtain the general solution.

1. Choosing $x_0 = 0, y_0 = 0$ for lower limits of integration and using Method 1, we obtain

$$F(x,y) = \int_0^x M(x,0)\, dx + \int_0^y N(x,y)\, dy = \int_0^x x^4\, dx + \int_0^y (2x - 3y^5)\, dy = \frac{1}{5}x^5 + 2yx - \frac{1}{2}y^6.$$

Hence the general solution is given implicitly by

$$\frac{1}{5}x^5 + 2yx - \frac{1}{2}y^6 = k.$$

2. Choosing again $x_0 = 0, y_0 = 0$ and using Method 2, we obtain

$$F(x,y) = \int_0^x M(x,y)\, dx + \int_0^y N(0,y)\, dy = \int_0^x (x^4 + 2y)\, dx + \int_0^y -3y^5\, dy = \frac{1}{5}x^5 + 2yx - \frac{1}{2}y^6.$$

The general solution is again defined by

$$\frac{1}{5}x^5 + 2yx - \frac{1}{2}y^6 = k.$$

3. Using Method 3, we integrate $M(x,y)$ with respect to x and hold y as a constant; we let $H(y)$ represent the constant of integration. Then

$$F(x,y) = \int (x^4 + 2y)\, dx + H(y) = \frac{1}{5}x^5 + 2xy + H(y).$$

Now, clearly $F_x = x^4 + 2y = M$. It remains to find $H(y)$ so that $F_y = N$. To this end, we set

$$2x + H'(y) = 2x - 3y^5,$$

which yields $H'(y) = -3y^5$ and hence $H(y) = -\frac{1}{2}y^6$. So once again we find that the general solution is given by

$$F(x,y) = \frac{1}{5}x^5 + 2yx - \frac{1}{2}y^6 = k.$$

4. This time we integrate $N(x,y)$ with respect to y and hold x as a constant; we let $K(x)$ represent the constant of integration. Then

$$F(x,y) = \int (2x - 3y^5)\, dy + K(x) = 2xy - \frac{1}{2}y^6 + K(x)$$

so that $F_y = N$. Now we choose K so that $F_x = M$. To this end, we let $2y + K'(x) = x^4 + 2y$, namely $K'(x) = x^4$. Integrating, we find $K(x) = \frac{1}{5}x^5$ and $F(x,y) = \frac{1}{5}x^5 + 2yx - \frac{1}{2}y^6 = k$.

We can also use (5.9) yielding

$$F(x,y) = \int_0^1 (as^4x^4 + bsy)x\, ds + \int_0^1 (bsx + cs^5y^5)y\, ds$$

$$= \int_0^1 (as^4x^5 + 2bsxy + cs^5y^6)\, ds$$

$$= \left[\frac{1}{5}as^5x^5 + bs^2xy + \frac{1}{6}cs^6y^6\right]_0^1 = \frac{1}{5}ax^5 + bxy + \frac{1}{6}cy^6$$

as before.　　　　　　　　　　　　　　　　　　　　　　　　　　　　　　□

Below we discuss some examples in detail. For convenience, we mostly prefer using the first two methods. Students may use any of the above methods that they find convenient, when solving exact equations. As we will see later, sometimes one method may be more convenient – either due to the nature of the problem or due to difficulty with integration. In the first two methods, the choice of (x_0, y_0) in the rectangle R, i. e. the lower limits of x and y, is important.

Example 5.5. Find the general solution of

$$\frac{2y^2 + 1}{x}\, dx + 4y \ln x\, dy = 0.$$

First we see that $M_y = \frac{4y}{x} = N_x$, so the equation is exact. Next, we see that we cannot take the lower limits to be 0, as we did in the example above, because $\ln x$ is not defined at $x = 0$. So, let us try a couple different choices for x_0 and y_0 and compare the answers.

$x_0 = 1 = y_0$. Then, if we use Method 2, since $\ln 1 = 0$, we will have only one integral to handle, that is we will have

$$F(x,y) = \int_1^x \frac{2y^2 + 1}{x}\, dx = (2y^2 + 1)\ln x - (2y^2 + 1)\ln 1 = (2y^2 + 1)\ln x$$

and hence the general solution is given by

$$(2y^2 + 1)\ln x = k.$$

$x_0 = 3$, $y_0 = 0$. This time, let us use the first method. Then we have

$$F(x,y) = \int_3^x \frac{1}{x}\, dx + \int_0^y 4y\ln x\, dy = (\ln x - \ln 3) + 2y^2 \ln x = c.$$

Notice that the answer is different from the previous case only by a constant. More precisely, we have $F(x,y) = (\ln x - \ln 3) + 2y^2 \ln x = k$, or $(2y^2 + 1)\ln x = k + \ln 3$. Since k is an arbitrary constant then so is $k + \ln 3$; thus we can let $k + \ln 3 = c$, where c is an arbitrary constant. So, our answer is equivalent to the answer in the first case. ☐

Example 5.6. Find the solution of $(xy^2 + x^2)\, dx + (y^2 + x^2 y)\, dy = 0$ passing through $(a, 0)$. Since

$$\frac{\partial}{\partial y}(xy^2 + x^2) = 2xy = \frac{\partial}{\partial x}(y^2 + x^2 y)$$

the equation is exact. Using option 1, we find

$$F(x,y) = \int_0^x x^2\, dx + \int_0^y (y^2 + x^2 y)\, dy$$

$$= \frac{1}{3}x^3 + \frac{1}{3}y^3 + \frac{1}{2}x^2 y^2 = c.$$

Substituting $x = a$ and $y = 0$, we obtain $c = \frac{a^3}{3}$. Therefore the solution to the initial value problem can be written as

$$2x^3 + 2y^3 + 3x^2 y^2 = 2a^3.$$ ☐

Remark 5.6. Similar to what we did in Chapter 3 for equations of the form $x' = f(t,x)$, here also it is possible to deduce some qualitative information about the general solution of (5.3) without solving it explicitly. Precisely, if (x_0, y_0) is such that $N(x_0, y_0) \neq 0$ and $M(x_0, y_0) = 0$ then (5.2) implies that the solution has horizontal tangent at (x_0, y_0), whereas

$$\frac{M(x_0, y_0)}{N(x_0, y_0)} > 0,\ (\text{resp.} < 0) \quad \Longrightarrow \quad \frac{d}{dx}y(x_0) = -\frac{M(x_0, y_0)}{N(x_0, y_0)} < 0,\ (\text{resp.} > 0).$$

Similarly, if $N(x_0, y_0) = 0$ and $M(x_0, y_0) \neq 0$ then (5.1) implies that the solution has vertical tangent at (x_0, y_0) and

$$\frac{N(x_0, y_0)}{M(x_0, y_0)} > 0,\ (\text{resp.} < 0) \quad \Longrightarrow \quad \frac{d}{dy}x(y_0) = -\frac{N(x_0, y_0)}{M(x_0, y_0)} < 0\ (\text{resp.} > 0).$$

5.4 Integrating factor

Now we discuss how one might convert a non-exact equation into an exact equation by multiplying it by a suitable nonzero function called an *integrating factor*.

In other words, we consider the equation

$$M(x,y)\,dx + N(x,y)\,dy = 0 \tag{5.1}$$

in the case when $M_y \neq N_x$. If, for any $\mu(x,y) \neq 0$, the equation

$$\mu(x,y) \cdot M(x,y)\,dx + \mu(x,y) \cdot N(x,y)\,dy = 0 \tag{5.10}$$

is exact, then $\mu(x,y)$ is said to be an *integrating factor* of (5.3).

For example, the equation $y\,dx - x\,dy = 0$ is not exact. But after multiplying it by the function y^{-2}, $y \neq 0$, the corresponding equation $\frac{1}{y}\,dx - \frac{x}{y^2}\,dy = 0$, $y \neq 0$, becomes exact since

$$\frac{\partial}{\partial y}\left(\frac{1}{y}\right) = \frac{\partial}{\partial x}\left(-\frac{x}{y^2}\right).$$

Let us point out that, as in the previous example, it might happen that an integrating factor exists only on a proper subset S' of the set S where M, N are defined. Obviously, in such a case, the equations (5.3) and (5.12) are equivalent on S'; see also Exercise 5.7 (2) in the sequel. This will always be understood in the examples below.

Moreover, the integrating factor need not be unique. For example, all of the functions

$$\frac{1}{xy}, \quad \frac{1}{x^2}, \quad \frac{1}{y^2}, \quad \frac{1}{x^2 + y^2}$$

are integrating factors of $y\,dx - x\,dy = 0$.

It is important to know that the procedure of solving equations by finding integrating factors is convenient when it can be applied, but it has very limited use and cannot be applied to all equations of the form $M(x,y)\,dx + N(x,y)\,dy = 0$, even if they appear to be very simple equations. It is convenient for certain types of equations; for example, when an equation happens to have either an integrating factor $\mu(x)$, which is a function of x alone, or an integrating function $\mu(y)$ which is a function of y alone.

As we have already seen, occasionally an equation can become exact if we simply multiply both sides by a common factor. Also, multiplying both sides of an equation by a function may change an exact equation into a nonexact equation. For example, the equation $y\,dx + x\,dy = 0$ is exact but if we multiply it by x, the new equation $xy\,dx + x^2\,dy = 0$ is not exact.

Case 1. Let us assume that (5.3) has an integrating factor which is a function of x, call it $\mu(x)$. Then multiplying the equation by $\mu(x)$, we have

$$\mu(x)M(x,y)\,dx + \mu(x)N(x,y)\,dy = 0.$$

In order for this equation to be exact, we must have

$$\frac{\partial}{\partial y}(\mu(x)M(x,y)) = \frac{\partial}{\partial x}(\mu(x)N(x,y)),$$

namely

$$\mu_y(x)M(x,y) + \mu(x)M_y(x,y) = \mu'(x)N(x,y) + \mu(x)N_x(x,y).$$

Since $\mu(x)$ is a function of x, then $\mu_y(x) = \partial\mu(x)/\partial y = 0$, and we get

$$\mu(x)M_y(x,y) = \mu'(x)N(x,y) + \mu(x)N_x(x,y)$$

or

$$\mu'(x)N(x,y) = [M_y(x,y) - N_x(x,y)]\mu(x). \tag{5.11}$$

To solve this equation for μ, we assume that

$$\psi \stackrel{\text{def}}{=} \frac{M_y(x,y) - N_x(x,y)}{N(x,y)}, \quad N \neq 0,$$

depends on x only. If this is the case, then from (5.11) we get

$$\frac{\mu'(x)}{\mu(x)} = \frac{M_y(x,y) - N_x(x,y)}{N(x,y)} \cdot = \psi(x) \tag{5.12}$$

which can be integrated, yielding

$$\mu(x) = e^{\int \psi(x)\,dx}$$

which is a (positive) integrating factor of (5.3).

Case 2. If we look for an integrating factor $\mu(y)$ which is a function of y only, we follow a similar procedure: assuming $M \neq 0$ and

$$\varphi \stackrel{\text{def}}{=} \frac{N_x(x,y) - M_y(x,y)}{M(x,y)}$$

is a function of y only, we find

$$\mu'(y) = \varphi(y) \cdot \mu(y),$$

which can be integrated, yielding a (positive) integrating factor

$$\mu(y) = e^{\int \varphi(y)\,dy}.$$

Example 5.7. (1) Find an integrating factor and solve $(y^2 - 3ye^x)\, dx + (y - e^x)\, dy = 0$.
Since

$$\frac{M_y - N_x}{N} = \frac{2y - 3e^x - (-e^x)}{y - e^x} = 2,$$

an integrating factor $\mu(x)$ exists and $\frac{\mu'}{\mu} = 2$. Therefore, $\mu(x) = e^{2x}$ is an integrating factor.
The given equation is changed to the exact equation $(e^{2x}y^2 - 3ye^{3x})\, dx + (ye^{2x} - e^{3x})\, dy = 0$.
Solving by Method 1, we have

$$\int\limits_0^y (ye^{2x} - e^{3x})\, dy = \frac{1}{2}y^2e^{2x} - ye^{3x} = c.$$

(2) Find an integrating factor and solve $y^3\, dx + (y^2x + 1)\, dy = 0$.
Since

$$\frac{N_x - M_y}{M} = \frac{y^2 - 3y^2}{y^3} = -\frac{2}{y},$$

an integrating factor $\mu = \mu(y)$ exists and $\mu'(y) = -\frac{2}{y} \cdot \mu(y)$. Thus $\mu(y) = \frac{1}{y^2}, y \neq 0$, and the
given equation is changed to the exact equation

$$y^3 \cdot \frac{1}{y^2}\, dx + (y^2x + 1) \cdot \frac{1}{y^2}\, dy = y\, dx + \left(x + \frac{1}{y^2}\right) dy = 0, \quad y \neq 0.$$

Solving it, we find

$$\int\limits_0^x 1 \cdot dx + \int\limits_1^y \left(x + \frac{1}{y^2}\right) dy = x + xy - x - \frac{1}{y} = xy - \frac{1}{y} = c, \quad y \neq 0. \qquad \square$$

5.5 Appendix

In this short appendix we discuss some further theoretical results on the existence and
uniqueness for exact equations.

In section 5.2 we have used the abstract results of Chapter 3 to deduce the local
existence and uniqueness of solutions to the ivp for exact equations. Below we state a
theorem which requires merely the continuity of M and N and is independent of those
abstract results. It requires the use of the implicit function theorem, see Chapter 1, Theorem 1.2:

Implicit function theorem. *Let $F(x, y)$ be a continuously differentiable function on $S \subseteq$
$\mathbb{R}^2$ and let $(x_0, y_0) \in S$ be given such that $F(x_0, y_0) = c_0$. If $F_y(x_0, y_0) \neq 0$, resp. $F_x(x_0, y_0) \neq$*

0, *there exists a neighborhood I of x_0, resp. neighborhood J of y_0, and a unique differentiable function $y = g(x)$ ($x \in I$), resp. a unique differentiable function $x = h(y)$ ($y \in J$), such that $g(x_0) = y_0$ and $F(x, g(x)) = c_0$ for all $x \in I$, resp. $h(y_0) = x_0$ and $F(h(y), y) = c_0$ for all $y \in J$.*

Theorem 5.2 (Local existence and uniqueness for exact equations). *Let M, N be continuous on $S \subseteq \mathbb{R}^2$ and suppose that the equation $M(x, y) dx + N(x, y) dy = 0$ is exact. Let $P_0 = (x_0, y_0) \in S$ be such that $N(x_0, y_0) \neq 0$, or $M(x_0, y_0) \neq 0$. Then the exact equation $M(x, y) dx + N(x, y) dy = 0$ has one and only one solution passing through P_0.*

Proof. Since $Mdx + Ndy = 0$ is exact, there exists a C^1 function $F(x, y)$, such that $F_x(x, y) = M(x, y)$ and $F_y(x, y) = N(x, y)$. If $N(x_0, y_0) \neq 0$, then $F_y(x_0, y_0) = N(x_0, y_0) \neq 0$ and we can apply the implicit function theorem to $F(x, y) = c_0 = F(x_0, y_0)$ at $P_0 = (x_0, y_0)$, yielding a unique differentiable function $y = g(x)$, defined in a neighborhood I of x_0 such that

$$F(x, g(x)) = c_0, \quad \forall x \in I, \quad g(x_0) = y_0.$$

Differentiating the preceding identity we find $F_x(x, g(x)) + F_y(x, g(x))\frac{dg(x)}{dx} = 0$, namely $M(x, g(x)) + N(x, g(x))\frac{dg(x)}{dx} = 0$, $x \in I$. This shows that $y = g(x)$ is a solution of (5.2). Since, in addition, $g(x_0) = y_0$, it follows that $y = g(x)$ is the unique solution of the ivp for (5.3) at P_0 we were looking for.

Similarly, if $M(x_0, y_0) \neq 0$ then $F_x(x_0, y_0) = M(x_0, y_0) \neq 0$ and the implicit function theorem yields a unique $x = h(y)$ such that $F(h(y), y) = c_0$, $h(y_0) = x_0$. Repeating the previous arguments it follows that $x = h(y)$ solves the ivp for (5.3) at P_0.

Notice that the result is local, in the sense that $g(x)$, resp. $h(y)$, is defined (in general) near x_0, resp. y_0. □

5.6 Exercises

1. Solve $(x^2 + 2y + e^x) dx + (2x - y^3) dy = 0$.
2. Solve $x^2 + ye^x + (y + e^x)y' = 0$.
3. Solve $(x^2 + 2y + 1) dx + (2x - y^3) dy = 0$.
4. Solve the initial value problem $(3x^2y + 2xy^2) dx + (x^3 + 2x^2y - 6) dy = 0, y(1) = -1$.
5. Solve $2y - e^x + (2x + \sin y)\frac{dy}{dx} = 0$.
6. Solve $(12x^5 - 2y) dx + (6y^5 - 2x) dy = 0$.
7. Solve $(y + \frac{1}{x}) dx + (x - \frac{1}{y}) dy = 0$.
8. Find the number a such that $(x^3 + 3axy^2) dx + (x^2y + y^4) dy = 0$ is exact and solve it.
9. Find numbers a and b such that $(xy + ay^3) dx + (bx^2 + xy^2) dy = 0$ is exact and solve it.
10. Solve $2xy^3 + 1 + (3x^2y^2)y' = 0, y(1) = 1$.
11. Solve $(y + 8x^3) dx + (x + 3y^2) dy = 0, y(1) = -1$.

12. Find the solution of $(x^2 - 1)\, dx + y\, dy = 0$ passing through $(-1, b)$ with $b > 0$ and show that it can be given in the form $y = y(x)$.

13. Solve $x^2(y^3 + 1)\, dx + y^2(x^3 - 1)\, dy = 0$.

14. Solve $y\, dx - 3x\, dy = 0$.

15. Solve $(y^3 + 1)\, dx + 3y^2\, dy = 0$.

16. Solve $\frac{2y^2 + 1}{x}\, dx + 4y \ln x\, dy = 0$.

17. Show that there exists an integrating factor $\mu = \mu(y)$ for the equation $(1 + f(y))\, dx + (xg(y) + y^2)\, dy = 0$, where f and g are some differentiable functions $f \neq -1$.

18. (A) Solve $(y + x)\, dx + dy = 0$ by finding an integrating factor $\mu(x)$.
    (B) Explain whether or not it also has an integrating factor that is a function of y only.

19. Solve $(2y + 1)\, dx + (x + \sqrt{y})\, dy = 0$, $y \geq 0$.

20. Solve $y(\cos x + \sin^2 x)\, dx + \sin x\, dy = 0$ in two ways, first as a separable equation and then as an exact equation.

21. Solve $2y\, dx + (x + \sqrt{y})\, dy = 0$, $(y \geq 0)$.

22. Solve $(y + \frac{1}{2}xy^2 + x^2y)\, dx + (x + y)\, dy = 0$.

23. Solve $[(1 + x)y + x]\, dx + x\, dy = 0$.

24. Solve $(3y + x)\, dx + x\, dy = 0$.

25. Solve $(y + xy + y^2)\, dx + (x + 2y)\, dy = 0$.

26. Does the equation $(x - 2y)\, dx + (xy + 1)\, dy = 0$ have an integrating factor that is a function of y only? Explain.

27. Prove that if you multiply an exact equation $M(x, y)\, dx + N(x, y)\, dy = 0$ by a differentiable function $f(x)$, then the new equation is still exact if and only if $f(x) = k$ is a constant.

28. Consider the exact equation $(xy^2 - 1)\, dx + (x^2y + 1)\, dy = 0$. Solve it by taking the lower limits as $(2, 1)$, i. e. $x_0 = 2$ and $y_0 = 1$. Then find the particular solution satisfying the initial condition $y(1) = -1$.

29. Find the value of c that will make $(x + ye^{xy})\, dx + (cxe^{xy})\, dy = 0$ exact and then solve it.

6 Second order linear differential equations

6.1 Preliminaries

A general second order linear differential equation has the form

$$a_2(t)\frac{d^2x}{dt^2} + a_1(t)\frac{dx}{dt} + a_0(t)x = k(t)$$

where $a_i(t)$, $i = 0,\ldots,2$, and $k(t)$ are continuous functions on an interval $I \subseteq \mathbb{R}$. Assuming that $a_2(t) \neq 0$, we can divide by a_2 and, letting $p_i(t) = \frac{a_i(t)}{a_2(t)}$, $h(t) = \frac{k(t)}{a_2(t)}$, we have

$$\frac{d^2x}{dt^2} + p_1(t)\frac{dx}{dt} + p_0(t)x = h(t).$$

It is convenient to introduce the differential operator L acting on functions $x \in C^2(I)$ by setting

$$L[x] = \frac{d^2x}{dt^2} + p_1(t)\frac{dx}{dt} + p_0(t)x.$$

With this notation, a second order linear differential equation can be written in a compact form as

$$L[x] = h. \tag{6.1}$$

If $h = 0$ the equation

$$L[x] = 0 \tag{6.2}$$

is called the _homogeneous_ linear second order equation associated to nonhomogeneous equation (6.1).

It is easy to check that

Lemma 6.1. _The differential operator L is linear, namely $L[c_1x_1 + c_2x_2] = c_1L[x_1] + c_2L[x_2]$, $c_1, c_2 \in \mathbb{R}$._

Proof. By differentiating and regrouping, we have

$$
\begin{aligned}
L[c_1x_1 + c_2x_2] &= (c_1x_1 + c_2x_2)'' + p_1(c_1x_1 + c_2x_2)' + p_0(c_1x_1 + c_2x_2) \\
&= c_1x_1'' + c_2x_2'' + c_1p_1x_1' + c_2p_1x_2' + p_0c_1x_1 + p_0c_2x_2 \\
&= c_1(x_1'' + p_1x_1' + p_0x_1) + c_2(x_2'' + p_1x_2' + p_0x_2) = c_1L[x_1] + c_2L[x_2]. \quad \square
\end{aligned}
$$

From the previous lemma we infer the following.

https://doi.org/10.1515/9783111185675-006

Lemma 6.2.

(a) *If x_1, x_2 are solutions of the homogeneous equation $L[x] = 0$, then any linear combi-nation $c_1x_1 + c_2x_2$ is also a solution of (6.2).*

(b) *If y is a solution of the non-homogeneous equation $L[y] = h$ and x is any solution of the associated homogeneous equation $L[x] = 0$, then $x + y$ is a solution of the non-homogeneous equation (6.1).*

Proof. (a) Using the linearity of L we find

$$L[c_1x_1 + c_2x_2] = c_1L[x_1] + c_2L[x_2] = 0,$$

since $L[x_1] = L[x_2] = 0$. This proves that $c_1x_1 + c_2x_2$ is a solution of (6.2).

(b) One has $L[x + y] = L[x] + L[y] = L[y] = h$, since $L[x] = 0$. □

The *initial value problem*, or *Cauchy problem* associated to $L[x] = h$ consists of find-ing a function $x(t)$ such that

$$\begin{cases} L[x] = h, \\ x(t_0) = y_0, \\ x'(t_0) = y_1, \end{cases} \tag{6.3}$$

where (y_0, y_1) is any point in $\mathbb{R}^2$ and $t_0 \in I$.

For example, the Cauchy problem associated to $x'' + p_1x' + p_0x = h$ involves seeking a solution $x(t)$ of the equation satisfying the two initial conditions $x(t_0) = y_0, x'(t_0) = y_1$.

The following theorem is an extension of the existence and uniqueness results proved for first order equations in Chapter 2.

Theorem 6.1. *Let p_0, p_1 be continuous on the interval $I \subseteq \mathbb{R}$. Then for any $t_0 \in I$ and $(y_0, y_1) \in \mathbb{R}^2$ the Cauchy problem (6.3) has one and only one solution, which is defined on I.*

From the preceding theorem we can deduce the counterpart of a result proved in Chapter 2 for first order linear equations; see Corollary 2.1(a) therein.

Corollary 6.1. *Let $x(t)$ be a solution of $L[x] = 0$. If there exists $t_0 \in I$ such that $x(t_0) = x'(t_0) = 0$, then $x(t)$ is identically zero.*

Proof. The function $x_0(t) \equiv 0$ solves the initial value problem $L(x) = 0$ and the ivp $x(t_0) = x'(t_0) = 0$. Therefore it follows from Theorem 6.1 above that $x(t) \equiv x_0(t) \equiv 0$, that is, $x(t)$ is the *trivial solution*. □

Remark 6.1. Contrary to the statement of Corollaries 2.1(b) and (c), Chapter 2, a solution of $L[x] = 0$ can vanish without being identically zero. For example, $x(t) = \sin t$ solves $x'' + x = 0$ and vanishes at $t = k\pi, k \in \mathbb{Z}$.

6.2 General solution of $L[x] = h$

In the sequel we will assume, for the sake of simplicity, that p_0, p_1 are continuous in $\mathbb{R}$. The case in which they are continuous on an interval $I \subset \mathbb{R}$ requires minor adjustments.

The *general solution (or integral)* of $L[x] = h$ is a family of smooth functions $\phi(t; c_1, c_2)$ such that:
(1) for any $c_1, c_2 \in R$, the function $\phi(t; c_1, c_2)$ is a solution of (6.1);
(2) for any solution $x(t)$ of (6.1) there exist $c_1, c_2 \in \mathbb{R}$ such that $x(t) = \phi(t; c_1, c_2)$.

In order to find the general solution of $L[x] = 0$, it is convenient to introduce the notion of linear dependence and independence of functions.

6.2.1 Linear dependence and independence

Definition 6.1.
(a) Two functions u_1, u_2 are said to be *linearly independent* provided

$$c_1 u_1(t) + c_2 u_2(t) = 0, \quad \forall t \in \mathbb{R} \quad \Longleftrightarrow \quad c_1 = c_2 = 0.$$

(b) u_1, u_2 are said to be *linearly dependent* provided there exist nonzero constants c_1, c_2 such that

$$c_1 u_1(t) + c_2 u_2(t) = 0, \quad \forall t \in \mathbb{R}.$$

Example 6.1. In order to prove that

$$u_1 = \left(t^2 + 2\right)^3 \quad \text{and} \quad u_2 = \left(t^2 + 1\right)^4$$

are linearly independent, we will show that there do not exist nonzero constants c_1 and c_2 such that $c_1 u_1(t) + c_2 u_2(t) \equiv 0$, or equivalently, that $c_1 u_1(t) + c_2 u_2(t) \equiv 0$ implies $c_1 = c_2 = 0$. So, suppose that

$$c_1\left(t^2 + 2\right)^3 + c_2\left(t^2 + 1\right)^4 \equiv 0.$$

If we let $t = 0$, we get $8c_1 + c_2 = 0$. Similarly, if we let $t = 1$, we get $27c_1 + 16c_2 = 0$. Solving the algebraic system

$$\begin{cases} 8c_1 + c_2 = 0, \\ 27c_1 + 16c_2 = 0. \end{cases}$$

we obtain $c_1 = c_2 = 0$. Or, alternatively, we simply notice that the determinant of the coefficients is $8 \times 16 - 27 \neq 0$ and hence $c_1 = c_2 = 0$.

Remark 6.2. Two functions u_1, u_2 are linearly dependent if and only if there is a constant λ such that $u_1(t) = \lambda u_2(t)$ for all t.

If $u_1(t) \equiv \lambda u_2(t)$ then it suffices to take $c_1 = 1, c_2 = -\lambda$ to see that $c_1 u_1(t) + c_2 u_2(t) \equiv 0$. If, on the other hand, there exist nonzero constants $c_1, c_2 \in \mathbb{R}$ such that $c_1 u_1(t) + c_2 u_2(t) \equiv 0$, then we divide by c_1 and obtain $u_1(t) = \lambda u_2(t)$ with $\lambda = -\frac{c_2}{c_1}$.

In particular, it follows that linearly dependent functions have common zeros.

To check linear dependence/independence it is convenient to introduce the *Wronskian W* of two differentiable functions u_1, u_2, defined as

$$W = W(u_1, u_2; t) = \begin{vmatrix} u_1(t) & u_2(t) \\ u_1'(t) & u_2'(t) \end{vmatrix} = u_1(t)u_2'(t) - u_2(t)u_1'(t).$$

We may alternatively use the notation $W(u_1, u_2)(t)$, or simply W when there is no risk of confusion.

The connection between the Wronskian and linear dependence/independence is shown in the following lemmas.

Lemma 6.3. *If two differentiable functions* u_1, u_2 *are linearly dependent, then* $W(u_1, u_2; t) = 0$ *for all* $t \in \mathbb{R}$.

Proof. Suppose that u_1, u_2 are linearly dependent and there exists a number t^* such that $W(u_1, u_2)(t^*) \neq 0$. Then since u_1 and u_2 are linearly dependent, there exist nonzero constants c_1, c_2 such that $c_1 u_1(t) + c_2 u_2(t) \equiv 0 \implies c_1 u_1'(t) + c_2 u_2'(t) \equiv 0$. Now consider the algebraic system

$$\begin{cases} c_1 u_1(t^*) + c_2 u_2(t^*) = 0, \\ c_1 u_1'(t^*) + c_2 u_2'(t^*) = 0. \end{cases}$$

The above system has nonzero solution c_1, c_2 only if the coefficient determinant is zero, i.e if $W(u_1, u_2)(t^*) = 0$, contradiction. Therefore, $W(u_1, u_2)(t) \equiv 0$. $\square$

For example, $u_1 = t^2, u_2 = t$ are linearly independent since $W(u_1, u_2; t) = t^2 - 2t^2 = -t^2$ is not identically zero.

Arguing by contradiction, the following lemma follows immediately from the preceding lemma.

Lemma 6.4. *If* $W(u_1, u_2; t_0) \neq 0$ *for some* $t_0 \in \mathbb{R}$ *then* u_1, u_2 *are linearly independent.*

Let us point out that, in general, the converse of Lemma 6.3 is not true: it could happen that $W(u_1, u_2; t) = 0$ for all $t \in \mathbb{R}$ even if u_1, u_2 are not linearly dependent. For example, let $u_1 = t^3$ and

$$u_2(t) = |t^3| = \begin{cases} t^3, & \text{for } t \geq 0, \\ -t^3, & \text{for } t \leq 0. \end{cases}$$

Then

$$W(u_1, u_2; t) = u_1(t)u_2'(t) - u_1'(t)u_2(t) = \begin{cases} t^3 \cdot 3t^2 - 3t^2 \cdot t^3 = 0 & \text{for } t \geq 0, \\ t^3 \cdot (-3t^2) - 3t^2 \cdot (-t^3) = 0 & \text{for } t \leq 0. \end{cases}$$

So, $W = 0$ for all t, but u_1, u_2 are obviously linear independent on $\mathbb{R}$. On the other hand the following lemma holds.

Lemma 6.5. *Let u_1, u_2 be solutions of $L[x] = 0$. If $W(t_0) = 0$ for some t_0, then u_1, u_2 are linearly dependent on $\mathbb{R}$.*

Proof. Let us consider the system

$$\begin{cases} c_1 u_1(t_0) + c_2 u_2(t_0) = 0, \\ c_1 u_1'(t_0) + c_2 u_2'(t_0) = 0, \end{cases}$$

whose determinant is $W(t_0)$. If $W(t_0) = 0$, this system has nonzero solution c_1, c_2. The function $x(t) = c_1 u_1(t) + c_2 u_2(t)$ solves $L[x] = 0$; see Lemma 6.2(a). Moreover, from the system we infer that $x(t_0) = x'(t_0) = 0$. From Corollary 6.1 it follows that $x(t) \equiv 0$, namely $c_1 u(t) + c_2 u_2(t) = 0$ for all $t \in \mathbb{R}$. Thus u_1, u_2 are linearly dependent on $\mathbb{R}$. ☐

From Lemmas 6.4 and 6.5 we infer

Lemma 6.6. *Two solutions u_1, u_2 of $L[x] = 0$ are linearly independent on $\mathbb{R}$ if and only if their wronskian is different from zero for all $t \in \mathbb{R}$.*

The following theorem shows an important property of the wronskian of solutions of $L[x] = 0$.

Theorem 6.2 (Abel's theorem). *If u_1, u_2 are solutions of $L[x] = x'' + p_1 x' + p_0 x = 0$, then their wronskian W satisfies the equation*

$$W' + p_1 W = 0$$

and is hence given by what is known as the Abel formula,

$$W(t) = ce^{-\int p_1(t)\, dt}, \quad c \in \mathbb{R}.$$

Proof. Differentiating $W = u_1 u_2' - u_1' u_2$ we obtain $W' = u_1' u_2' + u_1 u_2'' - u_1' u_2' - u_1'' u_2 = u_1 u_2'' - u_1'' u_2$. Recalling that u_1, u_2 are solution of $x'' + p_1 x' + p_0 x = 0$, we infer

$$u_1'' = -p_1 u_1' - p_0 u_1, \quad u_2'' = -p_1 u_2' - p_0 u_2$$

This in turn implies

$$W' = u_1 u_2'' - u_1'' u_2 = u_1(-p_1 u_2' - p_0 u_2) - u_2(-p_1 u_1' - p_0 u_1)$$
$$= -p_1 u_1 u_2' - p_0 u_1 u_2 + p_1 u_2 u_1' + p_0 u_1 u_2 = -p_1(u_1 u_2' - u_1' u_2) = -p_1 W.$$

Solving the first order linear equation $W' = -p_1 W$, the result follows. ☐

Corollary 6.2. *The wronskian* $W(u_1, u_2; t)$ *of two solutions* u_1, u_2 *of* $L[x] = 0$ *is either identically zero or it never vanishes.*

Example 6.2. Consider the functions $u_1 = t$, $u_2 = t^2$. Their wronskian $W = 2t^2 - t^2 = t^2$ is such that $W(0) = 0$, $W(t) \neq 0$ for $t \neq 0$. Thus they cannot be solutions of the same second order linear homogeneous equation on $\mathbb{R}$. ☐

6.2.2 Exercises on linear dependence and independence

1. Show that $x_1 = e^{2t}$, $x_2 = e^t$ are linearly independent.
2. Show that $x_1 = 2 \sin t$ and $x_2 = \sin 2t$ are linearly independent.
3. Show that $x_1 = (t^2 - 1)^5$ and $x_2 = (2 - 2t^2)^5$ are linearly dependent.
4. Show that $x_1 = \sin^2 t$ and $x_2 = \cos^2 t$ are linearly independent.
5. Show that $x_1 = 2t$, $x_2 = 3|t|$ are linearly independent on $(-\infty, 0)$ but linearly dependent on $\mathbb{R}$.
6. Let $y(t)$ be any function, $y(t) \neq 0$ for $t \in \mathbb{R}$. Show that if $x_1(t)$ and $x_2(t)$ are linearly independent, then so are $y(t)x_1(t)$ and $y(t)x_2(t)$.
7. Let f, g be linearly independent. Show that for any $a, b \in \mathbb{R}$, $a \neq 0$, $x_1 = af + bg$ and $x_2 = g$ are linearly independent.
8. Show that if $p \neq q$ then $u_1 = t^p$ and $u_2 = t^q$ cannot be solutions of the same second order equation $L[x] = 0$.
9. (a) Show that $f(t) = t^2$ and $g(t) = e^t$ are linearly independent.
   (b) Evaluate $W(f, g)(t)$, and then explain why f, g cannot be solutions of an equation of the form $L(x) = 0$, a, b constants.
10. Solve $W(t + 1, x(t)) = t + 1$, $x(0) = 1$.
11. Solve $W(t, x') = 1$, $x(1) = x'(1) = 2$.
12. Let u_1, u_2 be solutions of $L[x] = 0$ on I such that $u_1(t_0) = u_2(t_0) = 0$ for some $t_0 \in I$. Show that they are linearly dependent and hence all their zeros are in common.
13. Consider the equation $(t^2 + 1)x'' - tx' + e^t x = 0$. If $W(x_1, x_2)(1) = 2$ for a pair of linearly independent solutions x_1, x_2, find $W(x_1, x_2)(2)$.
14. Solve $W(t, x) = 1$, $x(1) = 2$.
15. Prove or disprove the statement that if $x_1(t)$ and $x_2(t)$ are linearly independent on $(-4, 3)$ then they are linearly independent on $(-2, 2)$.
16. Let x_1, x_2 be solutions of $x'' - x' + q(t)x = 0$, p, q continuous. If $W(x_1, x_2)(2) = 5$, find $W(x_1, x_2)(3)$.
17. Show that if x_1, x_2 are linearly independent then so are $y_1 = 2x_1 + 3x_2$ and $y_2 = 5x_1 - 4x_2$.

6.2.3 General solution of homogeneous equation

Definition 6.2. We say that two solutions u_1, u_2 of $L[x] = 0$ are *fundamental solutions* if they are linearly independent.

According to Lemma 6.6, u_1, u_2 are two fundamental solutions of $L[x] = 0$ whenever their wronskian $W \neq 0$. Of course, by Abel's theorem, to check that u_1, u_2 are two fundamental solutions, it suffices to show that $W(t_0) \neq 0$ for some t_0.

Theorem 6.3. *Let u_1, u_2 be a fundamental set of solutions of $L[x] = 0$. Then the general solution of $L[x] = 0$ is given by $x = c_1 u_1 + c_2 u_2$, $c_1, c_2 \in \mathbb{R}$.*

Proof. First, it follows that if $x = c_1 u_1 + c_2 u_2$, $c_i \in \mathbb{R}$, then $L[x] = c_1 L[u_1] + c_2 L[u_2] = 0$ since u_1 and u_2 are solutions. Let $x(t)$ be any solution of $L[x] = 0$. We want to show that there exist constants c_1, c_2 such that $x(t) = c_1 u_1 + c_2 u_2$. Let $t_0 \in \mathbb{R}$ be any fixed number and consider the system

$$\begin{cases} c_1 u_1(t_0) + c_2 u_2(t_0) = x(t_0), \\ c_1 u_1'(t_0) + c_2 u_2'(t_0) = x'(t_0). \end{cases}$$

The determinant of this system is simply the wronskian $W(u_1, u_2; t_0)$, which is different from zero since u_1, u_2 are linearly independent. Thus by the Cramer rule, this algebraic system has a unique solution $\bar{c}_1, \bar{c}_2$ (depending on t_0). The function

$$y(t) := \bar{c}_1 u_1(t) + \bar{c}_2 u_2(t)$$

is a solution of the equation such that $y(t_0) = x(t_0)$ and $y'(t_0) = x'(t_0)$. By uniqueness $y(t) = x(t)$. Thus we have found constants $\bar{c}_1, \bar{c}_2$ such that $x = \bar{c}_1 u_1 + \bar{c}_2 u_2$, proving the theorem. □

Remark 6.3. As pointed out above, $\bar{c}_1, \bar{c}_2$ depend on the choice of t_0. This is consistent with the fact that, given a solution $x = c_1 u_1 + c_2 u_2$, the constants c_1, c_2 are determined by the initial conditions $x(t_0) = x_0$, $x'(t_0) = x_1$.

Example 6.3. (a) Find the general solution of

$$x'' + k^2 x = 0, \quad k \neq 0.$$

Two solutions are given by $u_1 = \sin kt$, $u_2 = \cos kt$. Their wronskian is

$$W = u_1 u_2' - u_2 u_1' = -k \sin^2 kt - k \cos^2 kt = -k(\sin^2 kt + \cos^2 kt) = -k \neq 0.$$

Thus they are a pair of fundamental solutions and hence the general solution is $x = c_1 \sin kt + c_2 \cos kt$.

(b) Find the general solution of $x'' - k^2 x = 0$, $k \neq 0$. Two solutions are given by $u_1 = e^{kt}$, $u_2 = e^{-kt}$. Their wronskian is

$$W = u_1 u_2' - u_2 u_1' = e^{kt} \cdot (-ke^{-kt}) - e^{-kt} \cdot (ke^{kt}) = -2k \neq 0.$$

Thus they are a pair of fundamental solutions and hence the general solution is $x = c_1 e^{kt} + c_2 e^{-kt}$. □

6.2.4 General solution of nonhomogeneous equation

Theorem 6.4. *Let u_1, u_2 be two fundamental solutions of $L[x] = 0$ and let y be a particular solution of $L[y] = h$. Then the general solution of $L[x] = h$ is given by $x = c_1u_1 + c_2u_2 + y$.*

Proof. From Lemma 6.2 (b) it follows that $c_1u_1 + c_2u_2 + y$ is a solution of $L[x] = h$.

Conversely, let z be any solution of $L[z] = h$. Then $z-y$ satisfies $L[z-y] = L[z]-L[y] = h - h = 0$. Since $z - y$ is a solution of the homogeneous equation $L[x] = 0$, Theorem 6.3 yields $c_1, c_2 \in \mathbb{R}$ such that $z - y = c_1u_1 + c_2u_2$, namely $z = c_1u_1 + c_2u_2 + y$. □

According to the previous theorem, the general solution of $L[x] = h$ is obtained by combining the general solution of the corresponding homogeneous equation $L[x] = 0$ and any particular solution of the non-homogeneous equation.

In order to find a particular solution of $L[y] = h$ we may use the method of *variation of parameters* described below.

Variation of parameters. Given two linearly independent solutions u_1, u_2 of $L[x] = 0$, we seek two functions $v_1(t)$, $v_2(t)$ such that $y = v_1u_1 + v_2u_2$ is a solution of the non-homogeneous equation $L[y] = h$. We will now show that this can be accomplished if we can find two functions v_1 and v_2 such that

$$\begin{cases} v_1'u_1 + v_2'u_2 = 0, \\ v_1'u_1' + v_2'u_2' = h. \end{cases} \tag{6.4}$$

To this end, we first differentiate $y(t) = u_1v_1 + v_2u_2$, obtaining $y' = v_1u_1' + v_2u_2' + v_1'u_1 + v_2'u_2$. Our goal is to find two equations involving v_1' and v_2' so that we can solve for them and then integrate each to get v_1 and v_2. We get the first equation by an arbitrary choice of setting $v_1'u_1 + v_2'u_2 = 0$. This reduces y' to $y' = v_1u_1' + v_2u_2'$. In order to establish the second equation in (6.4), we differentiate y', obtaining $y'' = v_1'u_1' + v_1u_1'' + v_2'u_2' + v_2u_2''$. Hence setting $L[y] = h$, we have

$$\underbrace{v_1u_1'' + v_2u_2'' + v_1'u_1' + v_2'u_2'}_{y''} + p_1\underbrace{(v_1u_1' + v_2u_2')}_{y'} + p_0\underbrace{(v_1u_1 + v_2u_2)}_{y} = h.$$

Gathering all the terms with v_1 and all those with v_2 together and factoring v_1 and v_2, we obtain

$$v_1(u_1'' + p_1u_1' + p_0u_1) + v_2(u_2'' + p_1u_2' + p_0u_2) + v_1'u_1' + v_2'u_2' = h.$$

Since u_1 and u_2 are solutions of the homogeneous equation $L[x] = x'' + p_1x' + p_0x = 0$, the above equation reduces to

$$v_1'u_1' + v_2'u_2' = h,$$

which is the second equation in (6.4).

Since the determinant of the system (6.4) is the wronskian of u_1, u_2, which is different from zero by assumption, it follows that (6.4) has a unique solution $v_1'(t)$, $v_2'(t)$ given by

$$v_1' = -\frac{hu_2}{W}, \quad v_2' = \frac{hu_1}{W}.$$

Integrating, we find

$$v_1(t) = -\int \frac{h(t)u_2(t)}{W}\,dt, \quad v_2(t) = \int \frac{h(t)u_1(t)}{W}\,dt.$$

Choosing v_1, v_2 in this manner, a particular solution of $L[x] = h$ is given by $y(t) = v_1(t)u_1(t) + v_2(t)u_2(t)$.

Example 6.4. Find the general solution of

$$x'' + x = \tan t.$$

It can be easily verified that $u_1 = \sin t$ and $u_2 = \cos t$ are two linearly independent solutions. Hence the general solution of $x'' + x = 0$ is given by

$$x(t) = c_1 \sin t + c_2 \cos t.$$

In order to find the general solution of the given nonhomogeneous equation, we need a particular solution of the nonhomogeneous equation (see Theorem 6.4). Let us use the method of Variation of Parameters directly. We seek functions v_1, v_2 such that $x_p = v_1 u_1 + v_2 u_2 = v_1 \sin t + v_2 \cos t$ is a solution of $L(x) = \tan t$. This will be accomplished if we can solve the following equations.

$$\begin{cases} v_1' \sin t + v_2' \cos t = 0, \\ v_1' \cos t - v_2' \sin t_2 = \tan t. \end{cases} \tag{6.5}$$

Solving the above equations for v_1', v_2' and then integrating, we find $v_1 = -\cos t$ and $v_2 = -\ln|\sec t + \tan t| + \sin t$. Thus, the particular solution x_p is given by

$$x_p = u_1 v_1 + u_2 v_2 = -\sin t \cos t + \cos t[-\ln|\sec t + \tan t| + \sin t] = -\cos t \ln|\sec t + \tan t.$$

Finally, the general solution of the given nonhomogeneous equation is

$$x + x_p = c_1 \sin t + c_2 \cos t - \cos t \ln|\sec t + \tan|. \qquad \square$$

We will solve the next example by using the formulas

$$v_1(t) = -\int \frac{h(s)u_2(s)}{W} \, ds, \quad v_2(t) = \int \frac{h(s)u_1(s)}{W} \, ds$$

developed above.

Example 6.5. Find a particular solution of

$$x'' - x = e^t \cdot \frac{2t - 1}{t^2}, \quad t \neq 0.$$

It is easy to check that $u_1 = e^t$, $u_2 = e^{-t}$ are two linearly independent solutions of the associated homogeneous equation $x'' - x = 0$, with wronskian $W = \begin{vmatrix} e^t & e^{-t} \\ e^t & -e^{-t} \end{vmatrix} = -2$. Hence

$$v_1(t) = -\int \frac{e^t \cdot \frac{2t-1}{t^2} \cdot e^{-t}}{W} \, dt = \frac{1}{2} \int \frac{2t - 1}{t^2} \, dt = \ln|t| + \frac{1}{2t},$$

$$v_2(t) = \int \frac{e^t \cdot \frac{2t-1}{t^2} \cdot e^t}{W} \, dt = -\frac{1}{2} \int e^{2t} \cdot \frac{2t - 1}{t^2} \, dt = -\frac{1}{2} \frac{e^{2t}}{t}.$$

It follows that

$$y = \left(\ln|t| + \frac{1}{2t} \right) e^t - \frac{e^{2t}}{2t} \cdot e^{-t} = e^t \ln|t| + \frac{e^t}{2t} - \frac{e^t}{2t} = e^t \ln|t| \quad (t \neq 0)$$

is a particular solution of our equation. □

6.2.5 Appendix: reduction of order method

Suppose that we know one solution $u(t) \neq 0$ of $L[x] = 0$. Substituting $v = zu$, we find

$$L[v] = L[zu] = (zu)'' + p_1(zu)' + p_0 zu$$
$$= (z''u + z'u' + zu'' + z'u') + p_1(z'u + u'z) + p_0 zu$$
$$= z''u + 2z'u' + p_1 z'u + z(u'' + p_1 u' + p_0 u) = z''u + 2z'u' + p_1 z'u + zL[u].$$

Since $L[u] = 0$ it follows that $L[v] = z''u + 2z'u' + p_1 z'u$. Thus $L[v] = 0$ provided $z''u + 2z'u' + p_1 z'u = 0$. Setting $w = z'$ we find $uw' + (2u' + p_1 u)w = 0$, which is a first order linear equation. Dividing by $u \neq 0$ we get

$$w' + \left(p_1 + 2\frac{u'}{u} \right) w = 0.$$

Solving this first order linear equation by the integrating factor method we see that

$$w = ce^{-\int (p_1 + 2\frac{u'}{u})} = ce^{-\int p_1} \cdot e^{-2\int \frac{u'}{u}} = ce^{-\int p_1} \cdot e^{-\ln u^2} = cu^{-2} \cdot e^{-\int p_1}.$$

Since we need only one such a function w, let us take $c = 1$. Moreover, from $w = z'$ we infer that

$$z(t) = \int_0^t \frac{e^{-\int_0^s p_1}}{u^2(s)} ds \tag{6.6}$$

is such that $v = zu$ is a solution of $L[x] = 0$.

Next, let us show that u and v are linearly independent. We can evaluate the wronskian of u and v at $t = 0$, yielding

$$W(0) = u(0)v'(0) - u'(0)v(0) = u(0) \cdot [z'(0)u(0) + z(0)u'(0)] - u'(0)z(0)u(0).$$

Since $z(0) = 0$ and $z'(0) = \dfrac{e^{\int_0^0 p_1}}{u^2(0)} = \dfrac{1}{u^2(0)}$, we find $W(0) = u^2(0)z'(0) = 1$, proving that u and v are linearly independent.

The preceding discussion is summarized in the following proposition.

Proposition 6.1 (Reduction of order). *Let u be a solution of the second order equation $x'' + p_1 x' + p_0 x = 0$, with $u(t) \neq 0$. Then*

$$v(t) = u(t) \int \frac{e^{-\int p_1(t) dt}}{u^2(t)} dt$$

is another solution. Moreover, u and v are linearly independent.

Example 6.6. Consider the equation

$$x'' - \frac{1}{t}x' + \frac{1}{t^2}x = 0, \quad t > 0.$$

We can see that $u(t) = t$ is a solution. Using Proposition 6.1, we find another linearly independent solution

$$v(t) = t \int \frac{e^{-\int \frac{-1}{t} dt}}{t^2(t)} dt = t \ln t.$$

Let us apply the method of reduction of order directly in order to find a second linearly independent solution, without using the formula. Let $v(t) = tz(t)$. Then we have $v' = z't + z$ and $v'' = z''t + 2z'$. Substituting in the equation and simplifying, we get

$$z'' + \frac{1}{t}z' = 0.$$

If we let $z' = w$ we obtain the first order equation

$$w' + \frac{1}{t}w = 0.$$

Solving for w, we find $w = \frac{c}{t}$. Since we are only interested in finding one solution, we take $c = 1$ and $w = \frac{1}{t}$. This means that $z' = \frac{1}{t}$ and therefore we can take $z(t) = \ln t$; resulting in a second linearly independent solution $v(t) = t \ln t$. $\qquad\qquad\square$

The preceding method is called *reduction of order* since, if we know one solution $u(t) \neq 0$ of a second order linear equation $L[x] = 0$, the search for a second solution leads to solving a linear first order equation.

6.3 Equations with constant coefficients

6.3.1 The homogeneous case

Here we deal with homogeneous second order equations with constant coefficients,

$$L[x] = x'' + ax' + bx = 0, \quad a, b \in \mathbb{R}. \tag{6.7}$$

Letting $x = e^{\lambda t}$ and substituting in (6.7), we find $\lambda^2 e^{\lambda t} + a\lambda e^{\lambda t} + b e^{\lambda t} = 0$ and hence, dividing by $e^{\lambda t}$,

$$\lambda^2 + a\lambda + b = 0. \tag{6.8}$$

This shows that the function $x = e^{\lambda t}$ is a solution of (6.7) whenever λ is a root of the second order algebraic equation (6.8). Equation (6.8) is called the *characteristic or auxiliary equation* corresponding to the second order linear differential equation (6.7). Solving (6.8) we find

$$\lambda = \frac{-a \pm \sqrt{a^2 - 4b}}{2}.$$

It is convenient to distinguish among three cases, which will be discussed separately:

1. $a^2 - 4b > 0$: (6.8) has two real roots λ_1, λ_2, with $\lambda_1 \neq \lambda_2$;
2. $a^2 - 4b = 0$: (6.8) has a repeated real root $\lambda_1 = \lambda_2 = -\frac{a}{2}$;
3. $a^2 - 4b < 0$: (6.8) has two complex conjugate roots $\lambda_{1,2} = -\frac{a}{2} \pm \frac{i\beta}{2}$, with $\beta = \sqrt{4b - a^2} > 0$.

Case 1. The functions $u_i(t) = e^{\lambda_i t}$, $i = 1, 2$, are solutions of (6.7). Let us show that they are linearly independent on $\mathbb{R}$. To verify this, we evaluate the wronskian

$$W(u_1, u_2)(0) = \begin{vmatrix} u_1(0) & u_2(0) \\ u_1'(0) & u_2'(0) \end{vmatrix} = \begin{vmatrix} 1 & 1 \\ \lambda_1 & \lambda_2 \end{vmatrix} = \lambda_2 - \lambda_1 \neq 0,$$

and hence, by Lemma 6.6, u_1, u_2 are linearly independent and therefore u_1, u_2 is a pair of fundamental solutions. As a consequence, the general solution of (6.7) is given by

$$x = c_1 e^{\lambda_1 t} + c_2 e^{\lambda_2 t}.$$

Case 2. We have only one solution $u = e^{-\frac{a}{2}t}$. To find a second linearly independent solution, we can use the reduction of order method; see Proposition 6.1. Let us repeat the
calculations for this specific case. Setting $v = zu$ we find

$$v'' + av' + bv = z(u'' + au' + bu) + z''u + z'(2u' + au) = z''u + z'(2u' + au).$$

Since $u = e^{-\frac{a}{2}t}$ and $u' = -\frac{a}{2}e^{-\frac{a}{2}t} = -\frac{a}{2}u$ it follows that $2u' + au = 0$ and hence $v'' + av' + bv = z''u$. Therefore, $v = zu$ is a solution provided $z'' = 0$. Since we need one function, it
suffices to take $z = t$. As a consequence, $v = tu$ is another linearly independent solution,
and the general solution of (6.7) is given by

$$x = c_1 u(t) + c_2 tu(t) = (c_1 + c_2 t)u(t) = (c_1 + c_2 t) \cdot e^{-\frac{a}{2}t}.$$

Case 3. Consider the complex valued exponential functions $e^{(\alpha+i\beta)t}$. Recalling the Euler
formula $e^{i\beta t} = \cos\beta t + i\sin\beta t$, we see that

$$e^{(\alpha+i\beta)t} = e^{\alpha t} \cdot e^{i\beta t} = e^{\alpha t} \cdot (\cos\beta t + i\sin\beta t) = \underbrace{e^{\alpha t} \cdot \cos\beta t}_{:=u(t)} + i\underbrace{e^{\alpha t} \cdot \sin\beta t}_{:=v(t)} = u(t) + iv(t).$$

We know that if $\alpha = -\frac{1}{2}a$ and $\beta = \frac{1}{2}\sqrt{4b - a^2}$ then $z = u + iv$ is a complex valued solution
of (6.7), in the sense that $z'' + az' + bz = 0$. Substituting $z = u + iv$ we find

$$(u + iv)'' + a(u + iv)' + b(u + iv) = 0.$$

Grouping the terms involving u and those involving v, we obtain

$$(u'' + au' + bu) + i(v'' + av' + bv) = 0.$$

It follows that both the real part $u'' + au' + bu$ and the complex part $v'' + av' + bv$ vanish.
In other words $u'' + au' + bu = 0$ and $v'' + av' + bv = 0$, which means that both $u = e^{-\frac{a}{2}t} \cdot \cos\beta t$ and $v = e^{-\frac{a}{2}t} \cdot \sin\beta t$ are solutions of (6.7). Next, we claim that u and v are
linearly independent. Actually,

$$u' = -\frac{a}{2}e^{-\frac{a}{2}t} \cdot \cos\beta t - \beta e^{-\frac{a}{2}t} \cdot \sin\beta t = -\frac{a}{2}u - \beta v,$$
$$v' = -\frac{a}{2}e^{-\frac{a}{2}t} \cdot \sin\beta t + \beta e^{-\frac{a}{2}t} \cdot \cos\beta t = -\frac{a}{2}v + \beta u.$$

Therefore

$$W = uv' - u'v = u\left(-\frac{a}{2}v + \beta u\right) - v\left(-\frac{a}{2}u - \beta v\right) = \beta v^2 + \beta u^2.$$

Since $\beta \neq 0$ we have $W \neq 0$ and we conclude that u and v are linearly independent and
hence they are a pair of fundamental solutions of (6.7). As a consequence, the general

solution of (6.7) is given by

$$x = c_1 u(t) + c_2 v(t) = c_1 e^{-\frac{a}{2}t} \cdot \cos \beta t + c_2 e^{-\frac{a}{2}t} \cdot \sin \beta t = e^{-\frac{a}{2}t} \cdot (c_1 \cos \beta t + c_2 \sin \beta t).$$

6.3.2 Applications

Linear second order equations with constant coefficients arise in many applications. We discuss below two of them.

1. The harmonic oscillator. Consider a body of unit mass on the x axis at the free end of an elastic spring which is attached at the origin O. According to Hooke's law, the force F acting on the body is given by $F(x) = -\omega^2 x$, where x is the distance of the body from O and the minus sign is due to the fact that the force brings back the body to the initial position, acting oppositely to the motion of the body. In absence of friction, then Newton's second law (mass × acceleration = force) yields $x'' = -\omega^2 x$ or

$$x'' + \omega^2 x = 0. \tag{6.9}$$

This is referred to as the equation of the *free harmonic oscillator*.

The associated characteristic equation is $\lambda^2 + \omega^2 = 0$, whose solutions are $\lambda = \pm\sqrt{-\omega^2} = \pm i\omega$. Thus the general solution of (6.9) is given by

$$x(t) = c_1 \sin \omega t + c_2 \cos \omega t,$$

the superposition of $\sin \omega t$ and $\cos \omega t$, which models the motion of the elastic spring. Setting

$$(i) \quad A = \sqrt{c_1^2 + c_2^2}, \qquad (ii) \quad \theta = \arctan \frac{c_1}{c_2}, \quad (c_2 \neq 0),$$

we can also write

$$x(t) = A \sin(\omega t + \theta).$$

Actually

$$x = A \cdot \left(\frac{c_1}{\sqrt{c_1^2 + c_2^2}} \sin \omega t + \frac{c_2}{\sqrt{c_1^2 + c_2^2}} \cos \omega t \right).$$

Let us assume, e. g., that $c_1 > 0$ and $c_2 > 0$ (the other cases require minor changes). If θ is such that

$$\sin \theta = \frac{c_1}{\sqrt{c_1^2 + c_2^2}}$$

then

$$\cos\theta = \sqrt{1 - \sin^2\theta} = \sqrt{1 - \frac{c_1^2}{c_1^2 + c_2^2}} = \sqrt{\frac{c_2^2}{c_1^2 + c_2^2}} = \frac{c_2}{\sqrt{c_1^2 + c_2^2}}.$$

On the other hand, in the presence of friction, equation (6.9) has to be modified as

$$x'' + kx' + \omega^2 x = 0, \tag{6.10}$$

referred as the equation of the *damped harmonic oscillator*. The solution of (6.10) depends on the relationship between $k > 0$ and $\omega > 0$. Precisely,

1.1. $k > 2\omega$. The characteristic equation $\lambda^2 + k\lambda + \omega^2 = 0$ has two real distinct roots λ_1, λ_2. Then the general solution of the damped oscillator is

$$x = c_1 e^{\lambda_1 t} + c_2 e^{\lambda_2 t}.$$

Noticing that λ_1 and λ_2 are both negative, for $k > 0$, it follows that the solutions decay to zero as $t \to +\infty$: the friction force is sufficiently larger than the elastic force in such a way that the body does not oscillate but is attracted at O (overdamped response); see Fig. 6.1.

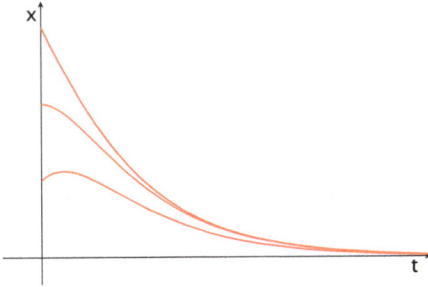

Figure 6.1: Overdamped response.

1.2. $k = 2\omega$. In this case, the characteristic equation has a double root $\lambda = -\frac{k}{2}$. As is well known, the general solution is given by

$$x = c_1 e^{-\frac{k}{2}t} + c_2 t e^{-\frac{k}{2}t} = e^{-\frac{k}{2}t} \cdot (c_1 + c_2 t).$$

Again, this solution decays to zero as $t \to +\infty$ without oscillations (critically damped response); see Fig. 6.2.

1.3. $k < 2\omega$. In this case, the characteristic equation has complex conjugate roots. As we know, the general solution is given by

$$x = e^{-\frac{k}{2}t}(c_1 \sin\beta t + c_2 \cos\beta t), \quad \beta = \sqrt{4\omega^2 - k^2}.$$

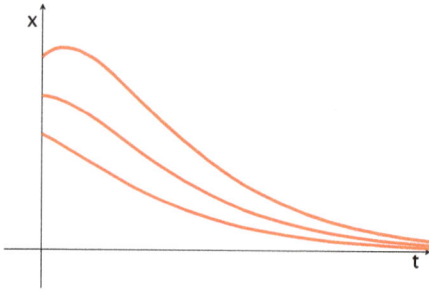

Figure 6.2: Critically damped response.

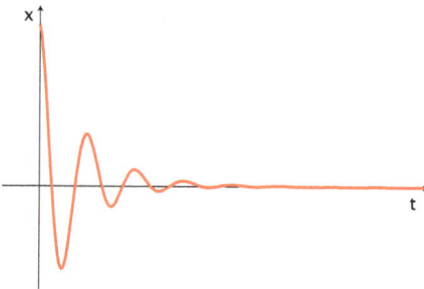

Figure 6.3: Underdamped response.

Now the solutions decay to zero at infinity but with an oscillating behavior (under-damped response); see Fig. 6.3.

2. RLC electric circuits. In an electric circuit with resistance R, inductance L, capacitance C and a source with constant voltage V, see Fig. 6.4, the intensity $x(t)$ of the current satisfies the equation

$$x'' + \frac{R}{L}x' + \frac{1}{LC}x = 0. \tag{6.11}$$

This equation is of the type (6.10) with $k = \frac{R}{L}$ and $\omega^2 = \frac{1}{LC}$.

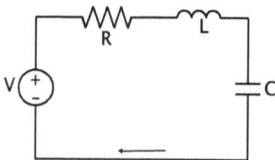

Figure 6.4: RLC circuit.

Repeating the previous calculations we find

2.1. if $k > 2\omega$, namely if $\frac{R}{L} > \frac{2}{\sqrt{LC}}$, or if $R > \frac{2\sqrt{L}}{C}$, we have an overdamped response.

2.2. if $k = 2\omega$, namely if $R = \frac{2\sqrt{L}}{C}$, we have a critically damped response.

2.3. if $k < 2\omega$, namely if $R < \frac{2\sqrt{L}}{C}$, we have an underdamped response.

Notice that, in any case, if there is a resistance $R > 0$, there is a decay of the current intensity in the circuit. If $R = 0$, i. e., if $k = 0$, the equation becomes the equation of the harmonic oscillations

$$x'' + \frac{1}{LC}x = 0,$$

whose general solution is given by the oscillating periodic functions

$$x = c_1 \sin \omega t + c_2 \cos \omega t = A \sin(\omega t + \theta), \quad \omega = \frac{1}{\sqrt{LC}}.$$

6.3.3 The non-homogeneous case

We know that in order to solve a non-homogeneous second order linear equation it suffices to add to the general solution of the associated homogeneous equation any particular solution of the non-homogeneous equation. If the equation has constant coefficients, this amounts to finding a particular solution of

$$x'' + ax' + bx = h(t). \tag{6.12}$$

One way to find a particular solution of (6.12) is to use the method of variations of parameters; see section 6.2.3.

In the case of an equation with constant coefficients, and when h has a specific form, it is simpler to use a method called the *method of undetermined coefficients*, which we now discuss. If $h(t) = A(t)e^{\lambda t}$, where $A(t)$ is a polynomial of degree m, or if $h(t)$ is a trigonometric polynomial, or else if it is a combination of the two, it is possible to find such a particular solution y which has the same form. Below we will deal with two specific examples. The arguments will indicate how to work in more general cases.

1. Let $h(t) = A(t)e^{\lambda t}$, where $A(t) = \sum_{r=0}^{m} a_r t^r$ is a polynomial of degree m.
   We look for a polynomial $Q(t)$ such that $y = Q(t)e^{\lambda t}$ is a solution of (6.12). Substituting $y' = \lambda Q e^{\lambda t} + Q' e^{\lambda t}$ and $y'' = \lambda^2 Q e^{\lambda t} + 2\lambda Q' e^{\lambda t} + Q'' e^{\lambda t}$ into the equation we find

$$[\lambda^2 Q e^{\lambda t} + 2\lambda Q' e^{\lambda t} + Q'' e^{\lambda t}] + a[\lambda Q e^{\lambda t} + Q' e^{\lambda t}] + bQ(t)e^{\lambda t} = Ae^{\lambda t}.$$

Canceling $e^{\lambda t}$ we get

$$[\lambda^2 Q + 2\lambda Q' + Q''] + a[\lambda Q + Q'] + bQ(t) = A$$

or

$$Q'' + (2\lambda + a)Q' + (\lambda^2 + a\lambda + b)Q = A,$$

which can be solved using the identity principle of polynomials. Notice that the degree of the unknown polynomial Q depends on λ.

1.1. If λ is not a root of the characteristic equation, namely if $\lambda^2 + a\lambda + b \neq 0$ then the degree of Q is m, the same as the degree of A.

1.2. If λ is a root of the characteristic equation (*resonant case*), namely if $\lambda^2 + a\lambda + b = 0$, we find $Q'' + [2\lambda + a]Q' = A$. This shows that, if $2\lambda + a \neq 0$, Q' has degree m and hence the degree of Q is $m + 1$.

1.3. If λ is a root of the characteristic equation and $2\lambda + a = 0$, then we find $Q'' = A$ and hence Q has degree $m + 2$.

Example 6.7. (a) Find a particular solution of $x'' + x = e^t$. We are not in the resonant case since $\lambda = 1$ is not a root of $\lambda^2 + 1 = 0$. Looking for $y = ke^t$ we find $y'' = ke^t$ and hence $ke^t + ke^t = e^t$ which yields $k = \frac{1}{2}$. Thus $y = \frac{1}{2}e^t$.

(b) Find a particular solution of $x'' + x' - 2x = t - 1$. Here $\lambda = 0$. The roots of the characteristic equation are -2 and 1 and hence we can seek a solution in the form $y = Q(t)$ with Q a polynomial of degree 1. Setting $Q == k_1 + k_2 t$ we find

$$Q' - Q = A \Rightarrow k_2 - 2(k_1 + k_2 t) = t - 1$$

Using the identity principle of polynomials it follows that

$$\begin{cases} k_2 - 2k_1 = -1, \\ -2k_2 = 1, \end{cases} \Rightarrow k_2 = -\frac{1}{2}, \quad k_1 = \frac{k_2 + 1}{2} = +\frac{1}{4}.$$

Thus $y = -\frac{1}{2}t + \frac{1}{4}$.

(c) Find a particular solution of $x'' + x' - 2x = (t - 2)e^t$. Here we are in the resonant case since $\lambda = 1$ is a root of the characteristic equation. Thus we seek for $Q = k_1 + k_2 t + k_3 t^2$, yielding

$$Q'' + 3Q' = P \Rightarrow 2k_3 + 3(k_2 + 2k_3 t) = t - 2 \Rightarrow \begin{cases} 2k_3 + 3k_2 = -2, \\ 6k_3 = 1. \end{cases}$$

Solving the second equation we get $k_3 = \frac{1}{6}$. Substituting into the first equation we find $\frac{1}{3} + 3k_2 = -2$ and hence $k_2 = -\frac{7}{9}$. The value of k_1 is arbitrary, and we can take $k_1 = 0$. Thus $Q = \frac{1}{6}t^2 - \frac{7}{9}t$ and a particular solution is given by $y = Qe^t = \left[\frac{1}{6}t^2 - \frac{7}{9}t\right]e^t$. □

2. The second case we deal with is when $h(t) = A(t)\cos\beta t + B(t)\sin\beta t$, where $A(t)$ and $B(t)$ are polynomials and $\beta \neq 0$ (if $\beta = 0$ we are in the preceding case).

We look for polynomials $P(t)$, $Q(t)$ such that $y = P\cos\beta t + Q\sin\beta t$ is a particular solution of $x'' + ax' + bx = h$. Repeating the previous calculations we find

$$y' = P'\cos\beta t - \beta P\sin\beta t + Q'\sin\beta t + \beta Q\cos\beta t$$
$$y'' = P''\cos\beta t - 2\beta P'\sin\beta t - \beta^2 P\cos\beta t + Q''\sin\beta t + 2\beta Q'\cos\beta t - \beta^2 Q\sin\beta t.$$

Hence

$$y'' + ay' + by = P''\cos\beta t - 2\beta P'\sin\beta t - \beta^2 P\cos\beta t + Q''\sin\beta t + 2\beta Q'\cos\beta t$$
$$- \beta^2 Q\sin\beta t + a(P'\cos\beta t - \beta P\sin\beta t + Q'\sin\beta t + \beta Q\cos\beta t)$$
$$+ b(P\cos\beta t + Q\sin\beta t).$$

Grouping the terms involving $\cos\beta t$ and those involving $\sin\beta t$, we find

$$y'' + ay' + by = (P'' - \beta^2 P + 2\beta Q' + aP' + a\beta Q + bP)\cos\beta t$$
$$+ (-2\beta P' + Q'' - \beta^2 Q - a\beta P + aQ' + bQ)\sin\beta t.$$

Rearranging the terms inside the parentheses,

$$y'' + ay' + by = (P'' + aP' + (b - \beta^2)P + 2\beta Q' + a\beta Q)\cos\beta t$$
$$+ (Q'' + aQ' + (b - \beta^2)Q - 2\beta P' - a\beta P)\sin\beta t.$$

Then the equation $y'' + ay + by = A\cos\beta t + B\sin\beta t$ yields

$$\begin{cases} P'' + aP' + (b - \beta^2)P + 2\beta Q' + a\beta Q = A, \\ Q'' + aQ' + (b - \beta^2)Q - 2\beta P' - a\beta P = B. \end{cases} \tag{6.13}$$

Inspection of the preceding system shows that:
2.1. If $b \neq \beta^2$, then the $\deg(P) = \deg(A)$ and $\deg(Q) = \deg(B)$; the same holds if $b = \beta^2$ but $a \neq 0$ (recall that $\beta \neq 0$).
2.2. If $b = \beta^2$ and $a = 0$, then $\deg(P) > \deg(A)$ and $\deg(Q) > \deg(B)$.

Remark 6.4. (a) The case in which $b = \beta^2$ and $a = 0$ is referred as the resonant case. Actually it arises whenever $\lambda = i\beta$ is a root of the characteristic equation, for one has $(i\beta)^2 + ai\beta + b = b - \beta^2 + ia\beta = 0$ if and only if $b - \beta^2 = 0$ and $a\beta = 0$. More precisely, we see that if $a = 0$ and $b = \beta^2$ system (6.13) becomes simply

$$\begin{cases} P'' + 2\beta Q' = A, \\ Q'' - 2\beta P' = B. \end{cases}$$

Setting $P = p_0 + p_1 t$, $Q = q_0 + q_1 t$, we find $2\beta q_1 = A$ and $-2\beta p_1 = B$, whereby $p_1 = -\frac{B}{2\beta}$, $q_1 = \frac{A}{2\beta}$. Thus

$$y = -\frac{B}{2\beta}\cos t + \frac{A}{2\beta}\sin t. \tag{6.14}$$

(b) It is worth pointing out that, in general, both P and Q can be different from zero even if A or B is zero.

Example 6.8. Find a particular solution of $x'' + x' = 2\sin t$. Here $a = 1, b = 0, A = 0$ and $B = 2$. Since $\beta = 1 \neq 0$ we have the non-resonant case and thus we can look for P, Q with degree 0, namely for $y = p\cos t + q\sin t$, $p, q \in \mathbb{R}$. Rather than substituting into the system (6.13), we prefer to repeat the calculations in this specific case, finding $y' = -p\sin t + q\cos t$ and $y'' = -p\cos t - q\sin t$. Thus $y'' + y' = 2\sin t$ yields $-p\cos t - q\sin t - p\sin t + q\cos t = 2\sin t$, hence $(q-p)\cos t - (p+q)\sin t = 2\sin t$ and the system

$$\begin{cases} q - p = 0, \\ p + q = -2. \end{cases}$$

Solving for p, q we find $p = q = -1$. Thus $y = -\cos t - \sin t$. □

Example 6.9 (The elastic spring subjected to an external sinusoidal force). We know, see subsection 6.3.1(1), that the free oscillations of a body of unit mass attached to an elastic spring satisfy the homogeneous equation $x'' + \omega^2 x = 0$. If the body is subjected to an external sinusoidal force $h = \sin kt$, Newton's law yields $x'' = -\omega^2 x + h$, namely

$$x'' + \omega^2 x = \sin kt. \tag{6.15}$$

According to the preceding discussion, we have to distinguish between $k = \omega$ or $k \neq \omega$.
(1) $k \neq \omega$. We are in the non-resonant case. Repeating the calculations above we find that the solution of (6.15) is given by

$$x = c_1\sin\omega t + c_2\cos\omega t + \frac{1}{\omega^2 - k^2}\sin kt$$

When $\omega \sim k$, namely when $\omega^2 - k^2$ is very small, we see the so-called *beating* phenomenon in which the solution is an oscillation with a periodic variation of intensity; see Fig. 6.5.
(2) $k = \omega$. We are in the resonant case. As in Remark 6.4(a) we can guess that a particular solution has the form $y = pt\cos kt + qt\sin kt$. Since $y'' = -2kp\sin kt + 2kq\cos kt - k^2pt\cos kt - k^2qt\sin kt = -2kp\sin kt + 2kq\cos kt - k^2(pt\cos kt + qt\sin kt) = -2kp\sin kt + 2kq\cos kt - k^2pt\cos kt - k^2y$, $y'' + k^2y = \sin kt$ yields

$$-2kp\sin kt + 2kq\cos kt = \sin kt \Rightarrow p = -\frac{1}{2k}, \quad q = 0.$$

Therefore $y = -\frac{1}{2k}t\cos kt$ and the solution of (6.15) (with $\omega = k$) is

$$x = c_1\sin kt + c_2\cos kt - \frac{1}{2k}t\cos kt.$$

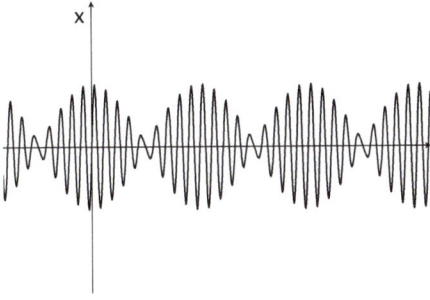

Figure 6.5: Beats.

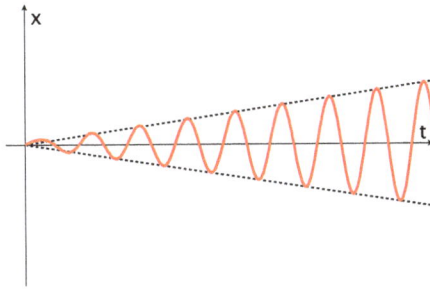

Figure 6.6: Plot of $c_1 \sin kt + c_2 \cos kt - \frac{1}{2k} t \cos kt$.

In this case we see that the solution is oscillating with increasing intensity, as shown in Fig. 6.6. This phenomenon is due to the presence of $\frac{1}{2k} t \cos kt$, which has an increasing amplitude. □

Remark 6.5. If $h = h_1 + \cdots + h_m$, a particular solution of $y'' + ay' + by = h$ can be found as $y = y_1 + \cdots + y_m$, where y_i solves $y_i'' + ay_i' + by_i = h_i$. For example, a particular solution of $y'' + y = e^t + \cos t$ is given by $y = y_1 + y_2$, where $y_1'' + y_1 = e^t$ and $y_2'' + y_2 = \cos t$. Using the results found in Examples 6.7(a) and 6.8(b) we infer that $y = \frac{1}{2} e^t + \frac{1}{2} \sin t$.

6.3.4 Exercises on equations with constant coefficients

1. Find the general solution of $x'' - 2x' - 4x = 0$.
2. Find the general solution of $x'' + 2x' + 10x = 0$.
3. Find the general solution of $x'' - x' + 2x = 0$.
4. Solve $x'' + 3x' = 0$, $x(0) = 1$, $x'(0) = -1$.
5. Solve $x'' - 2x' + x = 0$, $x(0) = 1$, $x'(0) = 2$.
6. Find the general solution of $9x'' + 6x' + x = 0$.
7. Solve the ivp $x'' + 4x' - 5x = 0$, $x(0) = 1$, $x'(0) = -1$.

8. Show that the nonconstant solutions of $x'' - kx' = 0, x(0) = 0, x'(0) = k$ are increasing or decreasing.

9. Let $x_a(t)$ be the solution of the ivp $x'' - x' - 2x = 0, x(0) = 0, x'(0) = a$. Find the $\lim_{t\to+\infty} x_a(t)$.

10. Find $k \in \mathbb{R}$ such that all solutions of $x'' + 2kx' + 2k^2x = 0$ tend to 0 as $t \to +\infty$.

11. Find the general solution of $x'' - x = e^t$.

12. Find $x(t)$ such that $x'' - 2x' = 0, \lim_{t\to-\infty} x(t) = 1$ and $x'(2) = 1$.

13. Solve $x'' + 2x' + 2x = 0$.

14. Show that all the solutions x of $x'' + 2x' + x = 0$ are such that $\lim_{t\to+\infty} x(t) = 0$.

15. Solve $x'' - 14x' + 49x = 0$.

16. Solve the bvp $x'' - 6x' + 9x = 0, x(-1) = 1, x(1) = 0$.

17. Solve $x'' - x = 2e^t - 3e^{-t}$.

18. Let $a \neq 0$. Show that the bvp $x'' + 3x' - 4x = 0, x(-a) = x(a) = 0$ has only the trivial solution.

19. Solve $x'' - x = te^t$.

20. Solve $x'' - x = t$.

21. Solve $x'' + x = t^2 + 1$.

22. Solve $x'' - 4x' + 3x = \sin t$.

23. Solve $x'' - 2x' + x = k$.

24. Solve $x'' + 4x = \sin \omega t$.

25. Solve $x'' + 4x = \sin t + \sin 2t$.

26. Find the general solution of $x'' - x = t^3 + 2t$.

6.4 Euler's equation

An equation of the form

$$at^2x'' + btx' + cx = 0, \quad a, b, c \in \mathbb{R}, \quad t > 0 \tag{6.16}$$

is called *Euler's equation*.

Euler's equations can be transformed into equations with constant coefficients simply by substituting $t = e^s$ or, equivalently, $s = \ln t$. We note that $s = \ln t$ implies $\frac{ds}{dt} = \frac{1}{t}$. Thus

$$\dot{x} \overset{\text{def}}{=} \frac{dx}{ds} = \frac{dx}{dt} \cdot \frac{dt}{ds} = t\frac{dx}{dt} = tx'.$$

This implies that

$$x' = \frac{1}{t}\dot{x}. \tag{6.17}$$

$$\ddot{x} = \frac{d^2x}{ds^2} = \frac{d}{ds}\left(\frac{dx}{ds}\right) = \frac{d}{ds}(tx') = \frac{d(tx')}{dt} \cdot \frac{dt}{ds} = (x' + tx'') \cdot t = t^2x'' + tx'.$$

Solving $\ddot{x} = t^2 x'' + tx'$ for x'' and using equation (6.17), we have

$$x'' = \frac{\ddot{x} - \dot{x}}{t^2}.$$ (6.18)

Thus the equation $at^2 x'' + btx' + cx = 0$ is transformed into

$$at^2 \cdot \frac{\ddot{x} - \dot{x}}{t^2} + bt \cdot \frac{1}{t}\dot{x} + cx = a\ddot{x} + (b - a)\dot{x} + cx = 0$$

which is a linear second order equation with constant coefficients. If $x(s)$ is a solution of this last equation, then $x(\ln t)$, $t > 0$, is a solution of (6.16).

Example 6.10. Solving Euler's equation $t^2 x'' + tx' + \omega^2 x = 0$ reduces to solving

$$\ddot{x} + \omega^2 x = 0.$$

The general solution is $x(s) = c_1 \sin(\omega s) + c_2 \cos(\omega s)$, the solution of the given Euler equation is $x(\ln t) = c_1 \sin(\omega \ln t) + c_2 \cos(\omega \ln t)$, $t > 0$. □

In order to solve a non-homogeneous Euler equation

$$at^2 x'' + btx' + cx = h(t)$$

we let $t = e^s$ ($s = \ln t$), which yields

$$a\ddot{x} + (b - a)\dot{x} + cx = h(e^s)$$

a non-homogeneous second order equation with constant coefficients.
    For example, the equation $t^2 x'' + tx' + \omega^2 x = t$ becomes

$$\frac{d^2 x}{ds^2} + \omega^2 x = e^s.$$

Searching for a particular solution y of this equation as ae^s we find $ae^s + \omega^2 \cdot ae^s = e^s$ which yields $a = \frac{1}{1+\omega^2}$. Then $y = \frac{1}{1+\omega^2}e^s$, and hence the general solution is

$$x(s) = c \sin(\omega s + \theta) + \frac{1}{1 + \omega^2}e^s.$$

Coming back to the variable $t = e^s$ we obtain

$$x(t) = c \sin(\omega \ln t + \theta) + \frac{1}{1 + \omega^2}t, \quad t > 0.$$

In Chapter 11 we will discuss another method to solve Euler's equation, by using solutions as power series.

6.4.1 Exercises on Euler equations

1. Solve the Euler equation $2t^2x'' - tx' + x = 0, t > 0$.
2. Solve $t^2x'' - tx' + x = 0$.
3. Solve $t^2x'' - 4tx' + 6x = 0$.
4. Solve $t^2x'' - 5tx' + 5x = 0$.
5. Solve

$$t^2x'' - 2tx' + 2x = 0, \quad x(1) = 1, \quad x'(1) = 0.$$

6. Solve

$$t^2x'' + tx' + x = 0, \quad x(1) = 1, \quad x'(1) = -1.$$

7. Consider the Euler equation

$$at^2x'' + btx' + cx = 0, \quad t > 0.$$

Show that if $a > 0$ and $c < 0$, then the general solution of the Euler equation, is of the form $c_1t^p + c_2t^q$.

7 Higher order linear equations

7.1 General results

This chapter can be viewed as an extention of the results in Chapter 6. Many of the extensions are fairly easy to comprehend and not so surprising. Let us start with the general definition of an nth order linear differential equation, where n is any natural number, $n \geq 1$.

A general *linear homogeneous differential equation of order n* has the form

$$\frac{d^n x}{dt^n} + p_{n-1}(t) \frac{d^{n-1} x}{dt^{n-1}} + \cdots + p_1(t) \frac{dx}{dt} + p_0(t)x = 0.$$

For convenience of notation, we let

$$L[x] \stackrel{\text{def}}{=} \frac{d^n x}{dt^n} + p_{n-1}(t) \frac{d^{n-1} x}{dt^{n-1}} + \cdots + p_1(t) \frac{dx}{dt} + p_0(t)x$$

so that

$$L[x] = 0 \tag{7.1}$$

represents a linear homogeneous differential equation of order n, while $L[x] = h$, where $h(t)$ is any nonzero function, is a linear *nonhomogeneous* differential equation of order n.

The *initial value problem*, or *Cauchy problem* for $L[x] = h$ consists of finding a function $x(t)$ such that

$$\begin{cases} L[x] = h, \\ x(t_0) = y_0, \\ x'(t_0) = y_1, \\ \cdots \\ x^{(n-1)}(t_0) = y_{n-1}, \end{cases} \tag{7.2}$$

where $(y_0, \ldots, y_{n-1})$ is any point in $\mathbb{R}^n$.

For example, if $n = 3$ the Cauchy problem for (6.1) consists of seeking a solution $x(t)$ of the equation satisfying the three initial conditions $x(t_0) = y_0, x'(t_0) = y_1, x''(t_0) = y_2$.

Below are listed some extensions of results from Chapter 2, which can be verified similarly.

- *If p_i, $1 \leq i \leq n$, is continuous on the interval $I \subseteq \mathbb{R}$, then for any $t_0 \in I$ and $(y_0, \ldots, y_{n-1}) \in \mathbb{R}^n$ the Cauchy problem (7.2) has one and only one solution, which is defined on I.*

The proof can be deduced from the existence and uniqueness results stated in Chapter 8.

https://doi.org/10.1515/9783111185675-007

As a consequence,

- Let $x(t)$ be a solution of $L[x] = 0$. If there exists $t_0 \in I$ such that $x(t_0) = x'(t_0) = \cdots = x^{(n-1)}(t_0) = 0$, then $x(t)$ is identically zero, called the trivial solution.

Next we list extensions of the main results for second order equations.

1. Any linear combination $\sum c_i x_i$ of solutions of (7.1) is a solution of (7.1).
2. The sum of a solution of $Lx = 0$ and a solution of $L[x] = h$ is a solution of $L[x] = h$.
3. Functions $u_i, 1 \leq i \leq n$, are *linearly independent* if $\sum c_i u_i = 0$ for all t implies that $c_i = 0, 1 \leq i \leq n$; otherwise they are called *linearly dependent*.
4. Solutions $u_i, i = 1, \ldots n$, of (7.1) are fundamental solutions if they are linearly independent.
5. $u_i, i = 1, \ldots, n$, are fundamental solutions of (7.1) if and only if their wronskian, defined as

$$ W = \begin{vmatrix} u_1(t) & \cdots & u_n(t) \\ u_1'(t) & \cdots & u_n'(t) \\ \cdots & \cdots & \cdots \\ u_1^{(n-1)}(t) & \cdots & u_n^{(n-1)}(t), \end{vmatrix} $$

is different from zero.
6. (Abel's theorem) The wronskian of n solutions of (7.1) satisfies $W' + p_{n-1}W = 0$ and hence is given by

$$ W(t) = ce^{-\int p_{n-1}(t)\,dt}, \quad c \in \mathbb{R}. $$

In particular, W is either identically zero or it never vanishes.
7. If $u_1, \ldots, u_n$ are fundamental solutions of $L[x] = 0$, then the general solution of $L[x] = 0$ is given by $x = c_1 u_1 + \cdots + c_n u_n, c_i \in \mathbb{R}$.

7.2 Higher order linear equations with constant coefficients

Let us consider the nth order linear homogeneous equation with constant coefficients

$$ L[x] = x^{(n)} + a_{n-1}x^{(n-1)} + \cdots + a_1 x' + a_0 x = 0, \quad a_i \in \mathbb{R}. \tag{7.3} $$

As in the case of second order equations, substituting $x = e^{\lambda t}$ in $L[x] = 0$ gives rise to the *characteristic polynomial*

$$ C(\lambda) = \lambda^n + a_{n-1}\lambda^{n-1} + \cdots + a_1\lambda + a_0. $$

Consider the (algebraic) equation $C(\lambda) = 0$, namely

$$\lambda^n + a_{n-1}\lambda^{n-1} + \cdots + a_1\lambda + a_0 = 0, \tag{7.4}$$

which is called the *characteristic equation* corresponding to (7.3). Let $\lambda_1 \le \lambda_2 \le \cdots \le \lambda_n$ denote n real roots, possibly repeated, of (7.4). Similar to the second order equation, we can check, by substitution that $L[e^{\lambda_i t}] = e^{\lambda_i t}C(\lambda_i) = 0$, and hence $x = e^{\lambda_i t}, 1 \le i \le n$, is a solution of $L[x] = 0$. Moreover, if $a \pm i\beta$ is a complex root of $C(\lambda) = 0$, then $u(t) = e^{at}\cos\beta t$ and $v(t) = e^{at}\sin\beta t$ is a pair of real valued linearly independent solutions of $L[x] = 0$.

Next we are going to find the general solution of $L[x] = 0$, focusing mainly on the cases $n = 3$ and $n = 4$. Rather than carrying out a general theoretical detailed analysis, we will give some examples, which will indicate the procedure to be followed in the general cases.

(a) $n = 3$ and $C(\lambda) = 0$ has three distinct (hence simple) real roots.

Evaluating the wronskian of $e^{\lambda_1 t}, e^{\lambda_2 t}, e^{\lambda_3 t}$ at $t = 0$, we find

$$W(0) = \begin{vmatrix} e^{\lambda_1 0} & e^{\lambda_2 0} & e^{\lambda_3 0} \\ \lambda_1 e^{\lambda_1 0} & \lambda_2 e^{\lambda_1 0} & \lambda_3 e^{\lambda_1 0} \\ \lambda_1^2 e^{\lambda_1 0} & \lambda_2^2 e^{\lambda_1 0} & \lambda_3^2 e^{\lambda_1 0} \end{vmatrix} = \begin{vmatrix} 1 & 1 & 1 \\ \lambda_1 & \lambda_2 & \lambda_3 \\ \lambda_1^2 & \lambda_2^2 & \lambda_3^2 \end{vmatrix}.$$

Multiplying the first row by $-\lambda_1$ and adding it to the second row, and then multiplying the first row by $-\lambda_1^2$ and adding it to the third row yields

$$W(0) = \begin{vmatrix} 1 & 1 & 1 \\ 0 & \lambda_1 - \lambda_2 & \lambda_1 - \lambda_3 \\ 0 & \lambda_1^2 - \lambda_2^2 & \lambda_1^2 - \lambda_3^2 \end{vmatrix} = (\lambda_1 - \lambda_2)(\lambda_1^2 - \lambda_3^2) - (\lambda_1 - \lambda_3)(\lambda_1^2 - \lambda_2^2) =$$

$$= (\lambda_1 - \lambda_2)(\lambda_1 - \lambda_3)(\lambda_1 + \lambda_3) - (\lambda_1 - \lambda_3)(\lambda_1 - \lambda_2)(\lambda_1 + \lambda_2)$$

$$= (\lambda_1 - \lambda_2)(\lambda_1 - \lambda_3)[(\lambda_1 + \lambda_3) - (\lambda_1 + \lambda_2)] = (\lambda_1 - \lambda_2)(\lambda_1 - \lambda_3)(\lambda_3 - \lambda_2).$$

Since the roots are distinct, we infer that $W(0) \ne 0$; and hence the general solution is given by

$$x(t) - c_1 e^{\lambda_1 t} + c_2 e^{\lambda_2 t} + c_3 e^{\lambda_3 t}.$$

(b) $n = 3$ and $C(\lambda) = 0$ has one simple root λ_1 and a double real root λ_2.

In this case we consider the functions $e^{\lambda_1 t}, e^{\lambda_2 t}, te^{\lambda_2 t}$. Let us show that, as in the case $n = 2$, $x = te^{\lambda_2 t}$ is also a solution of $L[x] = 0$. After evaluating $x' = e^{\lambda_2 t} + \lambda_2 te^{\lambda_2 t}, x'' = 2\lambda_2 e^{\lambda_2 t} + \lambda_2^2 te^{\lambda_2 t}, x''' = 3\lambda_2^2 e^{\lambda_2 t} + \lambda_2^3 te^{\lambda_2 t}$, we find

$$L[te^{\lambda_2 t}] = 3\lambda_2^2 e^{\lambda_2 t} + \lambda_2^3 te^{\lambda_2 t} + a_2(2\lambda_2 e^{\lambda_2 t} + \lambda_2^2 te^{\lambda_2 t}) + a_1(e^{\lambda_2 t} + \lambda_2 te^{\lambda_2 t}) + a_0 te^{\lambda_2 t}$$

$$= e^{\lambda_2 t} \cdot [3\lambda_2^2 + \lambda_2^3 t + 2a_2\lambda_2 + a_2\lambda_2^2 t + a_1 + a_1\lambda_2 t + a_0 t].$$

Rearranging

$$L[te^{\lambda_2 t}] = te^{\lambda_2 t} \cdot [\lambda_2^3 + a_2 \lambda_2^2 + a_1 \lambda_2 + a_0] + e^{\lambda_2 t} \cdot [3\lambda_2^2 + 2a_2 \lambda_2 + a_1]$$
$$= te^{\lambda_2 t} \cdot C(\lambda_2) + e^{\lambda_2 t} \cdot C'(\lambda_2).$$

Since λ_2 is a double root of $C(\lambda)$ we have $C(\lambda_2) = C'(\lambda_2) = 0$.[1] It follows that $L[te^{\lambda_2 t}] = 0$, as claimed.

Evaluating the wronskian of $e^{\lambda_1 t}, e^{\lambda_2 t}, te^{\lambda_2 t}$ we find

$$W(0) = \begin{vmatrix} 1 & 1 & 0 \\ \lambda_1 & \lambda_2 & 1 \\ \lambda_1^2 & \lambda_2^2 & 2\lambda_2 \end{vmatrix} = \begin{vmatrix} \lambda_2 & 1 \\ \lambda_2^2 & 2\lambda_2 \end{vmatrix} - \begin{vmatrix} \lambda_1 & 1 \\ \lambda_1^2 & 2\lambda_2 \end{vmatrix}$$
$$= 2\lambda_2^2 - \lambda_2^2 - (2\lambda_1 \lambda_2 - \lambda_1^2) = \lambda_2^2 - 2\lambda_1 \lambda_2 + \lambda_1^2 = (\lambda_2 - \lambda_1)^2 \neq 0.$$

Thus, $e^{\lambda_1 t}, e^{\lambda_2 t}, te^{\lambda_2 t}$ are linearly independent, and the general solution is given by

$$x = c_1 e^{\lambda_1 t} + c_2 e^{\lambda_2 t} + c_3 te^{\lambda_2 t}.$$

Similarly, if $n = 3$ and λ is a triple root, then the general solution is given by $x(t) = c_1 e^{\lambda t} + c_2 te^{\lambda t} + c_3 t^2 e^{\lambda t}$.

(c) $n = 3$ and $C(\lambda) = 0$ has one real root λ_1 and two complex roots $\alpha \pm i\beta$.

It follows from Euler's formula that $u = e^{\alpha t} \cos \beta t$ and $v = e^{\alpha t} \sin \beta t$ are solutions. Let us show that the solutions $e^{\lambda_1 t}, u = e^{\alpha t} \cos \beta t$ and $v = e^{\alpha t} \sin \beta t$ are linearly independent. One has: $u' = \alpha u - \beta v$, $v' = \alpha v + \beta u$ and $u'' = \alpha u' - \beta v'$, $v'' = \alpha v' + \beta u'$. Then $u(0) = 1$, $v(0) = 0$; $u'(0) = \alpha$, $v'(0) = \beta$; $u''(0) = \alpha^2 - \beta^2$, $v''(0) = 2\alpha\beta$ and hence

$$W(0) = \begin{vmatrix} e^{\lambda_1 0} & u(0) & v(0) \\ \lambda_1 e^{\lambda_1 0} & u'(0) & v'(0) \\ \lambda_1^2 e^{\lambda_1 0} & u''(0) & v''(0) \end{vmatrix} = \begin{vmatrix} 1 & 1 & 0 \\ \lambda_1 & \alpha & \beta \\ \lambda_1^2 & \alpha^2 - \beta^2 & 2\alpha\beta \end{vmatrix}$$
$$= \begin{vmatrix} \alpha & \beta \\ \alpha^2 - \beta^2 & 2\alpha\beta \end{vmatrix} - \begin{vmatrix} \lambda_1 & \beta \\ \lambda_1^2 & 2\alpha\beta \end{vmatrix}$$
$$= 2\alpha^2\beta - \alpha^2\beta + \beta^3 - 2\lambda_1\alpha\beta + \beta\lambda_1^2 = \beta \cdot (\lambda_1^2 - 2\lambda_1\alpha + \alpha^2 + \beta^2) = \beta[(\lambda_1 - \alpha)^2 + \beta^2] \neq 0.$$

Therefore $e^{\lambda_1 t}, u = e^{\alpha t} \cos \beta t$ and $v = e^{\alpha t} \sin \beta t$ are linearly independent and the general solution is given by

$$x = c_1 e^{\lambda_1 t} + c_2 e^{\alpha t} \cos \beta t + c_3 e^{\alpha t} \sin \beta t = c_1 e^{\lambda_1 t} + e^{\alpha t} (c_2 \cos \beta t + c_3 \sin \beta t).$$

---

[1] Recall that if λ is double root of a polynomial $C(\lambda)$ then $C'(\lambda) = 0$. More generally, if λ is a root with multiplicity $k > 1$ of $C(\lambda)$ then λ is a root with multiplicity $k - 1$ of $C'(\lambda)$.

(d) $n = 4$. We limit the discussion to the case in which $C(\lambda) = 0$ has two double complex roots $\alpha \pm i\beta$: the remaining cases can be handled exactly as the one for $n = 3$.

Setting again $u = e^{\alpha t} \cos \beta t$, $v = e^{\alpha t} \sin \beta t$, we can check that u, v, tu, tv provide 4 linearly independent solutions, since their wronskian is different from zero. We omit the tedious, though straight forward, calculations. It follows that the general solution of $L[x] = 0$ is given by

$$x = c_1 e^{\alpha t} \cos \beta t + c_2 e^{\alpha t} \sin \beta t + c_3 t e^{\alpha t} \cos \beta t + c_4 t e^{\alpha t} \sin \beta t$$
$$= e^{\alpha t}(c_1 \cos \beta t + c_2 \sin \beta t + c_3 t \cos \beta t + c_4 t \sin \beta t).$$

Example 7.1. (a) Solve $x''' - 2x'' - x' + 2x = 0$. The characteristic equation is $C(\lambda) = \lambda^3 - 2\lambda^2 - \lambda + 2 = 0$, whose solutions are $\lambda_1 = 1, \lambda_2 = -1, \lambda_3 = 2$. Thus the general solution is

$$x = c_1 e^t + c_2 e^{-t} + c_3 e^{2t}.$$

(b) Solve $x''' - 2x'' + x' = 0$. The characteristic equation is $C(\lambda) = \lambda^3 - 2\lambda^2 + \lambda = 0$, namely $\lambda(\lambda^2 - 2\lambda + 1) = \lambda(\lambda - 1)^2 = 0$, and hence the roots are $\lambda = 0$, simple, and $\lambda = 1$, double. It follows that the general solution is given by

$$x = c_1 + c_2 e^t + c_3 t e^t. \qquad \square$$

(c) Solve $x''' - x'' + x' - x = 0$. The characteristic equation is $C(\lambda) = \lambda^3 - \lambda^2 + \lambda - 1 = (\lambda_1 - 1)(\lambda^2 + 1) = 0$, whose solutions are $\lambda_1 = 1, \lambda_2 = i, \lambda_3 = -i$. Thus the general solution is

$$x = c_1 e^t + c_2 \cos t + c_3 \sin t. \qquad \square$$

(d) Solve $x'''' + 8x'' + 16x = 0$. The characteristic equation is $C(\lambda) = \lambda^4 + 8\lambda^2 + 16 = 0$, namely $(\lambda^2 + 4)^2 = 0$ whose roots are $\lambda = \pm 2i$, double. Since $\alpha = 0$ and $\beta = 2$, it follows that the general solution is given by

$$x = c_1 \cos 2t + c_2 \sin 2t + c_3 t \cos 2t + c_4 t \sin 2t. \qquad \square$$

Next, let us consider the general nonhomogeneous equation with constant coefficients in the form

$$x^{(n)} + a_{n-1} x^{(n-1)} + \cdots + a_1 x' + a_0 x = h(t), \quad a_i \in \mathbb{R}.$$

In order to find a particular solution of the nonhomogeneous equation $L[x] = h$, we may either use the method of *Variation of Parameters* or the method of *Undetermined Coefficients*, which are straight forward extensions of results from second order equations studied in Chapter 6. Variation of Parameters for n^{th} order equations involves solving n equations and n unknowns, instead of two equations and two unknowns, as is the case for second order equations. In particular, if $u_1, u_2, \ldots, u_n$ are n linearly independent solutions of the homogeneous equation $L[x] = 0$, then we seek n functions $v_1, v_2, \ldots, v_n$ such that $y = u_1 v_1 + u_2 v_2 + \cdots + u_n v_n$ is a solution of $L[x] = h$. This will be the case if $v_1', v_2', \ldots, v_n'$ are solutions of the following algebraic system:

$$\begin{cases} v_1' u_1 + v_2' u_2 + \cdots + v_n' u_n = 0, \\ v_1' u_1' + v_2' u_2' + \cdots + v_n' u_n' = 0, \\ \vdots \\ v_1' u_1^{(n-1)} + v_2' u_2^{(n-1)} + \cdots + v_n' u_n^{(n-1)} = h(t). \end{cases} \tag{7.5}$$

By way of demonstration, we solve the next example by both methods-Variation of Parameters and Undetermined Coefficients.

Example 7.2. Find the general solution of

$$x'''' - x'' = e^t.$$

First, we need to find the general solution of the homogeneous equation

$$x'''' - x'' = 0.$$

Solving the characteristic equation

$$\lambda^4 - \lambda^2 = \lambda^2(\lambda^2 - 1) = 0,$$

we obtain $\lambda_1 = 0$ (double root), $\lambda_2 = 1, \lambda_3 = -1$, and the general solution of the homogeneous equation is given by

$$x(t) = c_1 + c_2 t + c_3 e^t + c_4 e^{-t}.$$

Now we need a particular solution of the nonhomogeneous equation. Below we discuss two methods to accomplish this.

(A) **Method of Variation of Parameters.** In order to find a particular solution $y(t)$ of the nonhomogeneous equation, we let $y(t) = v_1 + v_2 t + v_3 e^t + v_4 e^{-t}$, where v_i', $1 \le i \le 4$, solve the algebraic system

$$\begin{cases} v_1' + v_2' t + v_3' e^t + v_4' e^{-t} = 0, \\ v_2' + v_3' e^t - v_4' e^{-t} = 0, \\ v_3' e^t + v_4' e^{-t} = 0, \\ v_3' e^t - v_4' e^{-t} = e^t. \end{cases} \tag{7.6}$$

Solving the above system, we find $v_1' = te^t, v_2' = -e^t, v_3' = \frac{1}{2}, v_4' = -\frac{1}{2}e^{2t}$. Integrating, we obtain $v_1 = te^t - e^t, v_2 = -e^t, v_3 = \frac{1}{2}t, v_4 = -\frac{1}{4}e^{2t}$. Thus the particular solution $y(t)$ is given by

$$y(t) = (te^t - e^t).1 - te^t + \frac{1}{2}te^t - \frac{1}{4}e^{2t}.e^{-t} = \frac{1}{2}te^t - \frac{5}{4}e^t.$$

Finally, the general solution of $L(x) = e^t$ is given by

$$y(t) = c_1 + c_2 t + c_3 e^t + c_4 e^{-t} + \frac{1}{2}te^t - \frac{5}{4}e^t.$$

Now, let us solve the above equation by the method of Undetermined Coefficients. It is a much easier and faster way to solve this problem. However, this is not always the case. This method is more of a guessing game. One has to think carefully and make a judicial initial choice. Both methods have advantages and disadvantages. The big advantage of the Variation of Parameters method is that it is consistent and, at least theoretically, it always works; which is not the case with the method of Undetermined Coefficients which depends greatly on the initial choice. For more detail, see the discussion and examples below.

(B) **Method of Undetermined Coefficients.** First we note that this is a case of resonant. So, Ae^t is not a good choice. Let us try $y(t) = Ate^t$. Then, $y' = Ae^t + Ate^t$, $y'' = 2Ae^t + Ate^t$, $y''' = 3Ae^t + Ate^t$, $y'''' = 4Ae^t + Ate^t$ and hence

$$y'''' - y'' = 4Ae^t + Ate^t - 2Ae^t - Ate^t = 2Ae^t.$$

Setting $2Ae^t = e^t$ yields $A = \frac{1}{2}$. Therefore, the particular solution is $y(t) = \frac{1}{2}te^t$.

Remark 7.1. We notice that in the above example, the answers that we obtained from the two different methods seem to be different. This is because in both cases we are asking for a particular solution of the nonhomogeneous equation while there are infinitely many such particular solutions. What is important is to know that each solution implies the other. First, starting with $x = \frac{1}{2}te^t$ as a solution of the nonhomogeneous equation, we note that e^t, and hence $-\frac{5}{4}e^t$, is a solution of the corresponding homogeneous equation. Therefore $y = \frac{1}{2}te^t - \frac{5}{4}e^t$ is a solution of the nonhomogeneous equation. Conversely, if we start with the solution $x = \frac{1}{2}te^t - \frac{5}{4}e^t$ then $y = \frac{1}{2}te^t - \frac{5}{4}e^t + \frac{5}{4}e^t = \frac{1}{2}te^t$ is also a solution, since it is the sum of a solution of the nonhomogeneous equation and a solution of the corresponding homogeneous equation.

In order to find a particular solution of the nonhomogeneous equation $L[x] = h$, we can use, e. g., the method of undetermined coefficients, with the following procedure:

In order to find a particular solution of $x''' + ax'' + bx' + cx = \rho e^{\lambda t}$, we let $y = re^{\lambda t}$ and find

$$y''' + ay'' + by' + cy = r\lambda^3 e^{\lambda t} + ar\lambda^2 e^{\lambda t} + br\lambda e^{\lambda t} + cre^{\lambda t} = \rho e^{\lambda t},$$

whereby $r\lambda^3 + ar\lambda^2 + br\lambda + cr = \rho$ or $r(\lambda^3 + a\lambda^2 + b\lambda + c) = \rho$. Notice that the term in parenthesis is just $C(\lambda)$. Thus solving $rC(\lambda) = \rho$ for r, we have to distinguish whether

$C(\lambda) = 0$ or not, namely whether λ is a root of the characteristic equation, or not. If $C(\lambda) \neq 0$ we find $r = p/C(\lambda)$ and hence

$$y = \frac{p}{C(\lambda)}e^{\lambda t}.$$

If $C(\lambda) = 0$, we have to take into account its multiplicity. Let us discuss the case in which λ is a simple root, which implies that $C(\lambda) = 0$, $C'(\lambda) \neq 0$. Setting $y = rte^{\lambda t}$ and repeating the calculations carried out several times, we find

$$y''' + ay'' + by' + cy = 3r\lambda^2 e^{\lambda t} + rt\lambda^3 e^{\lambda t} + ae^{\lambda t}(2r\lambda + rt\lambda^2) + be^{\lambda t}(r + rt\lambda) + crte^{\lambda t}.$$

Rearranging we get

$$y''' + ay'' + by' + cy = \left[rt\lambda^3 + art\lambda^2 + brt\lambda + crt + 3r\lambda^2 + 2ar\lambda + br \right] \cdot e^{\lambda t}$$
$$= [C(\lambda)rt + r(3\lambda + 2a\lambda + b)]e^{\lambda t} = [\underbrace{C(\lambda)}_{=0} rt + rC'(\lambda)] \cdot e^{\lambda t} = rC'(\lambda) \cdot e^{\lambda t}.$$

Thus, $L[y] = pe^{\lambda t}$ yields $rC'(\lambda) = p$. Since $C'(\lambda) \neq 0$ we find $r = \frac{p}{C'(\lambda)}$ and hence

$$y = \frac{p}{C'(\lambda)}te^{\lambda t}.$$

If $C'(\lambda) = 0$ we will choose $y = Q(t)e^{\lambda t}$ with a polynomial Q whose degree depends upon the multiplicity of λ. Specific examples are discussed below.

Example 7.3. (a) Solve $x''' - 3x'' + 5x' - 3x = 2e^{at}$ with (i) $a = 2$ and (ii) $a = 1$.
   Here $C(\lambda) = \lambda^3 - 3\lambda^2 + 5\lambda - 3 = 0$, and we note, by observation, that $\lambda = 1$ is a root. Then it follows that $\lambda = 1 \pm \sqrt{2}\,i$ are the other two solutions.
(i) This is the non-resonant case, for $C(2) = 3 \neq 0$. Hence $\frac{p}{C(2)} = \frac{2}{3}$ and $y = \frac{2}{3}e^{2t}$. Therefore the general solution is

$$x = c_1 e^t + e^{2t}(c_2 \cos \sqrt{2}\,t + c_3 \sin \sqrt{2}\,t) + \frac{2}{3}e^{2t}.$$

(ii) Since $a = 1$ is a simple root with $C'(1) = 2$, we find $\frac{p}{C'(1)} = \frac{2}{2} = 1$. Thus a particular solution is given by $y = te^t$ and the general solution is

$$x = c_1 e^t + e^{2t}(c_2 \cos \sqrt{2}\,t + c_3 \sin \sqrt{2}\,t) + te^t.$$

(b) Solve $x''' - x'' = t$. Solving $C(\lambda) = \lambda^3 - \lambda^2 = 0$ we find $\lambda = 1$ and $\lambda = 0$, double. Here $h = t \cdot e^{0 \cdot t}$. Since $\lambda = 0$ is a double root, we try a particular solution as $y = Q(t)e^{0 \cdot t} = Q(t)$, a polynomial. Then we find $y''' - y'' = Q''' - Q'' = t$. This shows that the degree of Q is 3. Setting $Q = k_3 t^3 + k_2 t^2 + k_1 t + k_0$, the equation $Q''' - Q'' = t$ yields

$$6k_3 - (6k_3t + 2k_2) = t \Rightarrow -6k_3t + 6k_3 - 2k_2 = 1 \Rightarrow k_3 = -\frac{1}{6}, \quad k_2 = -\frac{1}{2}$$

The coefficients k_0, k_1 are arbitrary and we can choose $k_0 = k_1 = 0$. Thus $y = -\frac{1}{2}t^2 - \frac{1}{6}t^3$ and the general solution is given by

$$x = c_1 + c_2t + c_3e^t - \frac{1}{2}t^2 - \frac{1}{6}t^3.$$

(c) Consider an elastic uniform beam of length L, modeled as a one-dimensional object. We denote by $u(x)$ the deflection of the beam in the vertical direction at some position x, by E the elastic Young modulus, by I is the second moment of area, and by $h(x)$ the distributed load on the beam; see Fig. 7.1.

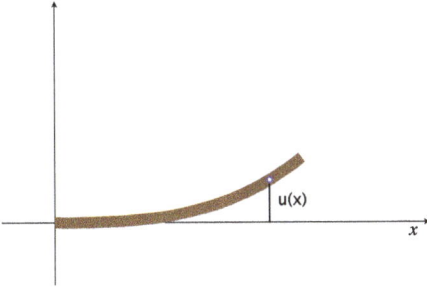

Figure 7.1: Deflection of the elastic beam.

Assuming that E, I are constant, it turns out that u satisfies the fourth order linear equation (notice that here the independent variable is denoted by x and the dependent variable by u)

$$EI\frac{d^4u}{dx^4} = h(x).$$

Since the corresponding characteristic equation $\lambda^4 = 0$ has $\lambda = 0$ as a root with multiplicity 4, the general solution is given by

$$u(x) = c_1 + c_2x + c_3x^2 + c_4x^3 + v(x)$$

where $v(x)$ is a particular solution, depending on h, E, I. Suppose that $h(x) = ax$. Then we can find a polynomial Q such that $v = Q(x)$. Substituting into the equation we find

$$EI\frac{d^4Q}{dx^4} = ax$$

Thus $Q(x)$ has degree 5. Letting $Q(x) = k_5x^5 + \cdots + k_1x + k_0$ we get

$$EI \cdot (5!) \cdot k_5x = ax.$$

Thus we can take $k_0 = \cdots = k_4 = 0$ and $k_5 = \frac{a}{(5!)EI}$. In conclusion, we obtain

$$u(x) = c_1 + c_2 x + c_3 x^2 + c_4 x^3 + \frac{a}{(5!)EI} x^5.$$

If we prescribe the initial condition $u(0) = u'(0) = u''(0) = u'''(0) = 0$ we find $c_1 = c_2 = c_3 = c_4 = 0$ and hence $u(x) = \frac{a}{(5!)EI} x^5$. For $x = L$ we get $u(L) = \frac{a}{(5!)EI} \cdot L^5$. Therefore measuring the deflection $u(L)$ at the end $x = L$ of the beam we can find the load coefficient a to be $a = \frac{(5!)EIu(L)}{L^5}$. □

7.2.1 Euler equations

Consider the general *Euler equation*

$$a_1 t^n x^{(n)} + a_2 t^{n-1} x^{(n-1)} + \cdots + a_{n+1} x = 0$$

where the $a_i, 1 \le i \le n + 1$, are constants. The method for solving such equations is similar to that introduced for second order Euler equations in Chapter 6. The difficulty is solving the characteristic equations. Below we demonstrate the method, for $n = 3$.
Let $t = e^s$. As we showed in Chapter 6, since $\frac{ds}{dt} = \frac{1}{t}$,

$$x' = \frac{dx}{dt} = \frac{dx}{ds}\frac{ds}{dt} = \frac{1}{t} \cdot \dot{x}$$

Thus $x' = e^{-s}\dot{x}$ and we infer

$$x'' = \frac{dx'}{dt} = \frac{d(e^{-s}\dot{x})}{dt} = \frac{d(e^{-s}\dot{x})}{ds} \cdot \frac{ds}{dt} = \frac{1}{t} \cdot (-e^{-s}\dot{x} + e^{-s}\ddot{x}) = \frac{\ddot{x} - \dot{x}}{t^2}$$

Similarly

$$x''' = \frac{dx''}{dt} = \frac{d(e^{-2s}(\ddot{x} - \dot{x}))}{dt} = \frac{d(e^{-2s}(\ddot{x} - \dot{x}))}{ds} \cdot \frac{ds}{dt}$$
$$= \frac{1}{t} \cdot [-2e^{-2s}(\ddot{x} - \dot{x}) + e^{-2s}(\dddot{x} - \ddot{x})] = \frac{1}{t^3} \cdot (\dddot{x} - 3\ddot{x} + 2\dot{x})$$

Thus the equation

$$a_1 t^3 x''' + a_2 t^2 x'' + a_3 t x' + a_4 x = 0$$

is transformed into

$$a_1(\dddot{x} - 3\ddot{x} + 2\dot{x}) + a_2(\ddot{x} - \dot{x}) + a_3\dot{x} + a_4 x = 0,$$

which is a third order linear equation with constant coefficients.

Example 7.4. Solve $t^3 x''' + 4t^2 x'' + 6tx' + 4x = 0$.

Making the substitution $t = e^s$ transforms the equation into

$$\dddot{x} - 3\ddot{x} + 2\dot{x} + 4(\ddot{x} - \dot{x}) + 6\dot{x} + 4x = \dddot{x} + \ddot{x} + 4\dot{x} + 4x = 0.$$

The corresponding characteristic equation is

$$\lambda^3 + \lambda^2 + 4\lambda + 4 = \lambda^2(\lambda + 1) + 4(\lambda + 1) = (\lambda + 1)(\lambda^2 + 4) = 0.$$

Therefore, roots of the characteristic equation are $\lambda_1 = -1$ and $\lambda_2 = \pm 2i$ and the solution is $x(s) = c_1 e^{-s} + c_2 \sin 2s + c_2 \cos 2s$. Since $t = e^s$ implies $s = \ln t$,

$$x(t) = x(\ln t) = c_1 \cdot \frac{1}{t} + c_2 \sin(\ln t^2) + c_3 \cos(\ln t^2). \qquad \square$$

7.3 Exercises

1. (A) Show that $t, -2t^2, t^3$ are linearly independent functions.
   (B) Show that e^{-t}, e^t, and e^{2t} are linearly independent.
2. Show that if $u_1, u_2, \ldots, u_k, 2 \le k < n$, are linearly dependent, then any larger set $u_1, u_2, \ldots, u_k, u_{k+1}, \ldots, u_n$ is linearly dependent.
3. Give an example to show that pairwise linear independence does not imply linear independence of the whole set of functions, i. e. any two functions in a set may be linearly independent without all the functions in the set being linearly independent.
4. Solve $x''' - 4x' = 0$.
5. Solve $x''' - 3x'' - x' + 3x = 0$.
6. Solve $x''' - 2x'' + x' - 2x = 0$.
7. Solve $x''' - 5x'' + 3x' + 9x = 0$.
8. Solve $x'''' - 3x'' + 2x = 0$.
9. Solve $x'''' - x' = 0$.
10. Solve $x'''' - 4x''' = 0$.
11. Solve $x''' - 25x' = e^t + 3\sin t$.
12. Solve $x''' - x = e^t$.
13. Solve $x'''' - 2x''' = -12$.
14. Solve the ivp $x''' + x'' = 0, x(0) = 1, x'(0) = 1, x''(0) = 0$.
15. Solve the ivp $x'''' - x = 0, x(0) = x'(0) = x''(1) = 0, x'''(0) = -1$.
16. Prove that the solutions of a third order equation cannot be all oscillatory.
17. Find a condition on p, q such that the solution of the ivp $x''' - x' = 0, x(0) = p, x'(0) = q, x''(0) = p$ tends to zero as $t \to +\infty$.
18. Show that for every $b \ne 0$ there is a one-parameter family of nontrivial solutions of $x'''' - b^2 x'' = 0$ which tend to zero at $t \to +\infty$.
19. Solve $x'''' - 2x''' + x'' = 0$.

20. Solve $x''' - x'' = t$ by the method of Variation of Parameters.
21. Solve $t^3 x''' - 3t^2 x'' + 6tx' - 6x = 0$.
22. Show that $t^3 x''' + 3t^2 x'' + tx' + x = 0$, $x(1) = x'(1) = x''(1) = 1$ has exactly one one-parametric solution that tends to 0, as $t \to \infty$, and two oscillating ones.
23. Solve $t^3 x''' + 2t^2 x'' + tx' - x = 0$.
24. Solve $t^3 x''' + 2t^2 x'' - 4tx' + 4x = 0$, $x(1) = x'(1) = 1$, $x''(1) = 0$.
25. Solve $t^3 x''' + 3t^2 x'' + tx' + x = 0$, $x(1) = x'(1) = x''(1) = 1$.
26. Solve $x''' - x'' + x' - x = 0$.
27. Solve $x'''' + 8x'' + 16x = 0$.
28. Evaluate the Wronskian $W(2, \sin^2 t, t^7 e^t, e^{t^2}, \cos^2 t, t^3, t^5 - 1)$.

8 Systems of first order equations

8.1 A brief review of matrices and determinants

An $m \times n$ *matrix* is an array of m rows and n columns of elements such as numbers, symbols, functions, etc. Here we will be mostly dealing with $n \times n$ matrices

$$
\mathbf{A} = \begin{pmatrix}
a_{11} & a_{12} & \cdots & a_{1n} \\
a_{21} & a_{22} & \cdots & a_{2n} \\
\vdots & & & \\
a_{n1} & a_{n2} & \cdots & a_{nn}
\end{pmatrix}
$$

involving real numbers.

Determinants. The determinant of a matrix $\mathbf{A}$ will be denoted either as $|\mathbf{A}|$ or $\det(\mathbf{A})$. The determinant of a matrix (a), consisting of a single element a, is defined as $|a| = a$. The determinant of an $n \times n$ matrix $\mathbf{A} = (a_{ij})$ is defined as

$$
|\mathbf{A}| = \begin{vmatrix}
a_{11} & a_{12} & \cdots & a_{1n} \\
a_{21} & a_{22} & \cdots & a_{2n} \\
\vdots & & & \\
a_{n1} & a_{n2} & \cdots & a_{nn}
\end{vmatrix} = a_{11}C_{11} + a_{12}C_{12} + \cdots + a_{1n}C_{1n},
$$

$C_{ij} = (-1)^{i+j}M_{ij}$, where M_{ij} is the determinant of the $n-1 \times n-1$ matrix obtained by crossing out the ith row and the jth column to which a_{ij} belongs. This defines the determinant of an $n \times n$ matrix, inductively, in terms of the smaller $n-1 \times n-1$ determinants C_{ij}. The smaller determinants M_{ij} and C_{ij} are called the *minors* and *cofactors* of a_{ij}, respectively.

According to the definition, since the determinant of a singleton matrix (a) is given by $|a| = a$, the determinant of a 2×2 matrix $\mathbf{A} = (a_{ij})$ is:

$$
|\mathbf{A}| = \begin{vmatrix}
a_{11} & a_{12} \\
a_{21} & a_{22}
\end{vmatrix} = a_{11}|a_{22}| - a_{12}|a_{21}| = a_{11}a_{22} - a_{12}a_{21}.
$$

Comment. It is useful to know that the evaluation of an $n \times n$ determinant by adding the products of the elements in the first row by their corresponding cofactors gives the same result as adding the products of the elements in any row or column by their corresponding cofactors. For example, if we evaluate the 2×2 matrix above along the second column, we will have

$$
(-1)^{(1+2)}a_{12}|a_{21}| + (-1)^{2+2}a_{22}|a_{11}| = -a_{12}a_{21} + a_{22}a_{11}
$$

which, as shown above, is the same as evaluating it along the first row.

https://doi.org/10.1515/9783111185675-008

Example 8.1. Evaluate the determinant

$$\begin{vmatrix} 1 & 2 & -1 \\ -2 & 1 & 0 \\ 1 & 1 & 1 \end{vmatrix}.$$

We note that in the third column and also in the second row, one of the elements is 0. So, it is convenient to evaluate the determinant either along the third column or the second row, since we will then have only two calculations instead of three. Let us do it both ways. First, choosing the third column, we have

$$(-1)^{(1+3)}(-1)\begin{vmatrix} -2 & 1 \\ 1 & 1 \end{vmatrix} + 0 + (-1)^{(3+3)}1.\begin{vmatrix} 1 & 2 \\ -2 & 1 \end{vmatrix} = -(-2-1)+(1+4) = 8.$$

Choosing the second row, we have

$$(-1)^{(1+2)}(-2)\begin{vmatrix} 2 & -1 \\ 1 & 1 \end{vmatrix} + (-1)^{(2+2)}(1)\begin{vmatrix} 1 & -1 \\ 1 & 1 \end{vmatrix} = 2(2+1)+(1+1) = 8.$$

Example 8.2. Evaluate the determinant

$$|A| = \begin{vmatrix} 3 & 0 & 0 & 5 \\ 0 & 1 & 1 & 2 \\ 1 & 0 & -1 & 3 \\ 0 & 0 & 0 & 3 \end{vmatrix}.$$

In this case, the last row is convenient since it has only one nonzero element. Let us go with the fourth row and then the first. Then

$$\begin{vmatrix} 3 & 0 & 0 & 5 \\ 0 & 1 & 1 & 2 \\ 1 & 0 & -1 & 3 \\ 0 & 0 & 0 & 3 \end{vmatrix} = (-1)^{4+4}3\begin{vmatrix} 3 & 0 & 0 \\ 0 & 1 & 1 \\ 1 & 0 & -1 \end{vmatrix} = 3\begin{vmatrix} 3 & 0 & 0 \\ 0 & 1 & 1 \\ 1 & 0 & -1 \end{vmatrix} = 3.(-1)^{1+1}3\begin{vmatrix} 1 & 1 \\ 0 & -1 \end{vmatrix} = -9.$$

We conclude the review of determinants by recalling some elementary and basic properties of determinants.

1. Multiplying a row (or column) of a determinant by a nonzero constant k multiplies the value of the determinant by k.
2. Adding a multiple of a row (or column) to another row (or column) does not change the value of the determinant.
3. Exchanging any two rows (or columns) of a determinant changes the sign of the determinant.
4. $|AB| = |A||B|$.
5. $|A| \neq 0 \iff$ both its rows and columns are linearly independent.

In order to illustrate property 5, we consider the determinant

$$
\begin{vmatrix}
4 & 7 & -1 \\
2 & 1 & 1 \\
1 & 3 & -1
\end{vmatrix}.
$$

The three rows of this determinant are linearly dependent since

$$2(1, 3, -1) + 1.(2, 1, 1) + (-1)(4, 7, -1) = (0, 0, 0).$$

Therefore, $|A| = 0$.

Matrices. An $n \times m$ *matrix* is an array of n rows and m columns of elements such as numbers, symbols, functions, etc. Here we will be mostly dealing with $n \times n$ matrices involving real numbers, but we start with general matrices first.

Matrices are designated by capital bold letters such as $\mathbf{A}$, $\mathbf{B}$, or $\mathbf{A} = (a_{ij})$, $\mathbf{B} = (b_{ij})$ etc. An $n \times 1$ matrix is called a *column vector* while a $1 \times n$ matrix is called a *row vector*. Vectors are usually designated by lower-case bold letters such as $\mathbf{u}, \mathbf{v}$, etc.

The *zero matrix*, designated as $\mathbf{0}$, is the matrix all of whose elements are zero. If $\mathbf{A} = (a_{ij})$ and $\mathbf{B} = (b_{ij})$ are $m \times n$ matrices, i. e. the same size, then $\mathbf{A} \pm \mathbf{B} = (a_{ij}) \pm (b_{ij}) = (a_{ij} \pm b_{ij})$, and $\mathbf{A} \pm \mathbf{0} = (a_{ij}) \pm \mathbf{0} = (a_{ij} \pm 0) = (a_{ij})$. If c is a real number then $c\mathbf{A} = (ca_{ij})$ and $c(\mathbf{A} + \mathbf{B}) = c\mathbf{A} + c\mathbf{B}$, $\mathbf{A}(\mathbf{BC}) = (\mathbf{AB})\mathbf{C}$ and $\mathbf{A}(\mathbf{B} + \mathbf{C}) = \mathbf{AB} + \mathbf{AC}$.

The *transpose* of a matrix $\mathbf{A}$, labeled as $\mathbf{A}^T$, is a matrix where the columns and rows of $\mathbf{A}$ are interchanged. For example, the transpose of

$$
\mathbf{A} = \begin{pmatrix}
3 & 6 & -2 \\
0 & 1 & -3 \\
1 & 2 & 5
\end{pmatrix},
$$

is

$$
\mathbf{A}^T = \begin{pmatrix}
3 & 0 & 1 \\
6 & 1 & 2 \\
-2 & -3 & 5
\end{pmatrix}.
$$

Before we discuss multiplication of matrices in general, let us recall that if $\mathbf{u}$ is a row vector and $\mathbf{v}$ is a column vector, then their dot product is a scalar, given by

$$
\mathbf{u}.\mathbf{v} = \begin{pmatrix} u_1 & u_2 & \cdots & u_n \end{pmatrix} \begin{pmatrix} v_1 \\ v_2 \\ \vdots \\ v_n \end{pmatrix} = u_1.v_1 + u_2.v_2 + \cdots + u_n.v_n.
$$

For example,

$$
\begin{pmatrix} 1 & 2 & 3 \end{pmatrix} \begin{pmatrix} 3 \\ 0 \\ 1 \end{pmatrix} = (1 \times 3) + (2 \times 0) + (3 \times 1) = 6.
$$

However, if $\mathbf{v}$ is a column vector and $\mathbf{u}$ is a row vector, then

$$\mathbf{vu} = \begin{pmatrix} v_1 \\ v_2 \\ \vdots \\ v_n \end{pmatrix} \begin{pmatrix} u_1 & u_2 & \cdots & u_m \end{pmatrix} = \begin{pmatrix} v_1 u_1 & v_1 u_2 & \cdots & v_1 u_m \\ v_2 u_1 & v_2 u_2 & \cdots & v_2 u_m \\ \vdots & & & \\ v_n u_1 & v_n u_2 & \cdots & v_n u_m \end{pmatrix}$$

which is an n by m matrix.

For example,

$$\begin{pmatrix} 3 \\ 0 \\ 1 \end{pmatrix} \begin{pmatrix} 1 & 2 \end{pmatrix} = \begin{pmatrix} 3 & 6 \\ 0 & 0 \\ 1 & 2 \end{pmatrix}.$$

Now, we discuss multiplication of matrices in general. Let $A = (a_{ij})$ and $B = (b_{ij})$ be two $n \times n$ matrices. In order to determine the matrix C such that $\mathbf{AB} = \mathbf{C}$, first of all $\mathbf{A}$ and $\mathbf{B}$ must be compatible in size, i. e. if $\mathbf{A}$ has n columns then $\mathbf{B}$ must have n rows. Then each element c_{ij} of $\mathbf{C}$ is obtained by multiplying the i-th row of $\mathbf{A}$ by the j-th column of $\mathbf{B}$, similar to the dot product of two vectors. For example,

$$c_{11} = (a_{11} \times b_{11}) + (a_{12} \times b_{21}) + \cdots + (a_{1n} \times b_{n1}).$$

In general,

$$c_{ij} = \sum_k a_{ik} b_{kj}.$$

The example below displays the sum of the rows in the first matrix multiplied by the columns of the second matrix in order to obtain the product of the two matrices.

$$\begin{pmatrix} 1 & 3 \\ -1 & 1 \\ 5 & 1 \end{pmatrix} \begin{pmatrix} 2 & 1 & 2 \\ -1 & 1 & 0 \end{pmatrix}$$

$$= \begin{pmatrix} (1 \times 2) + (3 \times -1) & (1 \times 1) + (3 \times 1) & (1 \times 2) + (3 \times 0) \\ (-1 \times 2) + (1 \times -1) & (-1 \times 1) + (1 \times 1) & (-1 \times 2) + (1 \times 0) \\ (5 \times 2) + (1 \times -1) & (5 \times 1) + (1 \times 1) & (5 \times 2) + (1 \times 0) \end{pmatrix}$$

$$= \begin{pmatrix} -1 & 4 & 2 \\ -3 & 0 & -2 \\ 9 & 6 & 10 \end{pmatrix}.$$

Notice that the first row of $\mathbf{AB}$ is sum of the products of the first row of $\mathbf{A}$ multiplied by all three columns of $\mathbf{B}$. The second row of $\mathbf{AB}$ is the sum of the products of the second row of $\mathbf{A}$ with the three columns of $\mathbf{B}$. The third row of $\mathbf{AB}$ is the sum of the products of the third row of $\mathbf{A}$ and the three columns of $\mathbf{B}$. It is important to note that multiplication of matrices is not commutative, i. e. generally $\mathbf{AB} \neq \mathbf{BA}$. For example,

$$\begin{pmatrix} 3 & 4 \\ 1 & -2 \end{pmatrix}\begin{pmatrix} 1 & -2 \\ 2 & 3 \end{pmatrix} = \begin{pmatrix} 11 & 6 \\ -3 & -8 \end{pmatrix} \neq \begin{pmatrix} 1 & -2 \\ 2 & 3 \end{pmatrix}\begin{pmatrix} 3 & 4 \\ 1 & -2 \end{pmatrix} = \begin{pmatrix} 1 & 8 \\ 9 & 2 \end{pmatrix}$$

Square matrices. From here on we concentrate on square matrices, i.e, $n \times n$ matrices, $n > 1$.

The $n \times n$ matrix

$$I = \begin{pmatrix} 1 & 0 & \dots & 0 & 0 \\ 0 & 1 & \dots & 0 & 0 \\ & \vdots & & & \\ 0 & 0 & \dots & 0 & 1 \end{pmatrix}$$

is called the identity matrix. It is a square matrix all of whose diagonal elements are 1 and the rest are all 0. It can be easily verified, that if A is any matrix of the same size as I, then $AI = IA = A$. For example,

$$\begin{pmatrix} -1 & 4 & 2 \\ -3 & 0 & -2 \\ 9 & 6 & 10 \end{pmatrix}\begin{pmatrix} 1 & 0 & 0 \\ 0 & 1 & 0 \\ 0 & 0 & 1 \end{pmatrix} = \begin{pmatrix} -1 & 4 & 2 \\ -3 & 0 & -2 \\ 9 & 6 & 10 \end{pmatrix}.$$

Inverse of a matrix. A matrix B is the inverse of a matrix A if $AB = BA = I$. The inverse of a matrix is unique. For, if B and C are both inverses of A, then $AB = I$ and $AC = I$ $\Longrightarrow AB = AC \Longrightarrow C(AB) = C(AC) \Longrightarrow (CA)B = (CA)C \Longrightarrow IB = IC \Longrightarrow B = C$.

A matrix has to be a square matrix in order to have an inverse, but not all square matrices have inverses. For example,

$$\begin{pmatrix} 1 & 2 \\ 2 & 4 \end{pmatrix}\begin{pmatrix} a & b \\ c & d \end{pmatrix} = \begin{pmatrix} 1 & 0 \\ 0 & 1 \end{pmatrix} \Longrightarrow a + 2c = 1, \quad 2a + 4c = 0$$

which is absurd. Therefore, the matrix

$$A = \begin{pmatrix} 1 & 2 \\ 2 & 4 \end{pmatrix}$$

has no inverse. Notice that the determinant of the matrix A is 0. This is true in general, i. e. if the determinant of a matrix is 0 then the matrix is has no inverse.

A matrix that has an inverse is called *nonsingular* or *invertible*, one that has no inverse is called *singular*. One can determine whether or not a matrix is singular by checking its determinant. If its determinant is nonzero then it is nonsingular, otherwise it is singular.

Methods for finding inverses of matrices. First, let us consider a nonsingular 2×2 matrix

$$A = \begin{pmatrix} a & b \\ c & d \end{pmatrix}.$$

We search for a matrix

$$X = \begin{pmatrix} x_1 & x_2 \\ y_1 & y_2 \end{pmatrix}$$

such that $AX = I$, which is equivalent to solving the algebraic system

$$\begin{cases} ax_1 + by_1 = 1, \\ cx_1 + dy_1 = 0, \\ ax_2 + by_2 = 0, \\ cx_2 + dy_2 = 1. \end{cases}$$

Solving the first pair of equations for x_1 and y_1 and the second pair of equations for x_2 and y_2, we obtain the matrix

$$X = A^{-1} = \begin{pmatrix} \frac{d}{ad-bc} & \frac{-b}{ad-bc} \\ \frac{-c}{ad-bc} & \frac{a}{ad-bc} \end{pmatrix}.$$

Thus it follows that if the determinant of A is zero, then it has no inverse. It is easy to remember how to obtain the inverse of a 2×2 matrix by simply replacing the diagonal elements a and d by d and a, respectively, and replacing the diagonal elements b and c by $-b$ and $-c$, respectively, then dividing all the elements of the new matrix by the determinant of A. For example, to find the inverse of

$$A = \begin{pmatrix} 1 & 2 \\ -1 & -3 \end{pmatrix}$$

we first find the determinant, which is $[1 \times (-3)] - [(2) \times (-1)] = -1$. Since the determinant is nonzero, then A has an inverse. To find the inverse, we replace the diagonal elements 1 and -3 by -3 and 1, respectively, and the diagonal elements -1 and 2 by 1 and -2, respectively, and then dividing all elements by the determinant of A, which is -1 in this case. Therefore,

$$A^{-1} = \begin{pmatrix} 3 & 2 \\ -1 & -1. \end{pmatrix}$$

Inverses in general. There are more than one method to find inverses of matrices. We prefer the method using cofactors since it is straightforward and does not depend on choosing among several options.

The inverse of an $n \times n$ nonsingular matrix $A = (a_{ij})$ is given by $C = (c_{ij})$, where $c_{ij} = \frac{C_{ji}}{|A|}$, where C_{ji} is the cofactor of a_{ji}.

Example 8.3. Find the inverse of

$$A = \begin{pmatrix} 1 & 1 & 1 \\ 0 & 2 & 1 \\ 1 & 0 & 3 \end{pmatrix}.$$

First, evaluating the determinant along the first column, we obtain $|A| = (6 - 0) + 0 + (-1) = 5$. Therefore, $C = A^{-1}$ exists. Let us first list the cofactors of $C = (c_{ij})$. We have:

$$C_{11} = (2 \cdot 3 - 1 \cdot 0) = 6, \quad C_{12} = -(0 \cdot 3 - 1 \cdot 1) = 1, \quad C_{13} = (0 \cdot 0 - 2 \cdot 1) = -2$$
$$C_{21} = -(1 \cdot 3 - 1 \cdot 0) = -3, \quad C_{22} = (1 \cdot 3 - 1 \cdot 1) = 2, \quad C_{23} = -(1 \cdot 0 - 1 \cdot 1) = 1$$
$$C_{31} = (1 \cdot 1 - 2 \cdot 1) = -1, \quad C_{32} = -(1 \cdot 1 - 1 \cdot 0) = -1, \quad C_{33} = (1 \cdot 2 - 1 \cdot 0) = 2$$

Now,

$$A^{-1} = (c_{ij}) = \left(\frac{C_{ji}}{|A|}\right) = \begin{pmatrix} \frac{6}{5} & \frac{-3}{5} & \frac{-1}{5} \\ \frac{1}{5} & \frac{2}{5} & \frac{-1}{5} \\ \frac{-2}{5} & \frac{1}{5} & \frac{2}{5} \end{pmatrix}.$$

8.2 General preliminaries

A system of first order differential equations (in normal form) has the form

$$\begin{cases} x_1' = f_1(t, x_1, \ldots, x_n), \\ x_2' = f_2(t, x_1, \ldots, x_n), \\ \cdots \\ x_n' = f_n(t, x_1, \ldots, x_n), \end{cases} \tag{8.1}$$

where the f_i are continuous functions defined for $(t, x_1, \ldots, x_n) \in A \subseteq \mathbb{R}^{n+1}$. For brevity, it is convenient to introduce vectors $\overline{x} = (x_1, \ldots, x_n), \overline{f} = (f_1, \ldots, f_n)$. With this notation (8.1) becomes

$$\overline{x}' = \overline{f}(t, \overline{x}).$$

A solution of this system is given by n differentiable functions in the form $\overline{x}(t) = (x_1(t), \ldots, x_n(t))$, defined on a common interval I such that:
(i) $(t, \overline{x}(t)) \in A$ for all $t \in I$;
(ii) $\overline{x}'(t) = \overline{f}(t, \overline{x}(t))$ for all $t \in I$, namely $x_i'(t) = f_i(t, x_1(t), \ldots, x_n(t))$, $i = 1, \ldots, n$, for all $t \in I$.

Remark 8.1. Consider the system

$$
\begin{cases}
x_1' = x_2, \\
x_2' = x_3, \\
\cdots \\
x_n' = f(t, x_1, \ldots, x_n).
\end{cases}
$$

By renaming $x_1 = x$, we obtain $x_2 = x_1'$, $x_3 = x_2' = x''$, ..., $x_n = x^{(n-1)}$. Therefore,

$$
x_n' = x^{(n)} = f(t, x', \ldots, x^{(n-1)}).
$$

Conversely, given a single equation $x^{(n)} = f(t, x', \ldots, x^{(n-1)})$, it can be written in the form of the above system simply by renaming $x_1 = x$, $x_2 = x_1'$, ..., $x_n = x_{n-1}'$. It is then easy to see that $x_n = x_{n-1}' = x_{n-2}'' = \cdots = x_{(n-(n-1))}^{(n-1)} = x^{(n-1)} \Rightarrow x_n' = x^{(n)} = f(t, x_1, \ldots, x_{n-1})$.

For example, the 2×2 system

$$
\begin{cases}
x_1' = x_2, \\
x_2' = -p_1 x_2 - p_0 x_1,
\end{cases}
$$

is equivalent to the second order equation $x'' = -p_1 x' - p_0 x$, or $x'' + p_1 x' + p_0 x = 0$. Similarly, the system

$$
\begin{cases}
x_1' = x_2, \\
x_2' = x_3, \\
x_3' = tx_1 x_2 + 2x_3,
\end{cases}
$$

is equivalent to the third order equation $x''' = txx' + 2x''$.

Given $t_0 \in I$ and a vector $\overline{x}_0 = (x_{10}, x_{20}, \ldots, x_{n0})$, with $(t_0, \overline{x}_0) \in A$, the Cauchy problem for (8.1) consists of finding a solution $\overline{x}(t)$ satisfying the initial conditions $\overline{x}(t_0) = \overline{x}_0$, namely

$$
x_1(t_0) = x_{10}, \quad x_2(t_0) = x_{20}, \quad \ldots, \quad x_n(t_0) = x_{n0}.
$$

Using vector notation, a Cauchy problem for (8.1) can be written as

$$
\overline{x}' = \overline{f}(t, \overline{x}), \quad \overline{x}(t_0) = \overline{x}_0. \tag{8.2}
$$

Roughly speaking, definitions and results for scalar equations generally hold true for systems. In particular, let us state the following existence and uniqueness theorem, stated for first order equations in Chapter 3.

Theorem 8.1. *Suppose that $\bar{f}$ is continuous in $A \subseteq \mathbb{R}^{n+1}$ and has continuous partial derivatives $\bar{f}_{x_i}$, $i = 1,\ldots,n$.[1] Let $(t_0, \bar{x}_0)$ be a given point in the interior of A. Then there exists a closed interval I containing t_0 in its interior such that the Cauchy problem (8.2) has one and only solution, defined in I.*

Furthermore, if $A = I \times \mathbb{R}^n$, with $I \subseteq \mathbb{R}$, and $\bar{f}_{x_i}$, $i = 1,\ldots,n$, are bounded, then the solution is defined on all of I.

8.3 Linear systems

If the n functions f_i are linear with respect to $x_1, \ldots, x_n$, (8.1) becomes

$$\begin{cases} x_1' = a_{11}(t)x_1 + \cdots + a_{1n}(t)x_n + h_1(t), \\ x_2' = a_{21}(t)x_1 + \cdots + a_{2n}(t)x_n + h_2(t), \\ \cdots \\ x_n' = a_{n1}(t)x_1 + \cdots + a_{nn}(t)x_n + h_n(t). \end{cases} \tag{8.3}$$

Using vector notation, (8.3) can be written as

$$\bar{x}' = A(t)\bar{x} + \bar{h}(t),$$

where $A, \bar{x}$ and $\bar{h}$ denote, respectively, the matrix

$$A(t) = \begin{pmatrix} a_{11}(t) & \cdots & a_{1n}(t) \\ a_{21}(t) & \cdots & a_{2n}(t) \\ \cdots & \cdots & \cdots \\ a_{n1}(t) & \cdots & a_{nn}(t) \end{pmatrix} \quad \text{and the vectors} \quad \bar{x} = \begin{pmatrix} x_1 \\ x_2 \\ \cdots \\ x_n \end{pmatrix} \quad \text{and}$$

$$\bar{h}(t) = \begin{pmatrix} h_1(t) \\ h_2(t) \\ \cdots \\ h_n(t) \end{pmatrix}.$$

If $\bar{h} = 0$, the system

$$\bar{x}' = A(t)\bar{x} \tag{8.4}$$

is called *homogeneous*, otherwise *non-homogeneous*.

For example, if $n = 2$ a linear non-homogeneous system has the form

$$\begin{cases} x_1' = a_{11}x_1 + a_{12}x_2 + h_1, * \\ x_2' = a_{21}x_1 + a_{22}x_2 + h_2. \end{cases}$$

---

1 $\bar{f}_{x_i}$ means $\frac{\partial f_i}{\partial x_j}$, $i,j = 1,\ldots,n$.

Assuming that a_{ij} and h are continuous on an interval $I \subseteq \mathbb{R}$, a direct application of Theorem 8.1 shows that for each $(t_0, \overline{x}_0) \in I \times \mathbb{R}^2$ the system $\overline{x}' = A(t)\overline{x} + \overline{h}(t)$ has one and only one solution satisfying the initial condition $\overline{x}(t_0) = \overline{x}_0$, defined on all of I.

Lemma 8.1. *Any linear combination of solutions $\overline{x}_1, \ldots, \overline{x}_n$ of the homogeneous system (8.4) is also a solution.*

Proof. We want to show that any linear combination $c_1\overline{x}_1 + \cdots + c_n\overline{x}_n$ is a solution. This follows easily, since

$$(c_1\overline{x}_1 + \cdots + c_n\overline{x}_n)' = c_1\overline{x}_1' + \cdots + c_n\overline{x}_n' = c_1 A(t)\overline{x}_1 + \cdots + c_n A(t)\overline{x}_n$$
$$= A(t)[c_1\overline{x}_1 + \cdots + c_n\overline{x}_n]. \qquad \square$$

Given n vector valued functions $\overline{u}_1(t), \ldots, \overline{u}_n(t)$, they are said to be *linearly independent* on I if $c_1\overline{u}_1(t) + \cdots + c_n\overline{u}_n(t) = 0$, for all $t \in I$, implies $c_1 = \cdots = c_n = 0$; otherwise they are said to be *linearly dependent* on I.

Taking into account the fact that the ith component of $c_1\overline{u}_1(t) + \cdots + c_n\overline{u}_n(t)$ is $c_1 u_{1i}(t) + \cdots + c_n u_{ni}(t)$ it follows that the identity $c_1\overline{u}_1(t) + \cdots + c_n\overline{u}_n(t) = 0$ is equivalent to the system

$$\begin{cases} c_1 u_{11}(t) + \cdots + c_n u_{n1}(t) = 0, \\ c_1 u_{12}(t) + \cdots + c_n u_{n2}(t) = 0, \\ \cdots \\ c_1 u_{1n}(t) + \cdots + c_n u_{nn}(t) = 0, \end{cases}$$

where $c_1, \ldots, c_n$ are arbitrary constants. If $\overline{u}_1(t), \ldots, \overline{u}_n(t)$ are linearly independent on I then the unique solution of the preceding system in c_i is $c_1 = \cdots = c_n = 0$. Thus $\overline{u}_1(t), \ldots, \overline{u}_n(t)$ are linearly independent if and only if

$$\begin{vmatrix} u_{11}(t) & \cdots & u_{n1}(t) \\ u_{12}(t) & \cdots & u_{n2}(t) \\ \cdots & \cdots & \cdots \\ u_{1n}(t) & \cdots & u_{nn}(t) \end{vmatrix} \neq 0, \quad \forall t \in I.$$

The above determinant is called the *wronskian* of $\overline{u}_1(t), \ldots, \overline{u}_n(t)$, written as $W(\overline{u}_1, \ldots, \overline{u}_n)(t)$. When no misunderstanding can arise, for brevity, we may simply write $W(t)$ or even W instead of $W(\overline{u}_1, \ldots, \overline{u}_n; t)$ or $W(\overline{u}_1, \ldots, \overline{u}_n)(t)$. Consequently:

- If $\overline{u}_1(t), \ldots, \overline{u}_n(t)$ *are linearly dependent on I then $W(t) = 0$ for all $t \in I$,*

which can be equivalently stated as

- If $\exists t_0 \in I$ *such that $W(\overline{u}_1, \ldots, \overline{u}_n)(t_0) \neq 0$, then $\overline{u}_1, \ldots, \overline{u}_n$ are linearly independent on I.*

The next theorem is an extension of Abel's formula for systems.

Theorem 8.2. *Let* $\bar{u}_1(t), \ldots, \bar{u}_n(t)$ *be solutions of* $\bar{x}' = A(t)\bar{x}$. *Then*

$$W(t) = W(t_0) \cdot e^{\int_{t_0}^{t} [a_{11}(s) + \cdots + a_{nn}(s)] \, ds}.$$

As a consequence, $W(t)$ is either identically zero, or it is never zero.

The sum

$$T_r A = \sum_{i=1}^{i=n} a_{ii}$$

is called the *trace* of the matrix A.

Let us verify this result for $n = 2$. Letting $\bar{u} = (u_1, u_2)$ and $\bar{v} = (v_1, v_2)$, we find

$$W = \begin{vmatrix} u_1 & v_1 \\ u_2 & v_2 \end{vmatrix} = u_1 v_2 - u_2 v_1$$

and

$$W' = u_1' v_2 + u_1 v_2' - u_2 v_1' - u_2' v_1.$$

Since $\bar{u}' = A\bar{u}$ and $\bar{v}' = A\bar{v}$ imply

$$\begin{cases} u_1' = a_{11} u_1 + a_{12} u_2, \\ u_2' = a_{21} u_1 + a_{22} u_2, \end{cases} \quad \text{and} \quad \begin{cases} v_1' = a_{11} v_1 + a_{12} v_2, \\ v_2' = a_{21} v_1 + a_{22} v_2, \end{cases}$$

we find

$$W' = (a_{11} u_1 + a_{12} u_2) v_2 + u_1 (a_{21} v_1 + a_{22} v_2) - (a_{11} v_1 + a_{12} v_2) u_2 - v_1 (a_{21} u_1 + a_{22} u_2).$$

Rearranging and simplifying, we obtain

$$W' = a_{11}(u_1 v_2 - v_1 u_2) + a_{22}(u_1 v_2 - v_1 u_2) = (a_{11} + a_{22}) W$$

which yields

$$W(t) = W(t_0) \cdot e^{\int_{t_0}^{t} [a_{11}(s) + a_{22}(s)] \, ds}. \qquad \square$$

For example, the linear scalar equation $x''' + p_2 x'' + p_1 x' + p_0 x = 0$ is equivalent to the system

$$\begin{cases} x_1' = x_2, \\ x_2' = x_3, \\ x_3' = -p_2 x_3 - p_1 x_2 - p_0 x_1. \end{cases}$$

The above system can be written as the vector equation $\bar{x}' = A\bar{x}$, where the matrix A is given by

$$A = \begin{pmatrix} 0 & 1 & 0 \\ 0 & 0 & 1 \\ -p_0 & -p_1 & -p_2 \end{pmatrix},$$

and hence $T_r A = a_{11} + a_{22} + a_{33} = a_{33} = -p_2$. Therefore $W = W(t_0) \cdot e^{-\int_{t_0}^{t} p_2(s)\,ds}$, which agrees with the result found by applying Abel's theorem to the given scalar equation.

If $\bar{u}_1(t), \ldots, \bar{u}_n(t)$ are solutions of $\bar{x}' = A(t)\bar{x}$, by Lemma 8.1 we know that $c_1\bar{u}_1(t) + \cdots + c_n\bar{u}_n(t)$ is a solution of $\bar{x}' = A(t)\bar{x}$. Furthermore, we have the following theorem.

Theorem 8.3. *Let $\bar{u}_1(t), \ldots, \bar{u}_n(t)$ be linearly independent solutions of $\bar{x}' = A(t)\bar{x}$. Then for any solution $\bar{u}(t)$ of $\bar{x}' = A(t)\bar{x}$ there exist $c_1, \ldots, c_n \in \mathbb{R}$ such that $\bar{u}(t) = c_1\bar{u}_1(t) + \cdots + c_n\bar{u}_n(t)$.*

The sum $c_1\bar{u}_1(t) + \cdots + c_n\bar{u}_n(t)$ is called *the general solution* of $\bar{x}' = A(t)\bar{x}$.

The proof is quite similar to the one for scalar equations (see Theorem 6.2 of Chapter 6), and it is omitted.

The preceding result can be rephrased by saying that the vector space of solutions of $\bar{x}' = A(t)\bar{x}$ is spanned by any n linearly independent solutions.

Remark 8.2. In order to find n linearly independent solutions, it is convenient to find a solution $\bar{u}$ satisfying the Cauchy problem $\bar{x}' = A\bar{x}$ subject to the n initial conditions

$$\bar{u}_1(0) = \bar{x}_1 = \begin{pmatrix} 1 \\ 0 \\ \cdots \\ 0 \end{pmatrix}, \quad \bar{u}_2(0) = \bar{x}_2 = \begin{pmatrix} 0 \\ 1 \\ \cdots \\ 0 \end{pmatrix}, \ldots, \bar{u}_n(0) = \bar{x}_n = \begin{pmatrix} 0 \\ 0 \\ \cdots \\ 1 \end{pmatrix}, \quad (8.5)$$

which implies

$$W(0) = \begin{vmatrix} 1 & 0 & \cdots & 0 \\ 0 & 1 & \cdots & 0 \\ \cdots & \cdots & \cdots & \cdots \\ 0 & 0 & \cdots & 1 \end{vmatrix} = 1.$$

Dealing with non-homogeneous systems, we have the following theorem.

Theorem 8.4. *The general solution of $\bar{x}' = A\bar{x} + \bar{h}$ is given by the sum of the general solution of the associated homogeneous system $\bar{x}' = A\bar{x}$ and any one particular solution of the non-homogeneous system.*

8.4 Linear systems with constant coefficients

If the a_{ij} do not depend on t, (8.3) becomes a linear system with constant coefficients,

$$\bar{x}' = A\bar{x} + \bar{h}(t), \quad \text{where } A = \begin{pmatrix} a_{11} & \cdots & a_{1n} \\ a_{21} & \cdots & a_{2n} \\ \cdots & \cdots & \cdots \\ a_{n1} & \cdots & a_{nn} \end{pmatrix}. \tag{8.6}$$

Recall that the general solution to the scalar equation $x' = ax$ is given by $x(t) = ce^{at}$. So in order to solve $\bar{x}' = A\bar{x}$, where A is a constant matrix, one might try to find a solution of the form $\bar{v}e^{\lambda t}$, where $\bar{v}$ is a constant vector and λ is a real or complex number. Substituting $\bar{x} = e^{\lambda t}\bar{v}$ in the equation $\bar{x}' = A\bar{x}$ leads to $\lambda e^{\lambda t}\bar{v} = Ae^{\lambda t}\bar{v}$, and hence $\lambda\bar{v} = A\bar{v}$, which can be written as

$$(A - \lambda I)\bar{v} = 0. \tag{8.7}$$

Now we recall that, for any number λ, the algebraic system (8.7) has a nontrivial solution $\bar{v}$ if and only if

$$\det(A - \lambda I) = 0. \tag{8.8}$$

Given a constant n-by-n matrix A, the determinant of $A - \lambda I$ is an nth order polynomial called the *characteristic polynomial* or *auxiliary polynomial of A*. Any real or complex number λ satisfying (8.8) is called an *eigenvalue* of A. Given an eigenvalue λ of A, any nonzero vector $\bar{v}$ satisfying equation (8.7) is called an *eigenvector of A corresponding to the eigenvalue λ.*

Remark 8.3. As pointed out in Remark 8.1, any linear n-th order equation

$$L[x] = x^{(n)} + a_{n-1}x^{(n-1)} + \ldots + a_1 x' + a_0 x = 0$$

with constant coefficients is equivalent to a linear system

$$\begin{cases} x_1' = x_2, \\ x_2' = x_3, \\ \cdots \quad \cdots \\ x_n' = -a_{n-1}x_n - \cdots - a_1 x_2 - a_0 x_1, \end{cases}$$

or $\bar{x}' = A\bar{x}$ with matrix

$$A = \begin{pmatrix} 0 & 1 & 0 & \cdots & 0 \\ 0 & 0 & 1 & \cdots & 0 \\ \cdots & \cdots & \cdots & \cdots & \cdots \\ -a_0 & -a_1 & -a_2 & \cdots & -a_{n-1} \end{pmatrix}.$$

By mathematical induction one can show that the eigenvalues of A are just the roots of the characteristic polynomial $C(\lambda) = \lambda^n + a_{n-1}\lambda^{n-1} + \cdots + a_1\lambda + a_0 = 0$, introduced in the previous chapter.

Let us check this claim for $n = 2, 3$. If $n = 2$ the characteristic polynomial of equation $x'' + a_1x' + a_0x = 0$ is $C(\lambda) = \lambda^2 + a_1\lambda + a_0$. The matrix of the equivalent system is

$$A = \begin{pmatrix} 0 & 1 \\ -a_0 & -a_1 \end{pmatrix}$$

and hence

$$\det(A - \lambda I) = \begin{vmatrix} -\lambda & 1 \\ -a_0 & -a_1 - \lambda \end{vmatrix} = \lambda(a_1 + \lambda) + a_0 = \lambda^2 + a_1\lambda + a_0 = C(\lambda).$$

In the case of the third order equation $x''' + a_2x'' + a_1x' + a_0x = 0$, the characteristic polynomial is $C(\lambda) = \lambda^3 + a_2\lambda^2 + a_1\lambda + a_0$. To find the eigenvalues of

$$A = \begin{pmatrix} 0 & 1 & 0 \\ 0 & 0 & 1 \\ -a_0 & -a_1 & -a_2 \end{pmatrix},$$

we evaluate

$$\det(A - \lambda I) = \begin{vmatrix} -\lambda & 1 & 0 \\ 0 & -\lambda & 1 \\ -a_0 & -a_1 & -a_2 - \lambda \end{vmatrix} = -\lambda \cdot [\lambda(\lambda + a_2) + a_1] - 1 \cdot a_0 = -(\lambda^3 + a_2\lambda^2 + a_1\lambda + a_0),$$

and thus $\det(A - \lambda I) = -C(\lambda)$.

Theorem 8.5. *Any second order system*

$$\begin{cases} x_1' = ax_1 + bx_2, \\ x_2' = cx_1 + dx_2, \end{cases}$$

can be transformed into the second order linear equation

$$x'' - (a + d)x' + (ad - bc)x = 0.$$

Moreover, roots of the characteristic equation

$$\lambda^2 - (a + d)\lambda + (ad - bc) = 0$$

coincide with the eigenvalues of the matrix

$$A = \begin{pmatrix} a & b \\ c & d \end{pmatrix}.$$

Proof. Differentiating the first equation, we find $x_1'' = ax_1' + bx_2' = ax_1' + b(cx_1 + dx_2) = ax_1' + bcx_1 + d(bx_2) = ax_1' + bcx_1 + d(x_1' - ax_1) = ax_1' + bcx_1 + dx_1' - adx_1 = (a + d)x_1' + (bc - ad)x_1$ whereby

$$x_1'' - (a + d)x_1' + (ad - bc)x_1 = 0.$$

By renaming $x_1 = x$, we obtain the required equation

$$x'' - (a + d)x' + (ad - bc)x = 0.$$

Simple calculation shows that $C(\lambda) = \det(A - \lambda I)$. □

Lemma 8.2. *The vector valued function $\bar{u}(t) = e^{\lambda t}\bar{v}$ is a solution of the system $\bar{x}' = A\bar{x}$ if and only if λ is an eigenvalue of A and $\bar{v}$ an eigenvector corresponding to the eigenvalue λ.*

Proof. Since $\bar{u}'(t) = \lambda e^{\lambda t}\bar{v}$ and $A\bar{u} = e^{\lambda t}A\bar{v}$ it follows that

$$\bar{u}'(t) = A\bar{u} \Leftrightarrow \lambda e^{\lambda t}\bar{v} = e^{\lambda t}A\bar{v} \Leftrightarrow A\bar{v} = \lambda\bar{v} \Leftrightarrow (A - \lambda I)\bar{v} = 0,$$

proving the lemma. □

In order to find the eigenvalues of A, we first start with equation (8.8) and solve for the eigenvalue λ. Then, for each value of λ, we set up equation (8.7) in order to find a corresponding eigenvector $\bar{v}$, and finally we have the corresponding solution $\bar{x} = \bar{v}e^{\lambda t}$.

One may recall from linear algebra, and it can easily be verified, that the eigenvalues of a triangular matrix are its diagonal elements. Thus we could have found the eigenvalues of A by simply noticing that A is triangular, without calculating the value of the determinant.

If $A\bar{v} = \lambda\bar{v}$ then, for any nonzero constant c, $Ac\bar{v} = \lambda c\bar{v}$, which means that $c\bar{v}$ is also an eigenvector of A. This is why when solving such equations one often has to make some arbitrary judicial choices in selecting $\bar{v}$. We demonstrate this in the examples below, but first

Theorem 8.6. *If $\lambda_1, \lambda_2, \ldots, \lambda_k$, $1 < k \leq n$ are k distinct eigenvalues of an n–n constant matrix A with corresponding eigenvectors $\bar{v}_1, \bar{v}_2, \ldots, \bar{v}_k$, then $\bar{v}_1, \ldots, \bar{v}_k$ are linearly independent.*

Lemma 8.3. *If λ_i and λ_j are any of the k distinct eigenvalues of A, then the corresponding eigenvalues $\bar{v}_i$ and $\bar{v}_j$ are linearly independent.*

Proof. Suppose that there exist constants c_1 and c_2 such that

$$c_1\bar{v}_i + c_2\bar{v}_j = 0. \tag{8.9}$$

Multiplying (8.9) by A on the left and recalling that $A\bar{v}_i = \lambda_i\bar{v}_i$ and $A\bar{v}_j = \lambda_j\bar{v}_j$, we obtain

$$c_1 A \bar{v}_i + c_2 A \bar{v}_j = 0 \quad \Longrightarrow \quad c_1 \lambda_i \bar{v}_i + c_2 \lambda_j \bar{v}_j = 0. \qquad (8.10)$$

Now, multiplying (8.9) by $-\lambda_i$ and then adding it to (8.10), we obtain

$$c_2 (\lambda_j - \lambda_i) \bar{v}_j = 0.$$

Since λ_i and λ_j are distinct and v_j is nonzero by definition, we must have $c_2 = 0$, reducing (8.9) to $c_1 \bar{v}_i = 0$. Since $\bar{v}_i$ is nonzero, we must have $c_1 = 0$. □

Proof of Theorem 8.6. For $k = 2$, the assertion of the theorem follows from Lemma 8.3. Next, we show that it is true for $k = 3$. Then it will become clear how the proof for the general case follows from a straightforward inductive argument. Let $\bar{v}_1, \bar{v}_2, \bar{v}_3$ be eigenvectors corresponding to distinct eigenvalues λ_1, λ_2 and λ_3, respectively. Suppose that

$$c_1 \bar{v}_1 + c_2 \bar{v}_2 + c_3 \bar{v}_3 = 0.$$

Multiplying this equation by A, we have $c_1 A \bar{v}_1 + c_2 A \bar{v}_2 + c_3 A \bar{v}_3 =$

$$c_1 \lambda_1 \bar{v}_1 + c_2 \lambda_2 \bar{v}_2 + c_3 \lambda_3 \bar{v}_3 = 0.$$

Similar to the proof of Lemma 8.3, we have

$$(c_1 \lambda_1 \bar{v}_1 + c_2 \lambda_2 \bar{v}_2 + c_3 \lambda_3 \bar{v}_3) - \lambda_1 (c_1 \bar{v}_1 + c_2 \bar{v}_2 + c_3 \bar{v}_3) = c_2 (\lambda_2 - \lambda_1) \bar{v}_2 + c_3 (\lambda_3 - \lambda_1) \bar{v}_3 = 0.$$

Since $\bar{v}_2$ and $\bar{v}_3$ are linearly independent, by Lemma 8.3, and the λ_i are distinct, it follows that $c_2 = c_3 = 0$. Therefore, $c_1 = 0$ and the proof is complete. □

Corollary 8.1. *If A has n distinct eigenvalues $\lambda_1, \lambda_2, \ldots, \lambda_n$, then the general solution of $\bar{x}' = A \bar{x}$ is given by*

$$c_1 e^{\lambda_1 t} \bar{v}_1 + c_2 e^{\lambda_2 t} \bar{v}_2 + \cdots + c_n e^{\lambda_n t} \bar{v}_n$$

where $\bar{v}_i$ is an eigenvector corresponding to λ_i, $1 \leq i \leq n$.

Proof. The proof follows from linear independence of the eigenvectors and is easy to verify. □

Example 8.4. Solve the system

$$\begin{cases} x_1' = x_1 - x_2, \\ x_2' = 2x_1 + 4x_2, \end{cases}$$

which is equivalent to

$$\bar{x}' = \begin{pmatrix} 1 & -1 \\ 2 & 4 \end{pmatrix} \bar{x} = 0.$$

Solution. Using (8.8), we have

$$\det(A - \lambda I) = \begin{vmatrix} 1-\lambda & -1 \\ 2 & 4-\lambda \end{vmatrix} = \lambda^2 - 5\lambda + 6 = 0.$$

Solving for λ, we have $\lambda = 2$, $\lambda = 3$.

Corresponding to $\lambda = 2$: In order to find an eigenvector corresponding to $\lambda = 2$, we set up (8.7) and solve for $\bar{v}$, obtaining

$$(A - \lambda I)\bar{v} = \begin{pmatrix} -1 & -1 \\ 2 & 2 \end{pmatrix}\begin{pmatrix} v_1 \\ v_2 \end{pmatrix} = 0.$$

This means that $-v_1 - v_2 = 0$ and $2v_1 + 2v_2 = 0$, which is actually only one equation in two unknowns. So, we can take v_1 to be any nonzero number and then take $v_2 = -v_1$. For convenience, let us choose $v_1 = 1$ and $v_2 = -1$ so that

$$\bar{v} = \begin{pmatrix} 1 \\ -1 \end{pmatrix},$$

which yields the solution

$$\bar{x}_1 = e^{2t}\begin{pmatrix} 1 \\ -1 \end{pmatrix} = \begin{pmatrix} e^{2t} \\ -e^{2t} \end{pmatrix}.$$

Corresponding to $\lambda = 3$: we set up (8.7) and solve for $\bar{v}$, obtaining

$$(A - \lambda I)\bar{v} = \begin{pmatrix} -2 & -1 \\ 2 & 1 \end{pmatrix}\begin{pmatrix} v_1 \\ v_2 \end{pmatrix} = 0.$$

This means that $-2v_1 - v_2 = 0$ and $2v_1 + v_2 = 0$. So, we can take v_1 to be any nonzero number and then take $v_2 = -2v_1$. For convenience, let us choose $v_1 = 1$ and $v_2 = -2$ so that

$$\bar{v} = \begin{pmatrix} 1 \\ -2 \end{pmatrix},$$

which yields the solution

$$\bar{x}_2 = e^{3t}\begin{pmatrix} 1 \\ -2 \end{pmatrix} = \begin{pmatrix} e^{3t} \\ -2e^{3t} \end{pmatrix}.$$

Therefore the general solution is given by

$$\bar{x}(t) = c_1\bar{x}_1 + c_2\bar{x}_2 = \begin{pmatrix} c_1 e^{2t} + c_2 e^{3t} \\ -c_1 e^{2t} - 2c_2 e^{3t} \end{pmatrix}. \qquad \square$$

If a pair of eigenvalues are complex conjugate, $\lambda = \alpha \pm i\beta$, as explained in Chapters 6 and 7, we use the Euler formula $e^{i\beta t} = \cos \beta t + i \sin \beta t$ to find real valued linearly independent solutions, as shown in the next example.

Example 8.5. Consider the system $\bar{x}' = A\bar{x}$ where

$$A = \begin{pmatrix} 1 & 4 \\ -1 & 1 \end{pmatrix}.$$

After solving $\det(A - \lambda I) = 0$ for λ, we obtain the complex eigenvalues $\lambda = 1 \pm 2i$. Setting $\bar{v} = \begin{pmatrix} v_1 \\ v_2 \end{pmatrix}$, $A\bar{v} = (1 + 2i)\bar{v}$ yields

$$\begin{cases} v_1 + 4v_2 = (1 + 2i)v_1, \\ -v_1 + v_2 = (1 + 2i)v_2, \end{cases} \Rightarrow \begin{cases} 4v_2 = 2iv_1, \\ -v_1 = 2iv_2. \end{cases}$$

We notice that the two equations are essentially the same, because if we multiply the second by $-2i$, we get the first. So, for convenience, we choose $v_1 = 2$, $v_2 = i$.

Thus we find the complex valued solution $\begin{pmatrix} 2 \\ i \end{pmatrix} e^{(1+2i)t}$. Using Euler's formula, $e^{(1+2i)t} = e^t(\cos 2t + i \sin 2t)$, we find

$$\bar{x} = \begin{pmatrix} 2 \\ i \end{pmatrix} e^{(1+2i)t} = \begin{pmatrix} 2e^t \cos 2t + 2ie^t \sin 2t \\ ie^t \cos 2t - e^t \sin 2t \end{pmatrix} = \begin{pmatrix} 2e^t \cos 2t \\ -e^t \sin 2t \end{pmatrix} + i \begin{pmatrix} 2e^t \sin 2t \\ e^t \cos 2t \end{pmatrix}.$$

Now we can take the two real solutions to be

$$\bar{x}_1 = \begin{pmatrix} 2e^t \cos 2t \\ -e^t \sin 2t \end{pmatrix}, \quad \bar{x}_2 = \begin{pmatrix} 2e^t \sin 2t \\ e^t \cos 2t \end{pmatrix}.$$

Evaluating their wronskian at $t = 0$ we find

$$W(0) = \begin{vmatrix} 2 & 0 \\ 0 & 1 \end{vmatrix} = 2 \neq 0.$$

Thus $\bar{x}_1, \bar{x}_2$ are linearly independent and $\bar{x} = c_1 \bar{x}_1 + c_2 \bar{x}_2$ is the general solution. □

Remark 8.4. A different method to find solutions of 2×2 autonomous systems

$$\begin{cases} x_1' = ax_1 + bx_2, \\ x_2' = cx_1 + dx_2, \end{cases}$$

is to use Theorem 8.5 and convert it to the second order equation

$$x'' - (a + d)x' + (ad - bc)x = 0.$$

Once we find the general solution x of this equation, we set $x_2 = \frac{1}{b}(x_1' - ax_1)$, where $x_1 = x$, so that $\bar{x}(t) = \begin{pmatrix} x_1(t) \\ x_2(t) \end{pmatrix}$ gives the general solution of the system. Notice that in order to solve for x_2, we assumed that $b \neq 0$. If $b = 0$, both x_1 ans x_2 can be found directly by solving the corresponding first order linear equations: first the homogeneous equation $x_1' = ax_1$ which yields $x_1 = ke^{at}$, $k \in \mathbb{R}$, and then the nonhomogeneous equation $x_2' = dx_2 + cx_1 = dx_2 + c \cdot (ke^{at})$.

As an exercise, let us reconsider the previous two examples.

Example 8.4: Consider the system

$$\bar{x}' = \begin{pmatrix} 1 & -1 \\ 2 & 4 \end{pmatrix} \bar{x} = 0,$$

which can be written as

$$\begin{cases} x_1' = x_1 - x_2, \\ x_2' = 2x_1 + 4x_2. \end{cases}$$

Using Theorem 8.5, it suffices to solve

$$x'' - 5x' + 6x = 0.$$

The roots of the characteristic equation $\lambda^2 - 5\lambda + 6 = 0$ are $\lambda = 2, 3$, and hence the general solution is $x = x_1 = c_1 e^{2t} + c_2 e^{3t}$. Since, from the first equation, $x_2 = x_1 - x_1' = -c_1 e^{2t} - 2c_2 e^{3t}$ again we find the previous vector valued solution,

$$\bar{x}(t) = \begin{pmatrix} c_1 e^{2t} + c_2 e^{3t} \\ -c_1 e^{2t} - 2c_2 e^{3t} \end{pmatrix}.$$

Example 8.5: The system

$$\begin{cases} x_1' = x_1 + 4x_2, \\ x_2' = -x_1 + x_2, \end{cases}$$

is equivalent to

$$x'' - 2x' + 5x = 0.$$

The roots of the characteristic equation $\lambda^2 - 2\lambda + 5 = 0$ are $\lambda = 1 \pm 2i$ and hence the general solution is $x = x_1 = e^t(c_1 \cos 2t + c_2 \sin 2t)$. Then

$$x_2 = \frac{1}{4}(x_1' - x_1) = \frac{1}{4}[2e^t(-c_1 \sin 2t + c_2 \cos 2t)] = \frac{1}{2}e^t(-c_1 \sin 2t + c_2 \cos 2t)$$

and the vector valued solution is

$$\bar{x}(t) = \begin{pmatrix} e^t(c_1 \cos 2t + c_2 \sin 2t) \\ \frac{1}{2}e^t(-c_1 \sin 2t + c_2 \cos 2t) \end{pmatrix},$$

as before (up to an irrelevant constant). □

The next example deals with a non-homogeneous system.

Example 8.6. Solve the system

$$\begin{cases} x_1' = x_1 - x_2 + e^t, \\ x_2' = 2x_1 + 4x_2 - 2e^t. \end{cases}$$

From the previous example, we know that the general solution of the associated homogeneous system is

$$\bar{x}(t) = c_1\bar{x}_1 + c_2\bar{x}_2 = \begin{pmatrix} c_1 e^{2t} + c_2 e^{3t} \\ -c_1 e^{2t} - 2c_2 e^{3t} \end{pmatrix}.$$

Let us find a particular solution $\bar{y}_p$ of the non-homogeneous system of the form $y_1 = ae^t$, $y_2 = be^t$. Substituting into the system we find

$$\begin{cases} ae^t = ae^t - be^t + e^t, \\ be^t = 2ae^t + 4be^t - 2e^t, \end{cases}$$

which yields the algebraic system

$$\begin{cases} a = a - b + 1, \\ b = 2a + 4b - 2. \end{cases}$$

Solving we find $b = 1$ and $a = -\frac{1}{2}$. Thus the general solution is

$$\bar{x}(t) = c_1\bar{x}_1 + c_2\bar{x}_2 + \bar{y}_p = \begin{pmatrix} c_1 e^{2t} + c_2 e^{3t} + e^t \\ -c_1 e^{2t} - 2c_2 e^{3t} - \frac{1}{2}e^t \end{pmatrix}. \qquad \square$$

The case in which A has repeated eigenvalues is more complicated. Here we are only giving an algorithm, without proofs. Summarizing the method, suppose that an eigenvalue λ is repeated k times. Then:

1. $\bar{v}_1$ is an eigenvector corresponding to λ, i. e., $(A - \lambda I)\bar{v}_1 = 0$, $\Rightarrow \bar{x}_1 = \bar{v}_1 e^{\lambda t}$.
2. $\bar{v}_2$ satisfies $(A - \lambda I)\bar{v}_2 = \bar{v}_1 \Rightarrow \bar{x}_2 = [\bar{v}_2 + t\bar{v}_1]e^{\lambda t}$.
3. $\bar{v}_3$ satisfies $(A - \lambda I)\bar{v}_3 = \bar{v}_2 \Rightarrow \bar{x}_3 = [\bar{v}_3 + t\bar{v}_2 + \frac{1}{2}t^2\bar{v}_1]e^{\lambda t}$.

$\vdots$

$k.$ $(A - \lambda I)\bar{v}_k = \bar{v}_{k-1} \Rightarrow \bar{x}_k = [\bar{v}_k + t\bar{v}_{k-1} + \frac{1}{2}t^2\bar{v}_{k-2} + \cdots + \frac{1}{(k-1)!}t^{k-1}\bar{v}_1]e^{\lambda t}$.

It follows that $\bar{x}_1, \ldots, \bar{x}_k$ are linearly independent.

We point out that, while $\bar{v}_1$ is an eigenvector corresponding to the eigenvalue λ, $\bar{v}_2, \bar{v}_3, \ldots, \bar{v}_k$ are not eigenvectors. We might refer to them as *extended* or *generalized* eigenvectors.

Some of the examples below can be uncoupled and solved directly, but we use eigenvalues and eigenvectors in order to demonstrate the methods.

Example 8.7. Solve

$$\bar{x}' = A\bar{x} = \begin{pmatrix} -1 & 0 & 0 \\ 1 & -1 & 2 \\ 0 & 0 & 3 \end{pmatrix} \bar{x}.$$

We have

$$\det(A - \lambda I) = \begin{vmatrix} -1 - \lambda & 0 & 0 \\ 1 & -1 - \lambda & 2 \\ 0 & 0 & 3 - \lambda \end{vmatrix} = 0.$$

Calculating the eigenvalues, we obtain $\lambda = -1, -1, 3$.
Corresponding to $\lambda = 3$, we have

$$(A - \lambda I)\bar{v} = \begin{pmatrix} -4 & 0 & 0 \\ 1 & -4 & 2 \\ 0 & 0 & 0 \end{pmatrix} \begin{pmatrix} v_1 \\ v_2 \\ v_3 \end{pmatrix} = 0$$

and hence $-4v_1 = 0$ and $v_1 - 4v_2 + 2v_3 = 0$. This means that we must have $v_1 = 0$ and $v_3 = 2v_2$. So, letting $v_1 = 0$, $v_2 = 1$, $v_3 = 2$, yields the solution

$$x_3(t) = \begin{pmatrix} 0 \\ e^{3t} \\ 2e^{3t} \end{pmatrix}.$$

Corresponding to $\lambda = -1$, we have

$$(A - \lambda I)\bar{v} = \begin{pmatrix} 0 & 0 & 0 \\ 1 & 0 & 2 \\ 0 & 0 & 4 \end{pmatrix} \begin{pmatrix} v_1 \\ v_2 \\ v_3 \end{pmatrix} = 0,$$

which means that $v_1 + 2v_3 = 0$ and $4v_3 = 0$. Therefore, we must have $v_1 = v_3 = 0$. We can choose v_2 arbitrarily; so, let $v_2 = 1$. Then we obtain

$$\bar{v}_1 = \begin{pmatrix} 0 \\ 1 \\ 0 \end{pmatrix}.$$

The corresponding solution is

$$\bar{x}_1(t) = \begin{pmatrix} 0 \\ e^{-t} \\ 0 \end{pmatrix}.$$

But we need three linearly independent solutions in order to find the general solution. We find a third linearly independent solution by employing the algorithm:

Find $\bar{v}_2$ such that $(A - \lambda I)\bar{v}_2 = \bar{v}_1$. Then we can choose a second solution as

$$\bar{x}_2(t) = (\bar{v}_1 + \bar{v}_2 t)e^{-t}.$$

To find $\bar{v}_2$, we set

$$(A - \lambda I)\bar{v}_2 = \bar{v}_1, i.e., \begin{pmatrix} 0 & 0 & 0 \\ 1 & 0 & 2 \\ 0 & 0 & 4 \end{pmatrix}\begin{pmatrix} v_1 \\ v_2 \\ v_3 \end{pmatrix} = \begin{pmatrix} 0 \\ 1 \\ 0 \end{pmatrix} = 0,$$

which yields $v_1 + 2v_3 = 1$ and $4v_3 = 0$, v_2 arbitrary. Let $v_1 = 1$, $v_2 = 0$, $v_3 = 0$. Then a second solution corresponding to $\lambda = -1$ can be obtained:

$$\bar{x}_2(t) = [\bar{v}_2 + t\bar{v}_1]e^{-t} = \left[\begin{pmatrix} 1 \\ 0 \\ 0 \end{pmatrix} + t\begin{pmatrix} 0 \\ 1 \\ 0 \end{pmatrix} \right]e^{-t} = \begin{pmatrix} e^{-t} \\ te^{-t} \\ 0 \end{pmatrix}.$$

It is easy to verify that the wronskian of $\bar{x}_1, \bar{x}_2, \bar{x}_3$ at $t = 0$ is $W(0) = -2$ and hence they form a fundamental set of solutions and the general solution is given by

$$\bar{x}(t) = c_1\bar{x}_1 + c_2\bar{x}_2 + c_3\bar{x}_3 = \begin{pmatrix} c_2 e^{-t} \\ c_1 e^{-t} + c_2 te^{-t} + c_3 e^{3t} \\ 2c_3 e^{3t} \end{pmatrix}.$$

Example 8.8. Solve

$$\bar{x}' = A\bar{x} = \begin{pmatrix} 1 & 0 & 1 \\ 1 & 1 & 0 \\ 0 & 0 & 1 \end{pmatrix}\bar{x}.$$

Setting $\det(A - \lambda I) = 0$ yields one value of λ repeated three times, $\lambda = 1, 1, 1$. Setting

$$(A - \lambda)\bar{v} = \begin{pmatrix} 0 & 0 & 1 \\ 1 & 0 & 0 \\ 0 & 0 & 0 \end{pmatrix}\begin{pmatrix} v_1 \\ v_2 \\ v_3 \end{pmatrix}$$

shows that $v_1 = v_3 = 0$ and

$$\bar{v}_1 = \begin{pmatrix} 0 \\ 1 \\ 0 \end{pmatrix}, \quad \bar{x}_1(t) = \begin{pmatrix} 0 \\ e^t \\ 0 \end{pmatrix}.$$

We need two more solutions. To find the first one, we find $\bar{v}_2$ such that $(A - \lambda I)\bar{v}_2 = \bar{v}_1$. So we set

$$\begin{pmatrix} 0 & 0 & 1 \\ 1 & 0 & 0 \\ 0 & 0 & 0 \end{pmatrix} \begin{pmatrix} v_1 \\ v_2 \\ v_3 \end{pmatrix} = \begin{pmatrix} 0 \\ 1 \\ 0 \end{pmatrix},$$

which implies that $v_3 = 0$, $v_1 = 1$. So, we take $\bar{v}_2$ as

$$\bar{v}_2 = \begin{pmatrix} 1 \\ 0 \\ 0 \end{pmatrix}, \quad \bar{x}_2 = \bar{v}_2 e^t + t e^t \bar{v}_1 = \begin{pmatrix} e^t \\ t e^t \\ 0 \end{pmatrix}.$$

To find a third solution, we set

$$\begin{pmatrix} 0 & 0 & 1 \\ 1 & 0 & 0 \\ 0 & 0 & 0 \end{pmatrix} \begin{pmatrix} v_1 \\ v_2 \\ v_3 \end{pmatrix} = \begin{pmatrix} 1 \\ 0 \\ 0 \end{pmatrix},$$

which implies that $v_1 = 0$, $v_3 = 1$. So we choose

$$\bar{v}_3 = \begin{pmatrix} 0 \\ 0 \\ 1 \end{pmatrix}, \quad \bar{x}_3 = \bar{v}_3 + t\bar{v}_2 + \frac{1}{2} t^2 \bar{v}_1 = \begin{pmatrix} t e^t \\ \frac{1}{2} t^2 e^t \\ e^t \end{pmatrix}.$$

The general solution is

$$c_1 \bar{x}_1 + c_2 \bar{x}_2 + c_3 \bar{x}_3 = \begin{pmatrix} c_2 e^t + c_3 t e^t \\ c_1 e^t + c_2 t e^t + \frac{1}{2} c_3 t^2 e^t \\ c_3 e^t \end{pmatrix}. \qquad \square$$

Remark 8.5. If A is triangular, another method to find solution of $\bar{x}' = A\bar{x}$ is to notice that the last equation is $x'_n = a_{nn} x_n$ and hence it can be solved independently, yielding $x_n = c_n e^{a_{nn} t}$. Substituting into the first $n - 1$ equations we find an $(n-1) \times (n-1)$ non-homogeneous system in the variables $x_1, \ldots, x_{n-1}$. The method might be convenient especially if $n = 2, 3$, as illustrated in the example below.

Let $n = 2$ and consider the system

$$\begin{cases} x' = 2x + y, \\ y' = y, \end{cases} \quad \text{or} \quad \begin{pmatrix} x' \\ y' \end{pmatrix} = \begin{pmatrix} 2 & 1 \\ 0 & 1 \end{pmatrix} \begin{pmatrix} x \\ y \end{pmatrix}.$$

It is obvious that $y' = y$ implies that we can take $y = ce^t$; now we wish to find x so that $c_1x + c_2y$ is the general solution. Solving $x' = 2x + ce^t$ as first order linear equation, we obtain

$$x = c_1 e^{2t} - ce^t.$$

Setting $c_2 = -c$ we find the general solution

$$\begin{pmatrix} x \\ y \end{pmatrix} = \begin{pmatrix} c_1 e^{2t} + c_2 e^t \\ -c_2 e^t \end{pmatrix}.$$

Of course, we could obtain the same result by solving, as in Theorem 8.5, the equivalent equation $x'' - 3x' + 2x = 0$.

8.5 Appendix: the exponential matrix

Given an $n \times n$ matrix A with entries a_{ij}, the *exponential matrix* e^A is an $n \times n$ matrix defined as follows.

1. Consider the iterated matrix $A^k = \underbrace{A \cdot A \cdots A}_{k\text{-times}}$. Setting

$$\|A\| = \max\{|A\bar{x}| : \bar{x} \in \mathbb{R}^n, |\bar{x}| = 1\},$$

one has

$$\|A\|^2 \le \sum a_{ij}^2 := C^2, \quad C \in \mathbb{R}.$$

Since $|A^k \bar{x}| \le \|A\|^k \cdot |\bar{x}|$, it follows

$$|A^k \bar{x}| \le \left(\sum a_{ij}^2 \right)^{k/2} = C^k, \quad \forall \bar{x} \in \mathbb{R}.$$

2. For each $\bar{x} \in \mathbb{R}^n$, consider the series

$$\sum_{k=0}^{\infty} \frac{A^k \bar{x}}{k!} = A^0 \bar{x} + A\bar{x} + \frac{A^2 \bar{x}}{2!} + \frac{A^3 \bar{x}}{3!} + \cdots,$$

whereby, since $A^0 = I$,

$$\sum_{k=0}^{\infty} \frac{A^k \bar{x}}{k!} = \bar{x} + A\bar{x} + \frac{A^2 \bar{x}}{2!} + \frac{A^3 \bar{x}}{3!} + \cdots$$

From the preceding step it follows that

$$\frac{|A^k \overline{x}|}{k!} \le \frac{c^k}{k!}.$$

Since the numerical series $\sum \frac{c^k}{k!}$ is convergent, by comparison we infer that the series $\sum_{k=0}^{\infty} \frac{A^k \overline{x}}{k!}$ is totally convergent for any $\overline{x} \in \mathbb{R}$.

3. We define the exponential matrix e^A by setting

$$e^A = \sum_{k=0}^{\infty} \frac{A^k}{k!}, \quad \text{and} \quad e^A[\overline{x}] = \sum_{k=0}^{\infty} \frac{A^k \overline{x}}{k!}.$$

If $n = 1$ and $A = a \in \mathbb{R}$ then $A^k = a^k$ and hence we find

$$e^A = \sum_{k=0}^{\infty} \frac{a^k}{k!} = 1 + a + \frac{a^2}{2!} + \frac{3^3}{3!} + \cdots,$$

which is simply the Taylor series for e^a. This explains the name "exponential matrix". If A is diagonal with entries a_{ii}, then $\frac{A^k}{k!}$ is given by

$$\frac{A^k}{k!} = \begin{pmatrix} \frac{a_{11}^k}{k!} & 0 & \cdots & 0 \\ 0 & \frac{a_{22}^k}{k!} & 0 & \cdots \\ \cdots & \cdots & \cdots & 0 \\ 0 & \cdots & 0 & \frac{a_{nn}^k}{k!} \end{pmatrix}.$$

It follows that e^A is a diagonal matrix with entries

$$e^A = \begin{pmatrix} e^{a_{11}} & 0 & \cdots & 0 \\ 0 & e^{a_{22}} & 0 & 0 \\ \cdots & \cdots & \cdots & \cdots \\ 0 & \cdots & 0 & e^{a_{nn}} \end{pmatrix}.$$

Among the properties of e^A we recall here the following one:

$$\frac{d}{dt} e^{tA} = A e^{tA},$$

which can be proved by differentiating term by term the totally convergent series $e^{tA}[\overline{x}] = \sum_{k=0}^{\infty} \frac{(tA)^k \overline{x}}{k!}$:

$$\frac{d}{dt} e^{tA}[\overline{x}] = \frac{d}{dt} \sum_{k=0}^{\infty} \frac{(tA)^k \overline{x}}{k!} = \frac{d}{dt} \left[\overline{x} + tA\overline{x} + \frac{t^2 A^2 \overline{x}}{2!} + \frac{t^3 A^3 \overline{x}}{3!} + \frac{t^4 A^4 \overline{x}}{4!} + \cdots \right]$$

$$= A\overline{x} + tA^2 \overline{x} + \frac{t^2 A^3 \overline{x}}{2!} + \frac{t^3 A^4 \overline{x}}{3!} + \cdots = A\overline{x} \cdot \left[\overline{x} + tA\overline{x} + \frac{t^2 A^2 \overline{x}}{2!} + \frac{t^3 A^3 \overline{x}}{3!} + \cdots \right]$$

$$= A\bar{x}\left[\sum_{k=0}^{\infty} \frac{t^k A^k \bar{x}}{k!}\right] = A\bar{x}\left[\sum_{k=0}^{\infty} \frac{(tA)^k \bar{x}}{k!}\right] = A\bar{x}e^{tA}[\bar{x}].$$

Thus, setting $\bar{x}(t) = e^{tA}[\bar{c}]$, $\bar{c} \in \mathbb{R}^n$, we find

$$\bar{x}' = \frac{d}{dt}e^{tA}[\bar{c}] = Ae^{tA}[\bar{c}] = A\bar{x}.$$

This shows that the general solution of $\bar{x}' = A\bar{x}$ can be written as

$$\bar{x}(t) = e^{tA}[\bar{c}], \quad \bar{c} \in \mathbb{R}^n.$$

8.6 Exercises

1. Solve

$$\begin{cases} x'' - x' - y = 0, \\ y' - x' = -x. \end{cases}$$

2. Use the eigenvalue method to find the general solution of

$$\begin{cases} x' = 4x + 6y, \\ y' = x + 3y. \end{cases}$$

3. Use Theorem 8.5 to find the general solution of

$$\begin{cases} x_1' = -3x_1 - 5x_2, \\ x_2' = x_1 + x_2. \end{cases}$$

4. Solve

$$\bar{x}' = \begin{pmatrix} 1 & 1 \\ 2 & 2 \end{pmatrix}\bar{x}.$$

5. Solve

$$\bar{x}' = \begin{pmatrix} 3 & 1 \\ -1 & 1 \end{pmatrix}\bar{x}, \quad \bar{x}(0) = \begin{pmatrix} 1 \\ -2 \end{pmatrix}.$$

6. Solve problem 2 by converting the system to a linear scalar equation suggested by Theorem 8.5.

7. Solve

$$\begin{cases} x_1' = x_1 + x_2 + 3e^t, \\ x_2' = 2x_1 + 2x_2 + e^t. \end{cases}$$

8. Find the inverse of

$$\begin{pmatrix} a & 0 & 0 \\ 0 & a & 0 \\ 0 & 0 & a \end{pmatrix}.$$

9. Solve

$$\begin{cases} x_1' = x_1 + \sin t, \\ x_2' = -x_2 + \cos t. \end{cases}$$

10. Show that any system

$$\begin{cases} x_1' = ax_1 + bx_2, \\ x_2' = cx_1 + dx_2. \end{cases}$$

has a nontrivial constant solution, $x_1 = k_1$, $x_2 = k_2$, if and only if the determinant

$$\begin{vmatrix} a & b \\ c & d \end{vmatrix} = 0.$$

11. Solve

$$\bar{x}' = \begin{pmatrix} -1 & 0 & 0 \\ 1 & -1 & 2 \\ 0 & 0 & 3 \end{pmatrix} \bar{x}.$$

12. First explain why the equation below has a nontrivial constant solution and then solve it.

$$\bar{x}' = \begin{pmatrix} 1 & -1 \\ 2 & -2 \end{pmatrix} \bar{x}.$$

13. Find the general solution of

$$\bar{x}' = \begin{pmatrix} 4 & -2 \\ 1 & 3 \end{pmatrix} \bar{x}.$$

14. Use the eigenvalue method to find the solution to the initial value problem

$$\bar{x}' = \begin{pmatrix} 1 & 1 \\ -1 & 1 \end{pmatrix} \bar{x}, \quad \bar{x}(0) = \begin{pmatrix} 1 \\ -1 \end{pmatrix}.$$

15. Find a fundamental set of solutions of the system below.

$$\bar{x}' = \begin{pmatrix} 5 & -8 \\ 1 & 1 \end{pmatrix} \bar{x}.$$

16. Show that

$$\bar{x}' = \begin{pmatrix} a & b \\ c & d \end{pmatrix} \bar{x}.$$

cannot have periodic solutions if b and c are either both positive or both negative.

17. Prove that the following vectors are linearly dependent:

$$\bar{x}_1 = \begin{pmatrix} 2t^2 \\ t \\ \sin^2 t \end{pmatrix}, \quad \bar{x}_2 = \begin{pmatrix} -3t^2 \\ 4t \\ 2\cos^2(t) \end{pmatrix}, \quad \bar{x}_3 = \begin{pmatrix} t^2 \\ 6t \\ 2 \end{pmatrix}.$$

18. Solve

$$\bar{x}' = \begin{pmatrix} 1 & -1 & 1 \\ 1 & 2 & -2 \\ 2 & 1 & -1 \end{pmatrix} \bar{x}.$$

19. Solve the Cauchy problem

$$\bar{x}' = \begin{pmatrix} 1 & 1 & 1 \\ -1 & 1 & -1 \\ 1 & -1 & 1 \end{pmatrix} \bar{x}, \quad \bar{x}(0) = \begin{pmatrix} 1 \\ -1 \\ 1 \end{pmatrix}.$$

20. Find a fundamental set of solutions for

$$\bar{x}' = \begin{pmatrix} 3 & 4 & 1 \\ 1 & 3 & -2 \\ 0 & 0 & 3 \end{pmatrix} \bar{x}.$$

9 Phase plane analysis

In this chapter we mainly deal with second order autonomous equations; see (9.1) below. The last section is devoted to general planar systems.

9.1 Preliminaries

Consider the second order autonomous equation

$$x'' + V'(x) = 0, \tag{9.1}$$

where $V : \mathbb{R} \mapsto \mathbb{R}$ is smooth (this will always be understood, even if it is not explicitly stated). The importance of equation (9.1) is due to the fact that Newton's second law, force = mass $\times$ acceleration, gives rise to such an equation.

Setting $y = x'$, (9.1) is equivalent to the first order autonomous planar system

$$\begin{cases} x' = y, \\ y' = -V'(x), \end{cases} \tag{V}$$

whose solutions are referred to as *orbits* or *trajectories*.

The plane (x, y), with $y = x'$, is called the *phase plane*.

From the general theory it follows that, given any $t_0 \in \mathbb{R}$ and any $(x_0, y_0) \in \mathbb{R}^2$, there exists a unique solution $x(t), y(t)$ of (V) satisfying the initial condition $x(t_0) = x_0$, $y(t_0) = y_0$. Given any $x_0, y_0 \in \mathbb{R}$ there exists a unique solution $x(t)$ of (9.1) satisfying the initial condition $x(t_0) = x_0, x'(t_0) = y_0$. In the sequel we will assume that these solutions are defined for all $t \in \mathbb{R}$.

Let us remark that, (9.1) being an autonomous equation, if $x(t)$ is a solution of (9.1), then so is $x(t + h)$, for any $h \in \mathbb{R}$.

A key ingredient in studying (V) is its total energy, defined by

$$E = E(x, y) = \frac{1}{2}y^2 + V(x) = \frac{1}{2}x'^2 + V(x).$$

Lemma 9.1. *E is constant along any solution of* (V).

Proof. Let $x = x(t), y = y(t)$ be any solution of (V). Then

$$\frac{d}{dt}E(x(t), y(t)) = E_x(x(t), y(t)) \cdot \frac{dx}{dt} + E_y(x(t), y(t)) \cdot \frac{dy}{dt} = V'(x(t)) \cdot \frac{dx}{dt} + y(t) \cdot \frac{dy}{dt}.$$

Thus (V) yields

$$V'(x(t)) \cdot y(t) + y(t) \cdot (-V'(x(t))) = 0, \quad \forall t \in \mathbb{R}$$

and hence $E(x(t), y(t))$ is constant. $\quad\square$

https://doi.org/10.1515/9783111185675-009

Motivated by the previous lemma, we say that (V) is a *conservative system*.

Remark 9.1. The conservative system (V) is a particular *hamiltonian system*, which in the planar case is a system as

$$\begin{cases} x' = H_y(x,y) \\ y' = -H_x(x,y) \end{cases} \tag{HS}$$

where $H : \mathbb{R}{\times}\mathbb{R} \mapsto \mathbb{R}$ is a continuously differentiable function. Actually, setting $H(x,y) = \frac{1}{2}y^2 + V(x)$ we find $H_y = y, H_x = V'(x)$ and hence (HS) becomes (V).

Let $x = x(t), y = y(t)$ be any solution of (HS). Then

$$\frac{d}{dt}H(x(t),y(t)) = H_x(x(t),y(t))x' + H_y(x(t),y(t))y'$$

$$= H_x(x(t),y(t)) \cdot H_y(x(t),y(t)) + H_y(x(t),y(t)) \cdot (-H_x(x(t),y(t))) \equiv 0.$$

Therefore the function $t \mapsto H(x(t),y(t))$ is identically constant, hence $H(x(t),y(t)) = c, \forall t \in \mathbb{R}$.

Remark 9.2. In the particular case in which (9.1) possesses constant solutions $x(t) \equiv \bar{x}, y(t) \equiv \bar{y}$, we find

$$E(x(t),y(t)) = E(\bar{x},\bar{y}) = V(\bar{x}), \quad \forall t \in \mathbb{R}.$$

Moreover $(\bar{x},\bar{y})$ is a *singular point* of the curve $\{E = V(\bar{x})\}$, since $E_x(\bar{x},\bar{y}) = 0$ and $E_y(\bar{x},\bar{y}) = \bar{y} = 0$ whereby $\nabla E(\bar{x},\bar{y}) = 0$. Notice that in this chapter $\bar{x}, \bar{y}$ denote real numbers, not vectors.

9.2 Phase plane analysis: some examples

System (V) is in general nonlinear and it might be quite difficult to find an explicit solution.

The *phase plane analysis* is an analytical approach that provides several types of information about the behavior of the solutions of (V), without solving the equation explicitly. This method can be seen as the counterpart of the qualitative analysis for first order nonlinear equations carried out in Chapter 3.

Recall that very few nonlinear second order equations can be solved explicitly. Therefore, it is important to be able to deduce from the equation some qualitative properties of the solutions, such as monotonicity, convexity, maxima and minima, asymptotic behavior, etc.

The starting point is the conservation of energy

$$\frac{1}{2}y^2 + V(x) = c \tag{9.2}$$

discussed earlier. It is important to distinguish the two different settings: the phase plane (x,y) where we draw the orbits of (V) and the plane (t,x) where we plot the solution $x = x(t)$ of (9.1). However, the two planes are closely related in a way that we are going to explain.

Given $c \in \mathbb{R}$, let us consider the level sets of E

$$E_c = \{(x,y) \in \mathbb{R}^2 : E(x,y) = c\} = \left\{(x,y) \in \mathbb{R}^2 : \frac{1}{2}y^2 + V(x) = c\right\}.$$

Solving $E = c$ for y we find

$$y = \pm\sqrt{2c - 2V(x)}.$$

In particular:
1. $E_c \neq \emptyset$ if and only if $V(x) \leq c$ for some x;
2. if $\overline{x}$ is such that $V(\overline{x}) = c$, then E_c reduces to the point $(\overline{x},0)$;
3. if $E_c \neq \emptyset$, then E_c is symmetric with respect to the x-axis.

Let us illustrate the phase plane analysis on a couple of examples.

Example 9.1. Given $V(x)$ and $c \in \mathbb{R}$, let us suppose that $E_c = E_{c,1} \cup E_{c,2}$ (in the sequel, to simplify notations, we will simply write E_1, E_2 instead of $E_{c,1}, E_{c,2}$) as indicated in Fig. 9.1, where
-   E_1 is the upper branch of $\frac{1}{2}y^2 + V(x) = c$, marked in red in Fig. 9.1, and
-   E_2 is the right branch of $\frac{1}{2}y^2 + V(x) = c$, marked in blue.

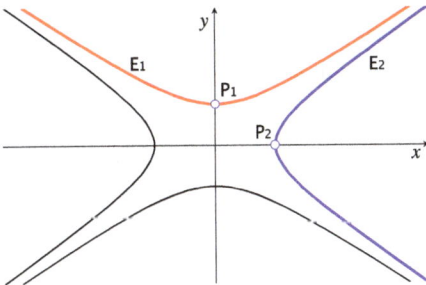

Figure 9.1: The branch E_1 in red and E_2 in blue.

Concerning the point $P_1 = (0,y_1) \in E_1$ there exists a unique solution $x(t), y(t)$ of $x'' + V(x) = 0$ with the initial condition $P_1 = (0,y_1)$, namely, such that $x(0) = 0, y(0) = x'(0) = y_1$. Moreover, from (9.2) it follows that

$$E(x(t),y(t)) = E(x(0),y(0)) = E(P_1) = c, \quad \forall t \in \mathbb{R} \quad \Longleftrightarrow \quad (x(t),y(t)) \in E_1, \quad \forall t \in \mathbb{R}.$$

On E_1 we have $y(t) \geq y(0) = y_1 > 0$, which means that $x' > 0$. Thus the solution $x(t)$ of $x'' + V(x) = 0$ is increasing. Moreover, $x''(t) = y'(t) > 0$ for $t > 0$, resp. $x''(t) = y'(t) < 0$ for $t < 0$, implies that $x(t)$ is a convex function for $t > 0$, resp. a concave function for $t < 0$. This information, together with the fact that E_1 is unbounded with respect to x, allows us to draw a qualitative graph of the solution $x(t)$; see Fig. 9.2, red curve.

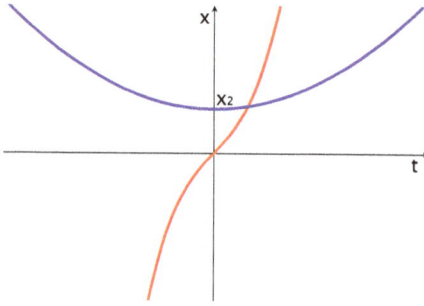

Figure 9.2: Plot of $x = x(t)$ in red and $x = \tilde{x}(t)$ in blue.

Similarly, the solution $\tilde{x}(t)$, $\tilde{y}(t)$ of $x'' + V'(x) = 0$, passing through $P_2 = (x_2, 0) \in E_2$, is such that $(\tilde{x}(t), \tilde{y}(t)) \in E_2$, $\forall t \in \mathbb{R}$. On E_2 we have $\tilde{x}(0) = x_2 > 0$ and $\tilde{x}(t) > x_2$ for all $t \neq 0$. Thus $\tilde{x}(t)$ has a minimum at $t = 0$. Notice that $\tilde{x}'(0) = \tilde{y}(0) = 0$, $\tilde{x}'(t) = \tilde{y}(t) > 0$ for $t > 0$ and $\tilde{x}'(t) = \tilde{y}(t) < 0$ for $t < 0$ imply that the point $(\tilde{x}(t), \tilde{y}(t))$ "moves" upwards on E_2. More precisely one has $\tilde{y}'(t) > 0$ and hence $\tilde{x}''(t) = \tilde{y}'(t) > 0$ so that $\tilde{x}(t)$ is a convex function. This information allows us to draw an approximate graph of the solution $\tilde{x}(t)$; see Fig. 9.2, blue curve. $\qquad\square$

Example 9.2. Suppose that in the phase plane the set $\{E(x,y) = c\}$ is the closed red curve in Fig. 9.3. Consider the arc A of this curve contained in the half plane $x \geq 0$ and the corresponding solution $x(t), y(t)$.

Starting from $(0, a)$ at $t = 0$, and moving along the arc A, we will first reach the point $(M, 0)$, at some $t_0 > 0$, and next the point $(0, -a)$ at some $\tau > t_0$ (actually, by symmetry

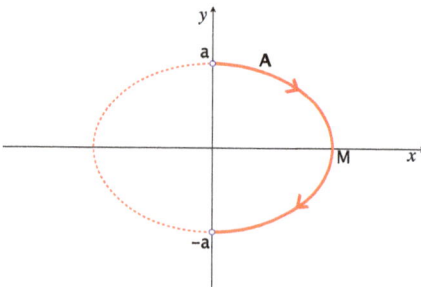

Figure 9.3: The arc A.

$\tau = 2t_0$; see the next section). The corresponding solution $x(t)$ of $x'' + V'(x) = 0$ is such that

1. $x(0) = x(\tau) = 0, x'(0) = a, x'(\tau) = -a$;
2. $0 \le x(t) \le M$ for $t \in [0, \tau]$ and $x(t_0) = M$, and is hence the maximum of $x(t)$;
3. $x'(t) = y(t) > 0$ for $0 \le t < t_0, x'(t) = y(t) < 0$ for $t_0 \le t < \tau$.

In particular, these features show that $x(t)$ is a positive solution of the boundary value problem $x'' + V'(x) = 0, x(0) = x(\tau) = 0$. In the next section we will evaluate τ and show that $\tau = 2t_0$. A possible graph of $x(t)$ is reported in Fig. 9.4. ☐

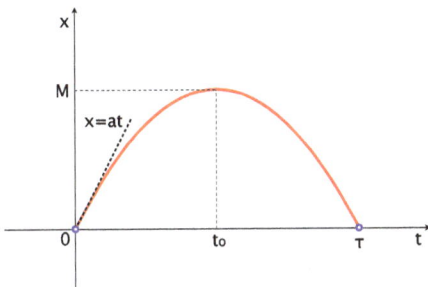

Figure 9.4: The solution $x(t)$ of the boundary value problem.

9.3 Periodic solutions

A solution $x(t), y(t)$ of (V) is periodic with period $T > 0$ if

$$x(t + T) = x(t), \quad y(t + T) = y(t), \quad \forall t \in \mathbb{R}.$$

Clearly, if $x(t), y(t)$ is T-periodic, then it is also kT-periodic for any integer $k \ne 0$. Usually, saying that T is the period, we understand that T is the *minimal* period.

In the sequel, by "periodic solution" we mean a *nontrivial* periodic solution of (V), namely a periodic solution which is not an equilibrium.

Let us again consider the level sets of E_c with equation

$$y = \pm\sqrt{2c - 2V(x)}.$$

It is clear that if $x = x(t), y = y(t)$ is a periodic solution of (V) then E_c is a closed bounded curve.

Next we consider the specific equation

$$x'' + x^3 = 0, \tag{9.3}$$

and show the typical argument that is used to find periodic solutions.

Here $V(x) = \frac{1}{4}x^4$ and $E = \frac{1}{2}y^2 + \frac{1}{4}x^4$. Thus $E = c$ becomes

$$\frac{1}{2}y^2 + \frac{1}{4}x^4 = c \quad \Longrightarrow \quad y = \pm\sqrt{2c - \frac{1}{2}x^4}.$$

Then:

1. for $c < 0$ the set E_c is empty;
2. for $c = 0$, E_0 reduces to the point $(0,0)$, which is the unique singular point of our equation.
3. for every $c > 0$, E_c is a smooth closed bounded curve which does not contain the singular point.

We refer to Fig. 9.5. Given a fixed $c > 0$, consider on E_c the point $(0, y_0)$. We want to show that the solution $x(t), y(t)$ with initial value $(0, y_0)$ is a periodic solution of (9.3).

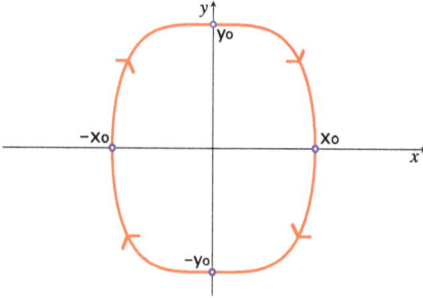

Figure 9.5: Plot of $\frac{1}{2}y^2 + \frac{1}{4}x^4 = c$, with $c > 0$.

Since $x'(0) = y(0) = y_0 > 0$, $x(t)$ enters in the first quadrant, where $x > 0$, $y > 0$. As $t > 0$ increases, the point $(x(t), y(t))$ moves on E_c: first it reaches $(x_0, 0)$ for some $t_0 > 0$, next it reaches $(0, -y_0)$ for some $t_1 > t_0$ and next $(-x_0, 0)$ for some $t_2 > t_1$. Finally, $(x(t), y(t))$ comes back to $(0, y_0)$, namely there exists $T > 0$ such that $(x(T), y(T)) = (0, y_0)$. We claim that this implies that $(x(t), y(t))$ is T-periodic. Actually, setting $\bar{x}(t) = x(t + T)$, $\bar{y}(t) = y(t + T)$ we notice that $\bar{x}(t), \bar{y}(t)$ is also a solution of (V), since (V) is autonomous. Moreover, $\bar{x}(0) = x(T) = 0$, $\bar{y}(0) = y(T) = y(0)$. Since $\bar{x}(t), \bar{y}(t)$ and $x(t), y(t)$ satisfy the same equation (V) and the same initial conditions, by uniqueness, $x(t) = \bar{x}(t), y(t) = \bar{y}(t)$ for all $t \in \mathbb{R}$, namely $x(t) = x(t + T), y(t) = y(t + T)$, for all $t \in \mathbb{R}$, proving the claim. It is possible to see that $t_1 = 2t_0$, $t_2 = 3t_0$ and $T = 4t_0$, as shown below.

First of all we evaluate t_0 such that $x(t_0) = x_0, y(t_0) = 0$. Since $y(0) = y_0$, the corresponding value of the energy is $c = \frac{1}{2}y_0^2$ and hence $y = \sqrt{y_0^2 - \frac{1}{2}x^4}$, where we have taken the sign $+$ since $y > 0$ for $0 < t \le t_0$. From $\frac{dx}{dt} = y$ we infer $dt = \frac{dx}{y}$ and thus $dt = \frac{dx}{\sqrt{y_0^2 - \frac{1}{2}x^4}}$, $(t \in [0, t_0])$. Integrating we find

$$t_0 = \int_0^{t_0} dt = \int_0^{x_0} \frac{dx}{y} = \int_0^{x_0} \frac{dx}{\sqrt{y_0^2 - \frac{1}{2}x^4}}.$$

If $t_1 > t_0$ is such that $x(t_1) = 0$, the same argument yields

$$t_1 - t_0 = \int_{t_0}^{t_1} dt = \int_{x_0}^{0} \frac{dx}{y} = \int_{x_0}^{0} \frac{dx}{-\sqrt{y_0^2 - \frac{1}{2}x^4}} = \int_0^{x_0} \frac{dx}{\sqrt{y_0^2 - \frac{1}{2}x^4}} = t_0,$$

whereby $t_1 = 2t_0$. In a similar way we can show that $t_2 = 3t_0$ and that $T = 4t_0$.

Furthermore, we have:

(a) for $0 < t < t_0$, $x(t) > 0$, $y(t) > 0$ and hence $x(t)$ is positive and increasing;

(b) for $t_0 < t < 2t_0$, $x(t) > 0$, $y(t) < 0$ and hence $x(t)$ is positive and decreasing;

(c) for $2t_0 < t < 3t_0$, $x(t) < 0$, $y(t) < 0$ and hence $x(t)$ is negative and decreasing;

(d) for $3t_0 < t < 4t_0 = T$, $x(t) < 0$, $y(t) > 0$ and hence $x(t)$ is negative and increasing.

More generally, using arguments similar to those carried out before, one can prove:

Theorem 9.1. *If E_c is a non-empty closed bounded curve which does not contain equilibria of* (V), *then E_c carries a periodic solution of* (V).

Corollary 9.1. *If V has a proper local minimum at x_0 then there exists $\epsilon > 0$ such that, for any c such that $V(x_0) < c < V(x_0) + \epsilon$, the equation $x'' + V'(x) = 0$ has a periodic solutions with energy c.*

Proof. It suffices to apply Theorem 9.1, noticing that, for any c with $V(x_0) < c < V(x_0) + \epsilon$, the set $\{E = c\}$ is a closed bounded curve with no singular point. $\square$

In the case of hamiltonian systems introduced in Remark 9.1, the counterpart of Theorem 9.1 is

Theorem 9.2. *If $H(x, y) = c$ is a non-empty smooth closed bounded curve which does not contain equilibria of* (HS), *then $H(x, y) = c$ carries a periodic solution of* (HS).

Remark 9.3. A typical case in which $\{E_c\}$ is a smooth closed bounded curve is when, as in Fig. 9.6(left):

(i) the set $\{V(x) \le c\}$ is a (not empty) bounded interval $I = [a, b]$, and

(ii) $V'(a) < 0$, $V'(b) > 0$.

To check this claim we will show that $E_c = \{\frac{1}{2}y^2 + V(x) = c\}$ is a bounded closed smooth curve, see Fig. 9.6 (right). Actually, solving $\frac{1}{2}y^2 + V(x) = c$ for y we find $y = \pm\sqrt{2(c - V(x))}$. The functions $y_1 = \sqrt{2(c - V(x))}$ and $y_2 = -\sqrt{2(c - V(x))}$ are defined on I and vanish at a, b. Moreover, a straight calculation yields $y_1'(a) = \frac{-V'(a)}{\sqrt{2(c-V(a))}}$, $y_2'(a) = \frac{V'(a)}{\sqrt{2(c-V(a))}}$ and $y_1'(b) = \frac{-V'(b)}{\sqrt{2(c-V(b))}}$, $y_2'(b) = \frac{V'(b)}{\sqrt{2(c-V(b))}}$. Therefore, if $V'(a) < 0$ and $V'(b) > 0$ we infer that

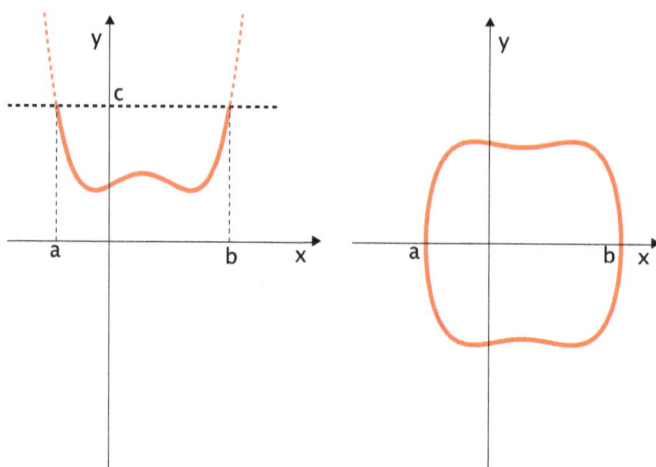

Figure 9.6: (Left) graph of $y = V(x)$; (right) plot of the curve E_c.

$y_1'(a) = +\infty, y_1'(b) = -\infty, y_2'(a) = -\infty, y_2'(b) = +\infty$. Thus y_1, y_2 glue together at $(a, 0), (b, 0)$ in such a way that E_c is a bounded closed smooth curve.

9.3.1 Remarkable examples

9.3.1.1 The nonlinear harmonic oscillator
As a first application, we consider the *nonlinear harmonic oscillator*

$$x'' + \omega^2 x - x^3 = 0 \tag{9.4}$$

which is equivalent to the system

$$\begin{cases} x' = y \\ y' = -\omega^2 x + x^3 \end{cases}$$

In this case $V(x) = \frac{1}{2}\omega^2 x^2 - \frac{1}{4}x^4$ has a proper local minimum at $x_0 = 0$ with $V(0) = 0$ and hence, by Corollary 9.1, $\exists \epsilon > 0$ such that the nonlinear harmonic oscillator has periodic solutions with energy c, for any $0 < c < \epsilon$. To evaluate ϵ, we argue as follows. Since $E = \frac{1}{2}y^2 + \frac{1}{2}\omega^2 x^2 - \frac{1}{4}x^4$, $E = c$ yields

$$y^2 + \omega^2 x^2 - \frac{1}{2}x^4 = 2c \quad \Longrightarrow \quad y = \pm\sqrt{\frac{1}{2}x^4 - \omega^2 x^2 + 2c}$$

The set E_c is symmetric both with respect to the x-axis and the y-axis. Moreover, the singular points are $(0, 0)$ and $(\pm\omega, 0)$. Some E_c are plotted in Fig. 9.7.

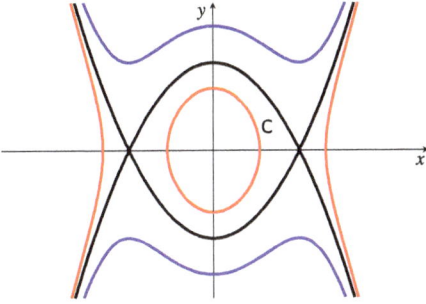

Figure 9.7: Phase plane of the nonlinear harmonic oscillator.

In particular, it is easy to check that, if $0 < c < \frac{1}{2}\omega^2$, the set E_c (marked in red) has a component C which is a closed bounded curve without singular points. Thus, by Theorem 9.1, it follows that the nonlinear harmonic oscillator has periodic solutions provided $0 < c < \frac{1}{2}\omega^2$.

9.3.1.2 Kepler's problem

As a second application, we consider the two-body Kepler problem describing the motions of two bodies of mass m_1, m_2 subjected to a gravitational force. Introducing the reduced mass $\mu = \frac{m_1 m_2}{m_1 + m_2}$, it is known (see, e. g., V. I. Arnold: *Mathematical methods in classical mechanics*, 2nd ed., Springer V., 1989) that the two-body problem is equivalent to a one-body problem for the reduced body of mass μ. Since the acting force is central, the angular momentum L is a constant of the motion. Let

$$U_{\text{eff}}(r) = \frac{L^2}{\mu r^2} - \frac{Gm_1 m_2}{r}, \quad r > 0,$$

denote the so-called *effective potential*, see Fig. 9.8.

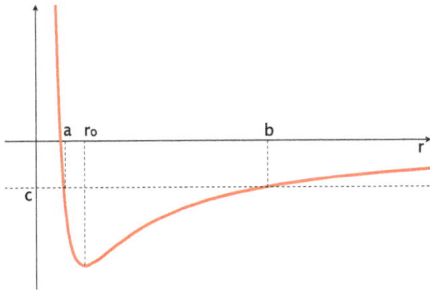

Figure 9.8: The effective potential U_{eff} of (K).

Then the motion of the reduced body is described by the equation

$$\mu\frac{d^2r}{dt^2} + U_{\text{eff}}(r) = 0, \tag{K}$$

where $r > 0$ is the distance of the reduced body to the origin. Notice that $U_{\text{eff}}(r)$ has a (global, negative) minimum at some $r_0 > 0$, see Fig. 9.8, and thus, according to Corollary 9.1, (K) possesses periodic solutions. More precisely, consider in the phase plane $(r, r'), r > 0$, the level sets of the energy

$$E_c = \left\{(r, r') : \frac{1}{2}\mu(r')^2 + U_{\text{eff}}(r) = c, \ r > 0\right\}$$

$$= \left\{(r, r') : \frac{1}{2}\mu(r')^2 + \frac{1}{2}\frac{L^2}{\mu r^2} - \frac{Gm_1m_2}{r} = c, \ r > 0\right\}.$$

Solving for r' we find

$$r' = \pm\sqrt{\frac{2}{\mu} \cdot (c - U_{\text{eff}}(r))}.$$

For any c such that $U_{\text{eff}}(r_0) < c < 0$ we can infer from Fig. 9.8 that $(r, r') \in E_c$ whenever $U_{\text{eff}}(r) < c < 0$, namely $a \le r \le b$; see Fig. 9.8. Thus the set E_c is a bounded closed curve with no singular point (the only singular point is $(r_0, 0)$) and hence it corresponds to a periodic solution of (K). It is possible to show that these periodic orbits are ellipses with $r_{\min} = a \le r \le r_{\max} = b$.

9.4 Van der Pol's equation

The van der Pol equation

$$x'' - \mu(1 - x^2)x' + x = 0 \tag{9.5}$$

is simply a damped oscillator $x'' + kx' + x = 0$ in which the damping term kx' depends on the position x. This equation also arises in electric circuits theory where it describes the current x in a vacuum tube.

The case we want to study is when $\mu > 0$. Since the results do not depend on the value of $\mu > 0$, we will take $\mu = 1$, to simplify notations.

Equation (9.5) is equivalent to the planar system

$$\begin{cases} x' = y, \\ y' = (1 - x^2)y - x, \end{cases}$$

but, due to the presence of the damping term $(1 - x^2)y$, this system is not conservative.

For this reason, setting $F(x) = -x + \frac{1}{3}x^3$ it is convenient to consider the following system, usually named the *Liénard system*:

$$\begin{cases} x' = y - F(x), \\ y' = -x. \end{cases} \tag{9.6}$$

Lemma 9.2. *The Liénard system is equivalent to* (9.5).

Proof. Differentiating the first equation we get $x'' = y' - F'(x)x'$ and hence the second equation yields $x'' = -x - F'(x)x'$ whereby $x'' + F'(x)x' + x = 0$, namely $x'' + (x^2 - 1)x' + x = 0$. □

We now deduce the behavior of the orbits of (9.6) carrying out in the *Liénard plane* (x, y) the arguments used before in the phase plane. One has to be careful and note that the Liénard plane is different from the phase plane (x, x').

The curve $y = F(x)$ and the axis $x = 0$ divide the Liénard plane into 4 regions

$$R_1 = \{(x,y) : x < 0, y > F(x)\}, \quad R_2 = \{(x,y) : x > 0, y > F(x)\},$$
$$R_3 = \{(x,y) : x > 0, y < F(x)\}, \quad R_4 = \{(x,y) : x < 0, y < F(x)\}.$$

Fig. 9.9 suggests that (9.6) possesses a periodic orbit corresponding to a periodic solution of van der Pol's equation (9.5).

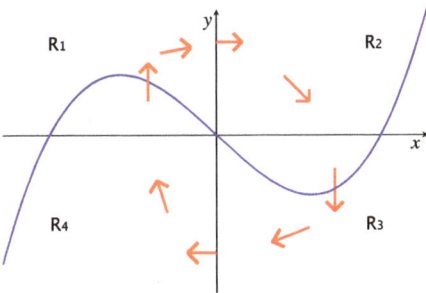

Figure 9.9: The Liénard plane: the curve in blue is $y = F(x)$; the arrows denote the vector field $(y - F(x), -x)$.

Precisely, one can show the following.

Theorem 9.3. *The van der Pol equation* (9.9) *has one and only one periodic solution.*

The complete proof of this result is rather long; we will limit ourselves to a sketch of the outlines of the main arguments.

Step 1. Given $z > 0$, let $x(t; z), y(t, z)$ be the solution of (9.6) satisfying the initial condition $x(0) = 0, y(0) = z$. This trajectory is such that $x' > 0, y' < 0$ as long as $(x(t; z), y(t, z)) \in R_1$;

see Fig. 9.9. Once it meets the curve $y = F(x)$, this orbit enters into the region R_2, where $x' < 0, y' < 0$ and reaches the axis $x = 0$ at some $z_1 < 0$, depending on z.

A similar argument shows that the trajectory $x(t; z), y(t, z)$ comes back to the y-axis at a certain $z_2 > 0$.

Step 2. Consider the map $g : \mathbb{R} \mapsto \mathbb{R}$ defined by setting $g(z) = z_2$. By the continuous dependence on the initial values, it follows that g is a continuous function. Moreover, it is possible to prove that g is decreasing and there exist $a, b > 0$ such that $g(a) > a$ whereas $g(b) < b$; see Fig. 9.10.

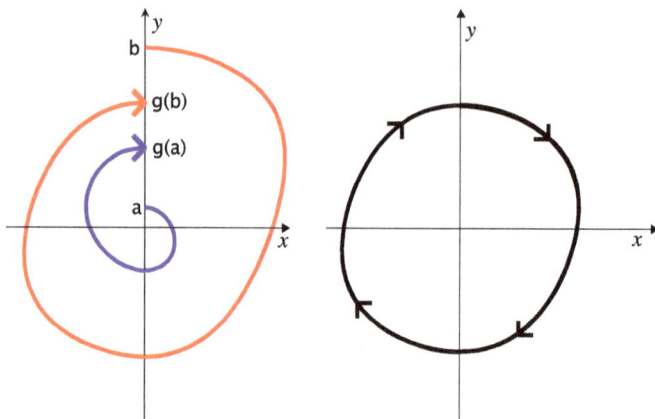

Figure 9.10: (left) The trajectories in the Liénard plane, with initial condition $(0, a)$ and $(0, b)$; (right) the periodic solution.

Therefore, by the intermediate value theorem it follows that there exists a unique $\tilde{z} \in (a, b)$ such that $g(\tilde{z}) = \tilde{z}$.

Step 3. The orbit $x(t, \tilde{z}), y(t, \tilde{z})$ with initial condition $(0, \tilde{z})$ is a closed trajectory and hence gives rise to a periodic solution of the van der Pol equation. There are no other periodic solutions since a periodic solution corresponds to a closed orbit of (9.6), which implies that $g(z) = z$ and hence $z = \tilde{z}$, by the uniqueness property seen in Step 2.

The previous arguments highlight the fact that that the periodic solution we have found behaves similar to an "attractor", in the sense that the other trajectories of (9.6) approach this periodic orbit as $t \to +\infty$: from inside if the initial condition is $(0, a)$, from outside if the initial condition is $(0, b)$.

That is why is this periodic solution is called a *limit cycle*, see Subsection 9.6.2 below. Figure 9.11 shows how the periodic orbit looks in the phase plane (x, x').

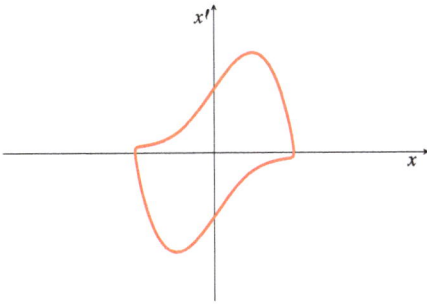

Figure 9.11: Plot of the periodic solution of van der Pol's equation in the phase plane (x, x').

9.5 Homoclinic and heteroclinic solutions

A solution $x(t)$ of (9.1) is *homoclinic to x^** if

$$\lim_{t \to \pm\infty} x(t) = x^*.$$

A solution $x(t)$ of (9.1) is a *heteroclinic from x_1 to x_2*, with $x_1 \neq x_2$, if

$$\lim_{t \to -\infty} x(t) = x_1, \quad \text{and} \quad \lim_{t \to +\infty} x(t) = x_2.$$

When we want to refer to solutions of (V) we will also say that $x(t), y(t)$ is a *homoclinic orbit*, respectively *heteroclinic orbit*, of (V).

To find homoclinics and heteroclinics we use a phase plane analysis. We will discuss some specific examples, but the arguments can easily be extended to more general cases.

Homoclinics

A typical situation in which homoclinics arise is the case in which the curve E_c contains a loop like the one drawn in Fig. 9.12.

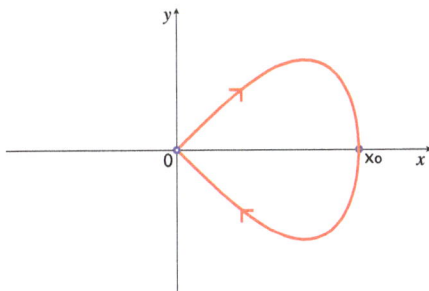

Figure 9.12: A loop with a homoclinic.

Let us prove that this loop corresponds to a homoclinic to $x^* = 0$. Consider the solution $x(t), y(t)$ of (V) with initial condition $(x_0, 0)$. From Fig. 9.12 we see that x_0 is the proper maximum of $x(t)$. Then the branch of the loop contained in the half plane $y < 0$ corresponds to a decreasing function $x = x_1(t)$ which satisfies the problem

$$\begin{cases} x''(t) + V'(x(t)) & = 0, \quad t \geq 0, \\ x(0) & = x_0 \end{cases}$$

Similarly, the branch of the loop contained in the half plane $y > 0$ corresponds to an increasing function $x = x_2(t)$ which satisfies the problem

$$\begin{cases} x''(t) + V'(x(t)) & = 0, \quad t \leq 0, \\ x(0) & = x_0 \end{cases}$$

Since $x_1(0) = x_2(0) = x_0$ and $x_1'(0) = x_2'(0) = 0$, the function obtained by gluing x_1 and x_2, namely

$$x(t) = \begin{cases} x_1(t) & \text{for } t \geq 0, \\ x_2(t) & \text{for } t \leq 0 \end{cases}$$

is continuously differentiable and gives rise to a solution of (9.1), since x_1 and x_2 are.
To show that $x(t)$ is homoclinic at $x^* = 0$ it remains to prove that

$$\lim_{t \to +\infty} x(t) = \lim_{t \to -\infty} x(t) = 0. \tag{9.7}$$

Recall that
(a) for $t > 0$ one has $x'(t) = x_1'(t) < 0$ and hence $x(t)$ is decreasing;
(b) for $t < 0$ one has $x'(t) = x_2'(t) > 0$ and hence $x(t)$ is increasing.

From (a–b) it follows that the limits $\lim_{t \to +\infty} x(t) = \lim_{t \to +\infty} x_1(t)$ and $\lim_{t \to -\infty} x(t) = \lim_{t \to -\infty} x_2(t)$ exist and are equal, respectively, to $\inf_{t > 0} x_1(t)$ and $\inf_{t < 0} x_2(t)$.
Finally, from the graph plotted in fig. 9.12, one could infer that

$$\inf_{t > 0} x_1(t) = \inf_{t < 0} x_2(t) = 0,$$

proving (9.7). ☐

Remark 9.4. Contrary to the periodic solution case, in the preceding example the curve $E(x, y) = 0$ has a cusp singularity at the point $(0, 0)$.

In general, we can show that

(†) *If* (V) *possesses a homoclinic orbit to* x^*, *then* $(x^*, 0)$ *is a singular point of the curve* $\{E(x, y) = V(x^*)\}$. ☐

Actually, if $x(t)$ is a homoclinic to x^*, then $y(t) = x'(t) \rightarrow 0$ as $t \rightarrow +\infty$. Since $y'(t) = -V'(x(t))$ we infer that $0 = \lim_{t\rightarrow+\infty} y'(t) = -V'(x^*)$, proving (†).

The point $(0, x^*)$ is called *homoclinic point*. Typically, at a homoclinic point the energy E has a saddle.

Example 9.3. Consider the equation $x'' = x - x^3$.
The energy is

$$E = \frac{1}{2}y^2 - \frac{1}{2}x^2 + \frac{1}{4}x^4.$$

First of all, we find the singular points since they are the only possible homoclinic points, as pointed out earlier. The singular points are the solutions of $\nabla E(x,y) = (0,0)$. Solving $E_x = -x + x^3 = 0$, $E_y = 0$ we find $x = 0$, $x = \pm 1$ and $y = 0$. It is easy to check that $(0,0)$ is a saddle with $E(0,0) = 0$, whereas $(\pm 1, 0)$ are absolute minima of E, with $E(\pm 1, 0) = -\frac{1}{2} + \frac{1}{4} = -\frac{1}{4}$. The orbit $\{E(x,y) = 0\}$ is the red curve in Fig. 9.13, which is a double loop passing through the singular point $(0,0)$.

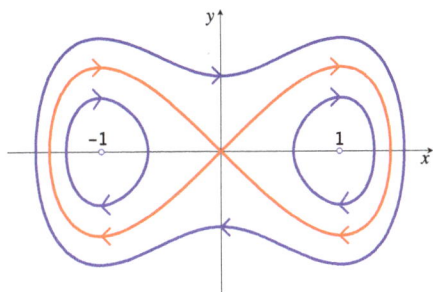

Figure 9.13: Plot of E_c for some values of c.

Repeating the arguments carried out before, we conclude that $x'' = x - x^3$ has two homoclinics to 0, one positive, corresponding to the loop on the right and one negative, corresponding to the loop on the left. In this case, a direct calculation shows that the homoclinics are the *solitary waves* $x(t) = \pm\frac{\sqrt{2}}{\cosh t}$.

For completeness, let us show that the phase plane is filled up by periodic orbits separated by the two homoclinics. For this, let us recall that the singular points $(\pm 1, 0)$ are minima of E with $E(\pm 1, 0) = -\frac{1}{4}$. As a consequence we see that E_c is empty if and only if $c < -\frac{1}{4}$. For $c = -\frac{1}{4}$ we find that $E_c = \{(-1, 0)\} \cup \{(1, 0)\}$, whereas for $c > -\frac{1}{4}$, $c \neq 0$ then E_c is a closed bounded curve (marked in blue in Fig. 9.13) which give rise to periodic solutions. ☐

Heteroclinics

As for homoclinics, heteroclinic orbits also arise in the presence of singular points. A typical situation is reported in Fig. 9.14, where the red curve is the graph of $\{E(x,y) = E(\pm x_1, 0) = V(\pm x_1)\}$.

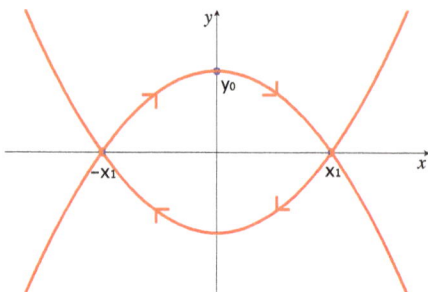

Figure 9.14: A loop with heteroclinics.

The two symmetric orbits joining $(-x_1, 0)$ and $(x_1, 0)$ are heteroclinics. To prove this claim it suffices to repeat the arguments carried out for homoclinics. On the upper branch of the loop we have $x'(t) = y(t) > 0$ and thus $x(t)$ is increasing. Since $-x_1 < x(t) < x_1$ we infer $\lim_{t\to-\infty} x(t) = -x_1$ and $\lim_{t\to+\infty} x(t) = x_1$ which means that $x(t)$ is a heteroclinic from $-x_1$ to x_1. On the lower branch we find that $-x(t)$ is a heteroclinic from x_1 to $-x_1$. Notice that $(-x_1, 0)$ and $(x_1, 0)$ are singular points, of the curve $\{E = V(\pm x_1)\}$.

More generally,

(‡) *if* (V) *possesses a heteroclinic from* x_1 *to* x_2, *then* $(x_1, 0)$ *and* $(x_2, 0)$ *are singular points of the curve* $\{E = V(x_1) = V(x_2)\}$.[1]

Example 9.4. We have seen that the nonlinear harmonic oscillator $x'' + \omega^2 x - x^3 = 0$ has periodic solutions. Now, let us show that it has also heteroclinics for suitable values of the energy

$$E = \frac{1}{2}y^2 + \frac{1}{2}\omega^2 x^2 - \frac{1}{4}x^4.$$

First of all, solving $E_x = \omega^2 x - x^3 = 0$ and $E_y = y = 0$ we find that the singular points are $(0,0)$ and $(\pm\omega, 0)$. Notice that $(0,0)$ is a local minimum of E and hence can be neither a homoclinic nor a heteroclinic point.

Let us consider $E(\pm\omega, 0) = -\frac{1}{4}\omega^4$. The level curve $\{E = -\frac{1}{4}\omega^4\}$ is quite similar to the one plotted in Fig. 9.14; see also Fig. 9.7. Then we find that the nonlinear harmonic oscillator has a pair of heteroclinics $\pm x(t)$ from $(\omega, 0)$ to $(-\omega, 0)$. ☐

---

1 Notice that $V(x_1) = V(x_2)$ since $(x_1, 0)$ and $(x_2, 0)$ belong to the same orbit.

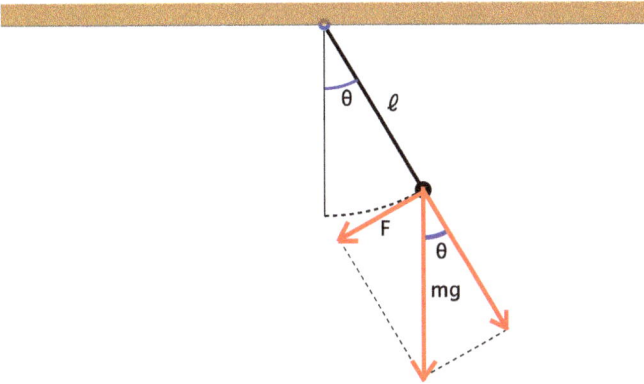

Figure 9.15: The mathematical pendulum.

Our last example deals with the *mathematical pendulum*.

Referring to Fig. 9.15, we see that the force that causes the movement of the bob is the component $F = mg \sin \theta$ of the gravity mg. Thus the Newton second law $ma = F$ yields $-ms'' = mg \sin \theta$, where $s = \ell\theta$ is the arclength on the circle of radius ℓ. Thus $s'' = \ell\theta''$ whereby $\ell\theta'' + g \sin \theta = 0$, and setting $x = \theta$ and taking $\frac{g}{\ell} = 1$,

$$x'' + \sin x = 0. \tag{MP}$$

Since $E(x,y) = \frac{1}{2}y^2 - \cos x$, the singular points are $(\pm k\pi, 0)$, $k \in \mathbb{Z}$.

Referring to Fig. 9.16 we infer:

- if $c = 1$ then E_c are the two curves marked in red passing through the singular points $(\pm\pi, 0)$, $(\pm3\pi, 0)$, etc. which correspond to heteroclinic solutions of (MP);
- if $-1 < c < 1$ then E_c are the closed curves marked in blue which correspond to periodic solutions of (MP);
- if $c > 1$ then E_c are the unbounded curves marked in violet: the branches contained in the half plane $y > 0$ which correspond to unbounded increasing solutions $x(t)$ of (MP), and the symmetric branches contained in the half plane $y < 0$ which correspond to unbounded decreasing solutions $-x(t)$ of (MP).

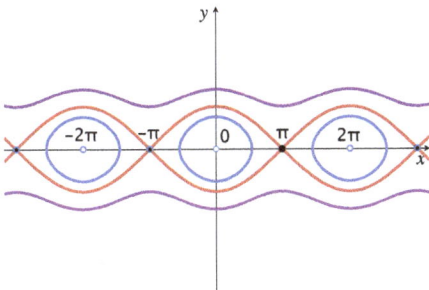

Figure 9.16: Plot of $\frac{1}{2}y^2 - \cos x = c$.

9.6 On planar autonomous systems

In this final section we deal with general planar autonomous systems as

$$\begin{cases} x' = f(x,y), \\ y' = g(x,y), \end{cases} \tag{9.8}$$

where $f, g : \mathbb{R}^2 \to \mathbb{R}$ are continuous.

We will only discuss two remarkable examples.

9.6.1 The Lotka–Volterra prey–predator system

Consider, for example, two species like sardines (the prey) and sharks (the predators) occurring in a bounded environment, like a portion of sea. Let $x(t)$ denote the number of prey at time t and $y(t)$ the number of predators at time t. We make a number of assumptions:

1. The prey have an unlimited food supply, while predators feed only on prey.
2. The prey grow up at a rate α depending on y only: $x' = a(y)x$. For $y = 0$, namely when there is no predator, prey grow exponentially since they have unlimited food supply. Hence $a(0) = a > 0$. When there are some predator $y > 0$, the growth rate of prey decreases with y because the larger the number of predators the greater is the difficulty for prey to prosper and reproduce: let us take $a(y) = a - by$.
3. The predators grow up at a rate β depending on x only: $y' = \beta(x)y$. For $x = 0$, namely when there is no prey, predators would go to extinction since predators do not have any food supply other than the prey. Thus $\beta(0) = -c < 0$. When there are some prey $x > 0$, the growth rate of predators increases with x since predators have more food supply: let us take $\beta(x) = -c + dx$.

Under these assumptions, we find the following system called *Lotka–Volterra system,* named after the two mathematicians A. Lotka and V. Volterra who first independently introduced this model:

$$\begin{cases} x' = x(a - by) = ax - bxy, \\ y' = y(dx - c) = -cy + dxy. \end{cases} \tag{LV}$$

Though (LV) is not of the same form as the systems discussed in the previous sections, we will see that it is a conservative system so that we can use arguments similar to those employed before.

9.6.1.1 Equilibria

The equilibrium solutions are found by setting $x' = y' = 0$ in (LV). Solving the resulting algebraic system

$$\begin{cases} x(a - by) = 0, \\ y(dx - c) = 0, \end{cases}$$

for x, y yields $(0, 0)$ and $P_0 = (x_0, y_0)$, with $x_0 = \frac{c}{d}, y_0 = \frac{a}{b}$.

9.6.1.2 Periodic solutions

Let

$$h(x, y) = dx + by - c \ln x - a \ln y, \quad x > 0, \ y > 0.$$

Lemma 9.3. *Let $x(t)$, $y(t)$ be a solution of* (LV). *Then $h(x(x), y(t))$ is identically constant.*

Proof. In view of (LV), we find

$$\frac{d}{dt} h(x(x), y(t)) = dx' + by' - \frac{cx'}{x} - \frac{ay'}{y}$$

$$= d(ax - bxy) + b(-cy + dxy) - \frac{c(ax - bxy)}{x} - \frac{a(-cy + dxy)}{y}$$

$$= adx - bdxy - bcy + bdxy - ac + bcy + ac - adx = 0.$$

Thus $h(x(x), y(t)) \equiv \text{const.}$ □

Lemma 9.4. $P_0 = (x_0, y_0)$ *is the proper global minimum of h.*

Proof. The equation $\nabla h = 0$, namely $h_x = h_y = 0$, yields

$$\begin{cases} d - \frac{c}{x} = 0, \\ b - \frac{a}{y} = 0, \end{cases}$$

whose solution is $x_0 = \frac{c}{d}, y_0 = \frac{a}{b}$. Evaluating the hessian matrix $h''(x_0, y_0)$ we find

$$h''(x_0, y_0) = \begin{pmatrix} h_{xx}(x_0, y_0) & h_{xy}(x_0, y_0) \\ h_{yx}(x_0, y_0) & h_{yy}(x_0, y_0) \end{pmatrix} = \begin{pmatrix} \frac{c}{x_0^2} & 0 \\ 0 & \frac{a}{y_0^2} \end{pmatrix},$$

which is positive definite. Thus P_0 is the global minimum of h. □

Theorem 9.4. *Let $h_0 = h(P_0)$. Then for every $y > h_0$ the system* (LV) *has a periodic solution $x(t)$, $y(t)$ such that $h(x(t), y(t)) = y$*

Proof. Since P_0 is the global minimum of h, $\forall y > h_0$ the set $h(x, y) = y$ is a non-empty closed curve around P_0. Notice that P_0 is the unique equilibrium of (LV). Then, repeating the arguments carried out to prove the existence of periodic solutions of (V), we can show that the curve $h(x, y) = y$ corresponds to a periodic solution of (LV). Some level curves $h = y > h_0$ are plotted in Fig. 9.17. □

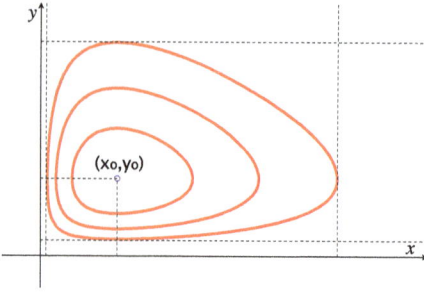

Figure 9.17: Plot of $h(x, y) = y > h_0$.

Next let us consider the mean values

$$\bar{x} = \frac{1}{T} \int_0^T x(t)\, dy, \quad \bar{y} = \frac{1}{T} \int_0^T y(t)\, dy.$$

Theorem 9.5. *If $x(t), y(t)$ is a T-periodic solution of (LV), then $\bar{x} = x_0 = \frac{c}{d}$ and $\bar{y} = y_0 = \frac{a}{b}$.*

Proof. The second equation in (LV) $y' = y(dx - c)$ yields $\frac{y'}{y} = dx - c$ whereby

$$\int_0^T \frac{y(t)'}{y(t)}\, dt = \int_0^T (dx(t) - c)\, dt = d \int_0^T x(t)\, dt - cT.$$

On the other hand,

$$\int_0^T \frac{y(t)'}{y(t)}\, dt = \ln y(T) - \ln y(0) = 0,$$

since $y(t)$ is T-periodic. Thus

$$\int_0^T (dx(t) - c)\, dt = d \int_0^T x(t)\, dt - cT = 0 \Rightarrow \frac{1}{T} \int_0^T x(t)\, dy = \frac{c}{d} = x_0.$$

The proof of the second equality is quite similar. □

The quantities $\bar{x}, \bar{y}$ are important in applications because they provide the average size of the prey and predator population.

9.6.1.3 Behavior of periodic solutions

In order to understand the behavior of a generic T-periodic solution $x(t), y(t)$ of (LV) as a function of t, let us focus on the corresponding closed orbit plotted in Fig. 9.18.

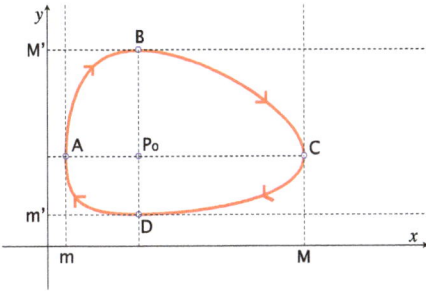

Figure 9.18: A generic T-periodic solution of (LV).

First of all we see that $x(t), y(t)$ share the following properties:

- $0 < m \leq x(t) \leq M$ for all t;
- $0 < m' \leq y(t) \leq M'$ for all t.

These first results can be interpreted as follows. The larger the predator population becomes the more difficult it becomes to find food, and hence the population size will remain bounded form above. Also, the number of prey cannot overcome a threshold because as they increase they provide a greater quantity of food for predators who would prosper. On the other hand, the number of prey cannot be zero either because, being the only food supply of predators, the latter would die out. Similarly, if predators become very few then, for example, few sardines would be eaten by sharks and their number would increase, providing more food for predators.

Moreover, let us select on this orbit four specific points: A, B, C, D as in Fig. 9.18. Without loss of generality, we can assume that $(x(0), y(0)) = A$ (otherwise we shift time). Let $0 < t_1 < t_2 < t_3$ be the times at which the orbit meets B, C, D, respectively. Notice that $x(0)$ is the minimum for $x(t), y(t_1)$ is the maximum for $y(t), x(t_2)$ is the maximum for $x(t)$ and $y(t_2)$ the minimum for $y(t)$. Thus $x'(0) = y'(t_1) = x(t_2) = y(t_3) = 0$. Then (LV) yields $a - by(0) = 0, c - dx(t_1) = 0, a - by(t_2) = 0$ and $c - dx(t_3) = 0$, whereby $y(0) = y(t_2) = \frac{a}{b} = y_0$ and $x(t_1) = x(t_3) = \frac{c}{d} = x_0$.

Repeating the arguments carried out in the phase plane analysis, we see that the solution $x(t), y(t)$ has the following features:

- for $0 < t < t_1$, namely along the arc $AB, x > \frac{c}{d}, y < \frac{a}{b}$ imply $x'(t) > 0, y'(t) > 0$: both prey $x(t)$ and predators $y(t)$ increase;
- for $t_1 < t < t_2$, namely along the arc $BC, x > \frac{c}{d}, y > \frac{a}{b}$ imply $x'(t) > 0, y'(t) < 0$: prey $x(t)$ increase whereas predators $y(t)$ decrease;
- for $t_2 < t < t_3$, namely along the arc $CD, x < \frac{c}{d}, y > \frac{a}{b}$ imply $x'(t) < 0, y'(t) < 0$: both prey $x(t)$ and predators $y(t)$ decrease;
- for $t_3 < t < T$, namely along the arc $DA, x < \frac{c}{d}, y < \frac{a}{b}$ imply $x'(t) < 0, y'(t) > 0$: prey $x(t)$ decrease and predators $y(t)$ increase.

The Lotka–Volterra model has shown to be consistent since, at a first approximation, the properties of the solutions of (LV) agree in a rather positive way with the concrete statistical data.

9.6.2 Limit cycles

Here we give a flavor of the *Poincaré–Bendixon theory* dealing with the existence of *limit cycles* which are particular periodic solutions of planar systems

$$\begin{cases} x' = f(x,y), \\ y' = g(x,y). \end{cases} \tag{9.8}$$

By definition, a *limit cycle* of (9.8) is an isolated periodic solution of (9.8) such that the nearby trajectories spiral towards and/or away from it, see Fig. 9.20. As anticipated before, an example of limit cycle is the periodic solution of van der Pol equation (see Fig. 9.19), whereas any periodic solution of conservative systems is not a limit cycle, since the nearby trajectories do not tend to such a solution, being themselves periodic.

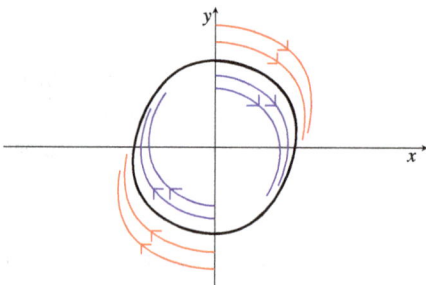

Figure 9.19: Example of limit cycle.

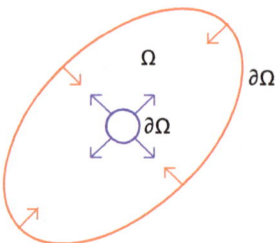

Figure 9.20: Example of a set $\Omega \subset \mathbb{R}^2$ where assumption (b) holds.

Consider the planar systems (9.8) and suppose that there is a set $\Omega \subset \mathbb{R}^2$ such that
(a) Ω is bounded and does not contain equilibria;

(b) on the boundary $\partial\Omega$ of Ω the vector field (f, g) is directed inwards Ω; see Fig. 9.20 (by this we mean that the scalar product of (f, g) with the outer unit normal $\bar{\nu}$ to $\partial\Omega$ is < 0).

Under these assumptions we can state a result which is a particular case of a more general theorem by *Poincaré–Bendixon*.

Theorem 9.6. *If (a)–(b) hold, then Ω contains a limit cycle of (9.8).*

To give an idea of the proof of Theorem 9.6 let us suppose that Ω is an annulus like the set plotted in Fig. 9.20. Roughly, any trajectory with initial point on $\Omega \cup \partial\Omega$ remains in Ω for all $t > 0$, since to escape out of Ω the trajectory should meet the boundary $\partial\Omega$ where the vector field (f, g) is directed inwardly; see assumption (b). Since Ω is bounded, by compactness these trajectories must converge in a suitable sense. If any of these trajectories converge to a point, this limit should be an equilibrium of (9.8) contained in Ω, which is impossible by assumption (a). Then one can prove that (9.8) possesses a closed orbit (hence a periodic solution) such that the nearby trajectories look like spirals approaching the closed orbit from outside and from inside, which is therefore a limit cycle.

For example, Theorem 9.6 might be used to find the periodic solution of van der Pol equation.

For a broader treatment of the Poincaré–Bendixon theory, we refer, e. g., to the aforementioned book by H. Amann.

9.7 Exercises

1. Using the phase plane analysis, show that the solution of $x'' - 3x^2 = 0$, $x(0) = 1$, $x'(0) = 0$, is positive, has a minimum at $t = 0$ and is convex. Sketch a qualitative graph of the solution. [Hint: see Example 9.1]
2. Using the phase plane analysis, show that the solution of $x'' - 4x^3 = 0$, $x(0) = 0$, $x'(0) = -1$ is increasing, convex for $t < 0$ and concave for $t > 0$. Sketch a qualitative graph of the solution.
3. Show that the solution of $x'' + x + 4x^3 = 0$, $x(0) = 1$, $x'(0) = 0$, is periodic.
4. Show that the solution of $x'' - x + 4x^3 = 0$, $x(0) = 1$, $x'(0) = 0$, is periodic.
5. Knowing that the boundary value problem $x'' + x^3 = 0$, $x(-1) = x(1) = 0$, has a solution $x(t)$ such that $M = \max_{[-1,1]} x(t) = 3$, find $x'(-1)$.
6. Let $x_a(t)$ be the solution of the nonlinear harmonic oscillator $x'' + \omega^2 x - x^3 = 0$, $x_a(0) = 0$, $x'_a(0) = a$. Find $a > 0$ such that $x_a(t)$ is periodic.
7. Find a, c such that the the equilibrium of the Lotka–Volterra system

$$\begin{cases} x' = ax - 3xy \\ y' = -cy + xy \end{cases}$$

is $(2, 3)$.

8. Let $x_a(t), y_a(t)$ be the solution of the Hamiltonian system

$$\begin{cases} x' = ax + y, \\ y' = -x - ay, \end{cases}$$

such that $x(0) = 0, y(0) = 1$. Find a such that $x_a(t), y_a(t)$ is periodic.
9. Show that $x'' = x - x^5$ has a positive and a negative homoclinic to $x^* = 0$ and find its maximum, resp. minimum, value.
10. Find k such that the equation $x'' = k^2x - x^3$ has a homoclinic $x(t)$ to 0 such that $\max_{\mathbb{R}} x(t) = 2$.
11. Show that $x'' = x - x^4$ has one and only one positive homoclinic to $x^* = 0$.
12. Show that for $k < 0$ the equation $x'' = kx - x^3$ has no homoclinic to 0.
13. Show that $x'' + V'(x) = 0$ cannot have homoclinics if V' does not change sign.
14. Show that for $p \geq 1$ the equation $x'' + x^p = 0$ has no heteroclinic.
15. Show that if $x'' + V'(x) = 0$ has a homoclinic to 0, then V cannot have a minimum nor a maximum at $x = 0$.
16. Find a such that $x'' + 2x - 2x^3 = 0, x(0) = 0, x'(0) = a$, is a heteroclinic.
17. Find the integers $k > 1$ such that $x'' + x - x^k = 0$ has a heteroclinic.
18. * Let $T = T(a)$ be the period of the periodic solution of the nonlinear harmonic oscillator $x'' + \omega^2 x - x^3 = 0$ such that $x(0) = 0, x'(0) = a > 0$. Show that $\lim_{a\to 0} T(a) = \frac{2\pi}{\omega}$.
19. * Describe the behavior of the solution of the Kepler problem with energy (i) $c = U_{\text{eff}}(r_0) = \min U_{\text{eff}}(r)$, (ii) $c = 0$ and (iii) $c > 0$.
20. * Show that there exists a unique $\lambda > 0$ such that the boundary value problem $x'' + \lambda x^3 = 0, x(0) = x(\pi) = 0$, has a positive solution.

10 Introduction to stability

Notation. If $\bar{x} = (x_1, \ldots, x_n), \bar{y} = (y_1, \ldots, y_n) \in \mathbb{R}^n$, $(\bar{x} \mid \bar{y}) = \sum x_i y_i$ denotes the Euclidean inner product and $\|\bar{x}\|$ denotes its Euclidean norm: $\|\bar{x}\| = (\bar{x} \mid \bar{x})^{1/2} = \sqrt{x_1^2 + \cdots + x_n^2}$. For $r > 0$, we also set $B(r) = \{\bar{x} \in \mathbb{R}^n : \|\bar{x}\| < r\}$.

10.1 Definition of stability

Let $f_i(t, x_1, \ldots, x_n)$, $i = 1, 2, \ldots, n$, be continuously differentiable functions on $\mathbb{R}^n$. Given $p = (p_1, \ldots p_n) \in \mathbb{R}^n$, let $\bar{u}(t) = (u_1(t), \ldots, u_n(t))$ be the unique solution of the ivp

$$x_i' = f_i(t, x_1, \ldots, x_n), \quad x_i(0) = p_i, \quad i = 1, 2, \ldots, n. \tag{10.1}$$

We write $\bar{u}(t; p)$ in order to stress the fact that the solution $\bar{u}$ depends on p. It can be proved that $\bar{u}$ depends continuously on p in the sense that

Theorem 10.1. *Under the preceding assumptions, for each $\epsilon > 0$ and $T > 0$ there exists $\delta > 0$ such that*

$$\max_{[0,T]} \|\bar{u}(t; p) - \bar{u}(t; q)\| < \epsilon$$

provided $\|p - q\| < \delta$.

The proof in the case when $n = 1$ and f is linear has been given in Remark 2.4 in Chapter 2.

Stability is concerned with the dependence of δ on T as $T \to +\infty$. We will limit our study to deal with the stability of equilibria. Let $p^* = (p_1^*, \ldots, p_n^*) \in \mathbb{R}^n$ be an equilibrium of (10.1), so that $f_i(p_1^*, \ldots, p_n^*) = 0$ for $i = 1, \ldots, n$. Since (10.1) is autonomous, up to a translation, we can assume without loss of generality that $p^* = 0$.

Definition 10.1 (Stability). We say that $p^* = 0$ is stable if $\forall \epsilon > 0$, there exists $\delta > 0$ such that $\max_{t \geq 0} \|\bar{u}(t; p)\| < \epsilon$ for all $\|p\| < \delta$. We say that $p^* = 0$ is unstable if it is not stable.

Roughly, if an equilibrium is stable, then the solution $\bar{u}(t; p)$ remains close to $p^* = 0$ for all time $t \geq 0$, provided p is sufficiently close to $p^* = 0$.

A more restrictive stability is *asymptotic stability*:

Definition 10.2 (Asymptotic stability). We say that $p^* = 0$ is asymptotically stable if it is stable and $\exists \delta > 0$ such that $\lim_{t \to +\infty} \bar{u}(t; p) = p^* = 0$ for all $\|p\| < \delta$.

Thus asymptotic stability requires not only that $\bar{u}(t; p)$ remain near the equilibrium for $t > 0$ sufficiently large, but also that $\bar{u}(t; p)$ converge to the equilibrium as $t \to +\infty$. In the sequel we will see that there are equilibria which are stable but not asymptotically stable.

https://doi.org/10.1515/9783111185675-010

The main purpose of the following analysis is to establish the stability of equilibria by a general method, without solving the system explicitly.

10.2 Stability of linear systems

The stability of $p^* = 0$ of the $n \times n$ linear homogeneous system

$$\bar{u}' = A\bar{u}, \quad \bar{u} = \begin{pmatrix} u_1 \\ u_2 \\ \cdots \\ u_n \end{pmatrix}, \quad A = \begin{pmatrix} a_{11} & \cdots & a_{1n} \\ a_{21} & \cdots & a_{2n} \\ \cdots & \cdots & \cdots \\ a_{n1} & \cdots & a_{nn} \end{pmatrix}, \tag{10.2}$$

depends on the sign of the eigenvalues of A and can be established in a rather direct way.

Theorem 10.2. *If all the eigenvalues of A have negative real parts then the origin $p^* = 0$ is asymptotically stable. More precisely, $\forall p \in \mathbb{R}^n$ one has $\bar{u}(t; p) \to 0$ as $t \to +\infty$.*

Proof. We give the proof in the case when A is diagonal and the eigenvalues are simple.
(i) We note that in this case all of the eigenvalues are real. Thus $A = \text{diag}\{\lambda_1, \lambda_2, \dots, \lambda_n\}$ and the system splits into n independent equations $x_i' = \lambda_i x_i$, $(i = 1, \dots, n)$, whose solutions are given by $u_i(t) = c_i e^{\lambda_i t}$. Then $\lambda_i < 0$, for all $i = 1, \dots, n$, implies that $u_i(t) \to 0$ as $t \to +\infty$.
(ii) A can have some pairs of complex conjugate eigenvalues $\alpha \pm i\beta$. Each of these eigenvalues gives rise to the solution $e^{\alpha t}(c_1 \cos \beta t + ic_2 \sin \beta t)$, which tends to 0 as $t \to +\infty$ provided $\alpha < 0$. In both cases, the result is independent of the initial condition p. □

On the other hand, we have the following.

Theorem 10.3. *If one eigenvalue of A is positive or has positive real part then the origin $p^* = 0$ is unstable.*

The preceding results rule out the case in which A has an eigenvalue with zero real part.

Example 10.1. Let $n = 2$ and

$$A = \begin{pmatrix} a & -1 \\ 1 & a \end{pmatrix}.$$

The eigenvalues of A are $a \pm i$. Thus: if $a < 0$ the origin is asymptotically stable, whereas if $a > 0$ we have instability. If $a = 0$ the preceding theorems do not give any information and we have to consider the system directly, which is the same as

$$\begin{cases} x_1' = -x_2, \\ x_2' = x_1. \end{cases}$$

Solving we find $x_1'' = -x_2' = -x_1$, or $x_1'' + x_1 = 0$ whereby $x_1(t) = c_1 \sin t + c_2 \cos t$ and $x_2(t) = -x_1'(t) = -c_1 \cos t + c_2 \sin t$. If $x_1(0) = p_1$ and $x_2(0) = p_2$, we find $c_2 = p_1$ and $c_1 = -p_2$. Thus the corresponding solution $\bar{u}(t; p)$ has components

$$u_1 = -p_2 \sin t + p_1 \cos t, \quad u_2 = p_2 \cos t + p_1 \sin t.$$

Thus

$$u_1^2 + u_2^2 = p_2^2 \sin^2 t + p_1^2 \cos^2 t - 2p_1 p_2 \sin t \cos t + p_1^2 \sin^2 t + p_2^2 \cos^2 t + 2p_1 p_2 \sin t \cos t$$
$$= p_2^2 (\sin^2 t + \cos^2 t) + p_1^2 (\sin^2 t + \cos^2 t) = p_1^2 + p_2^2,$$

which means that $\|\bar{u}(t, p)\|^2 = \|p\|^2$. It is clear that this implies stability since the point $(u_1(t : p), u_2(t; p))$ remains on the circle of radius $\|p\|$. On the other hand, $(0, 0)$ is not asymptotically stable since neither $u_1(t, p)$ nor $u_2(t, p)$ tends to zero as $t \to +\infty$, no matter what $p \neq 0$ is. □

10.2.1 Stability of 2×2 linear systems

For 2×2 linear homogeneous systems

$$\begin{cases} x' = ax + by, \\ y' = cx + dy, \end{cases} \tag{10.3}$$

it is possible to give a complete classification of the stability property of the equilibrium $p^* = (0, 0)$.

Equation (10.3) can be written in vector form as

$$u' = Au, \quad \text{where } u = \begin{pmatrix} x \\ y \end{pmatrix}, \quad A = \begin{pmatrix} a & b \\ c & d \end{pmatrix}.$$

We assume in the sequel that A is nonsingular, so that $p^* = (0, 0)$ is the unique equilibrium of (10.3).

Let us recall the following important properties of matrices. Given a nonsingular matrix A there exists a matrix J, called the *Jordan canonical normal form* of A, such that:
(a) J is similar to A, that is, there exists a nonsingular matrix X such that $XJ = AX$ or $J = X^{-1}AX$.
(b) J is upper triangular and has the same eigenvalues as A.

Let v be a solution of $v' = Jv$. Setting $u = Xv$ we find

$$u' = Xv' = XJv = A\underset{=u}{\underline{Xv}} = Au.$$

Moreover, since $\|u\| = \|Xv\| \le k \cdot \|u\|$, with k = the norm of X, we see that u and v have the same qualitative properties. In particular u is stable, asymptotically stable, unstable if and only if v is so. Therefore, the stability analysis of $u' = Au$ is equivalent to the stability analysis of $v' = Jv$.

Suppose that A is non-singular (the singular case will be discussed later) and denote by λ_1, λ_2 its eigenvalues (real and distinct, equal, or complex conjugate). In this case the unique equilibrium of (10.3) is $x = 0, y = 0$. Let J be its Jordan canonical normal form. Then there are four possible cases:

1. λ_1 and λ_2, $\lambda_1 \ne \lambda_2$, are real numbers and $J = \begin{pmatrix} \lambda_1 & 0 \\ 0 & \lambda_2 \end{pmatrix}$;
2. $\lambda_1 = \lambda_2 \in \mathbb{R}$ and $J = \begin{pmatrix} \lambda_1 & 0 \\ 0 & \lambda_1 \end{pmatrix}$;
3. $\lambda_1 = \lambda_2 \in \mathbb{R}$ and $J = \begin{pmatrix} \lambda_1 & 1 \\ 0 & \lambda_1 \end{pmatrix}$;
4. $\lambda = \alpha \pm i\beta$ with $\beta \ne 0$ and $J = \begin{pmatrix} \alpha & -\beta \\ \beta & \alpha \end{pmatrix}$.

Case 1. Solving $u' = Ju$ we find $x = c_1 e^{\lambda_1 t}, y = c_2 e^{\lambda_2 t}$. Thus, as we already know, if $\lambda_1 < 0$ and $\lambda_2 < 0$ we have asymptotic stability and the origin is called a *stable node*, whereas if one eigenvalue is positive we have instability.

Let us find solutions of the form $y = y(x)$. If either $c_1 = 0$ or $c_2 = 0$ we deal with one of the axes. Otherwise, we first solve $x = c_1 e^{\lambda_1 t}$ for t yielding

$$\lambda_1 t = \ln\left|\frac{x}{c_1}\right| \Rightarrow t = \frac{1}{\lambda_1}\ln\left|\frac{x}{c_1}\right| = \ln\left|\frac{x}{c_1}\right|^{1/\lambda_1}.$$

Then we obtain

$$y = c_2 e^{\lambda_2 t} = c_2 e^{\lambda_2 \cdot \ln\left|\frac{x}{c_1}\right|^{1/\lambda_1}} = c_2 e^{\ln\left|\frac{x}{c_1}\right|^{\lambda_2/\lambda_1}} = c_2 \cdot \left|\frac{x}{c_1}\right|^{\lambda_2/\lambda_1}.$$

Relabelling the constants, we find

$$y = c|x|^{\lambda_2/\lambda_1}, \quad c \ne 0.$$

These solutions are plotted in Fig. 10.1: the curves in red correspond to $\lambda_2/\lambda_1 > 1$, those in blue to $\lambda_2/\lambda_1 < 1$.

If both λ_1 and λ_2 are positive we have an *unstable node*. The phase plane portrait of the trajectories is similar to that plotted in Fig. 10.1, with the direction of the arrows reversed.

If the two eigenvalues have different signs, say $\lambda_1 < 0 < \lambda_2$, then $y = c|x|^{\lambda_2/\lambda_1}, c \ne 0$, which behaves like a hyperbola. In this case the origin is called a *saddle point*. If the initial condition implies that $x(t) = 0$ then we deal with solutions on the y-axis, and $|y(t)| = |c|e^{\lambda_2 t} \to +\infty$ as $t \to +\infty$. If the initial condition implies that $y(t) = 0$ then we deal with solutions on the x axis, and $|x(t)| = |c|e^{\lambda_1 t} \to 0$ as $t \to +\infty$. The saddle case is shown in Fig. 10.2.

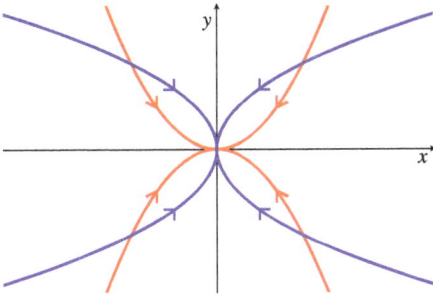

Figure 10.1: Stable node when $\lambda_1 \neq \lambda_2, \lambda_1 < 0, \lambda_2 < 0$.

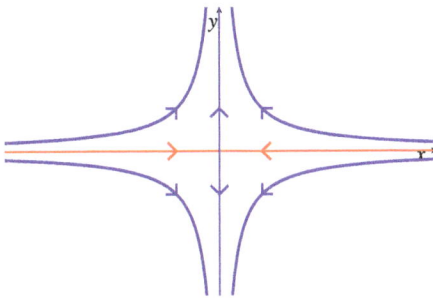

Figure 10.2: Saddle.

Case 2. This is similar to Case 1. Letting $\lambda = \lambda_1 = \lambda_2 \neq 0$ we find $x = c_1 e^{\lambda t}, y = c_2 e^{\lambda t}$. If either $c_1 = 0$ or $c_2 = 0$, we are led to solutions on the axes; otherwise, setting $c = \frac{c_2}{c_1}$, we find

$$y = cx$$

and the phase space portrait of the trajectories consists of straight lines passing through the singular point $(0,0)$. The origin is still a node, stable if $\lambda < 0$, unstable if $\lambda > 0$.

Case 3. Solving the second equation of $u' = Ju$, namely $y' = \lambda y$, we find $y = c_2 e^{\lambda t}$. Then the first equation, namely $x' = \lambda x + y$ becomes $x' = \lambda x + c_2 e^{\lambda t}$, which is a first order linear non-homogeneous equation whose solution is given by $x = (c_1 + c_2 t)e^{\lambda t}$. We see, as one might expect, that if $\lambda < 0$ we have asymptotic stability, otherwise instability. The origin is still called a stable or unstable node (see Fig. 10.3).

Here we can find the solution as $x = x(y)$. Let us assume that $c_2 \neq 0$, otherwise, $y = 0$ and $x = c_1 e^{\lambda t}$. Then

$$e^{\lambda t} = \frac{y}{c_2} \Rightarrow t = \frac{1}{\lambda} \cdot \ln\left|\frac{y}{c_2}\right|.$$

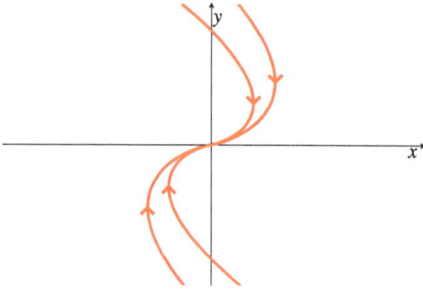

Figure 10.3: Stable node when $\lambda_1 = \lambda_2 < 0$ and $J = \begin{pmatrix} \lambda_1 & 1 \\ 0 & \lambda_1 \end{pmatrix}$.

Thus

$$x = (c_1 + c_2 t)e^{\lambda t} = \left(c_1 + c_2 \frac{1}{\lambda} \cdot \ln\left|\frac{y}{c_2}\right| \right) \cdot \frac{y}{c_2}$$

Case 4. The system becomes

$$\begin{cases} x' = ax - \beta y, \\ y' = \beta x + ay. \end{cases}$$

It is convenient to use polar coordinates $x = r\cos\theta$, $y = r\sin\theta$. We have $x' = r'\cos\theta - r\sin\theta \cdot \theta'$ and $y' = r'\sin\theta + r\cos\theta \cdot \theta'$. Then, substituting into the system we find

$$\begin{cases} r'\cos\theta - r\sin\theta \cdot \theta' = ar\cos\theta - \beta r\sin\theta, \\ r'\sin\theta + r\cos\theta \cdot \theta' = \beta r\cos\theta + ar\sin\theta. \end{cases}$$

Multiplying the first equation by $\cos\theta$ and the second by $\sin\theta$ and summing up, we find, using simple algebra, that $r'(\cos^2\theta + \sin^2\theta) = ar(\cos^2\theta + \sin^2\theta) \Rightarrow r' = ar$. Similarly, multiplying the first equation by $-\sin\theta$ and the second by $\cos\theta$ and summing up, we find $r(\cos^2\theta + \sin^2\theta) \cdot \theta' = \beta r(\cos^2\theta + \sin^2\theta) \Rightarrow \theta' = \beta$. In other words, the given system splits into the independent equations

$$r' = ar, \quad \theta' = \beta,$$

yielding $r = c_1 e^{at}$ and $\theta = \beta t + c_2$, which is a spiral. Clearly if $a < 0$ we have asymptotic stability since $r(t) \to 0$ as $t \to +\infty$, whereas if $a > 0$ we have instability since $r(t) \to +\infty$ as $t \to +\infty$. Of course, these results agree with those found in Theorems 10.2 and 10.3. In this case the origin is called (stable or unstable) *focus or spiral*. See Figs. 10.4–10.5.

If $a = 0$ we find $r = c_1$. If $c_1 > 0$ they are just circles centered at the origin and radius $\sqrt{c_1}$. It follows that when $a = 0$ the origin is stable but not asymptotically stable. In this case the origin is called a *center*. See Fig. 10.6 (the direction of the arrows depends on the sign of β).

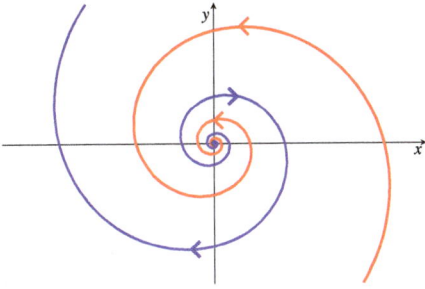

Figure 10.4: Stable (red) and unstable (blue) focus ($\beta > 0$).

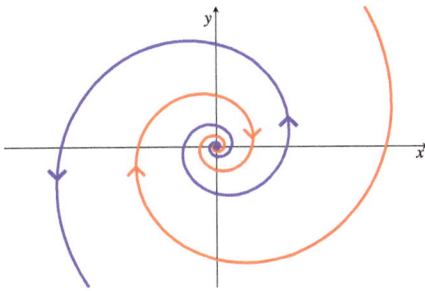

Figure 10.5: Stable (red) and unstable (blue) focus ($\beta < 0$).

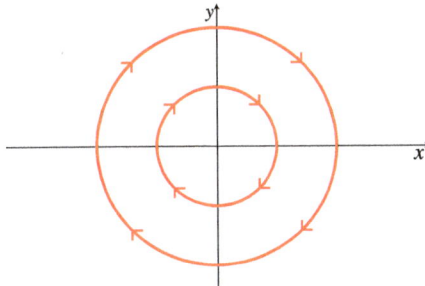

Figure 10.6: Center with $\beta > 0$.

Summarizing:

1. A has real eigenvalues $\lambda_{1,2}$:

$$\lambda_1 \cdot \lambda_2 > 0 \Rightarrow \text{Node} \begin{cases} \lambda_1, \lambda_2 < 0 & \text{asymptotically stable,} \\ \lambda_1, \lambda_2 > 0 & \text{unstable,} \end{cases}$$

$$\lambda_1 \cdot \lambda_2 < 0 \Rightarrow \text{Saddle (unstable).}$$

2. A has a complex conjugated eigenvalue $\alpha \pm i\beta$:

$$\alpha \neq 0 \Rightarrow \text{Focus} \quad \begin{cases} \alpha < 0 & \text{asymptotically stable,} \\ \alpha > 0 & \text{unstable,} \end{cases}$$

$$\alpha = 0 \Rightarrow \text{Center (stable).}$$

Remark 10.1. We know that the stability/instability of the origin for $u' = Au$ is the same as that of $v' = Jv$, where $J = X^{-1}AX$ is the canonical normal form of A. The behavior of u is the same as the behavior of v, up to the change of coordinates $u = Xv$.

For example, if $(0,0)$ is a stable center, the solutions might be ellipses rather than circles. Or, if $(0,0)$ is a saddle, the solutions might be hyperbolas as the one plotted in Fig. 10.7. With respect to the graph reported in Fig. 10.2, the hyperbolas and the asymptotes are rotated by an angle of $\pi/4$.

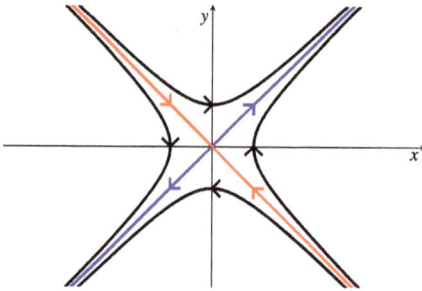

Figure 10.7: The stable (in red) and the unstable (in blue) manifold.

The asymptotes are called (linear) *stable and unstable manifolds* of $u' = Au$. In particular, a point p belongs to the stable manifold if and only if $\bar{u}(t;p) \to 0$ as $t \to +\infty$.

10.2.2 Stability of linear homogeneous equations with constant coefficients

We say that the trivial solution $x = 0$ of the linear homogeneous equation $L[x] = x^{(n)} + a_{n-1}x^{(n-1)} + \cdots + a_1x' + a_0x = 0$ is stable, resp. asymptotically stable, unstable, if 0 is stable, resp. asymptotically stable, unstable, for the equivalent system $\bar{x}' = A\bar{x}$, where

$$\begin{cases} x_1 = x_2, \\ x_2' = x_3, \\ \cdots \\ x_n' = -a_{n-1}x^{(n-1)} - \cdots - a_1x' - a_0x. \end{cases}$$

In particular $x = 0$ is asymptotically stable if

$$\lim_{t \to +\infty} x(t,p) = \lim_{t \to +\infty} x'(t,p) = \cdots = \lim_{t \to +\infty} x^{(n-1)}(t,p) = 0$$

for the initial condition of $p = (p_0, p_1, \ldots, p_{n-1})$ being sufficiently close to $p^* = 0$.

Theorem 10.4. *If all the roots of the characteristic equation of $L[x] = 0$ have negative real parts, then $x = 0$ is asymptotically stable. If at least one root has a positive real part, then $x = 0$ is unstable.*

Proof. We know that the solutions of $L[x] = 0$ are a superposition of terms as $t^m \cdot e^{\lambda t}$. Then it is clear that we have asymptotic stability provided all the eigenvalues have negative real part, whereas we have instability if one eigenvalue has positive real part.

An alternative proof may be given by noticing that the roots of the characteristic equation of $L[x] = 0$ are the eigenvalues of the matrix A of the equivalent first order system. Then the result follows from Theorems 10.2 and 10.3. □

Example 10.2. Let us consider the equilibrium $x = 0$ of the damped harmonic oscillator $x'' + kx' + \omega^2 x = 0$ with $\omega \neq 0$. Since the roots of the characteristic equations are given by $\lambda_{1,2} = \frac{1}{2} \cdot (-k \pm \sqrt{k^2 - 4\omega^2})$, we infer:

1. if $k > 0$ we have asymptotic stability. Actually, either
   (1a) $k^2 - 4\omega^2 \geq 0$, in which case the eigenvalues are real and negative (repeated if $k^2 - 4\omega^2 = 0$) and we have a node, or
   (1b) $k^2 - 4\omega^2 < 0$, in which case the eigenvalues are complex conjugate with negative real part and we have a focus;
2. if $k = 0$ then $\lambda = \pm i\omega$ and we have a center;
3. if $k < 0$ we have instability. □

10.3 Stability of conservative systems

A system of first order autonomous equations

$$x_i' = f_i(x_1, \ldots, x_n), \quad i = 1, \ldots, n, \tag{10.4}$$

is said to be *conservative* if there exists a real valued function $E(x_1, \ldots, x_n)$ such that $E(u_1(t), \ldots, u_n(t)) \equiv$ const. for every solution $u_1(t), \ldots, u_n(t)$ of (10.4). The function E is called the *energy* of the system.

For example the 2D system

$$\begin{cases} x' = y, \\ y' = -U'(x), \end{cases}$$

corresponding to the second order equation $x'' + U'(x) = 0$, is conservative since $E(x,y) = \frac{1}{2}y^2 + U(x)$ is constant along its solutions, as we saw in Chapter 9. More generally, any gradient system as $x_i'' + U_{x_i}(x_1, \ldots, x_n) = 0$ or, using vector notation, $\bar{x}'' + \nabla V(\bar{x}) = 0$ is conservative with energy $E(x_1, \ldots, x_n) = \frac{1}{2}\|\bar{x}'\|^2 + U(\bar{x})$.

Another example of a conservative system is the Lotka–Volterra system discussed in Chapter 9.

Theorem 10.5. *Let $p^* = 0$ be an equilibrium of the conservative system* (10.4). *Suppose that 0 is a strict local minimum of the energy E, namely $\exists r > 0$ such that $E(\bar{x}) > E(p^*)$, $\forall \bar{x} \in B(r)$. Then $p^* = 0$ is a stable equilibrium (but not asymptotically stable).*

Proof. We first introduce some notations. For $p \in \mathbb{R}^n$, we set $S_p = \{\bar{x} \in B(r) : E(\bar{x}) < E(p)\}$.

Notice that if $E(p) > E(p^*)$ is sufficiently small then S_p is a non-empty bounded neighborhood of 0, since 0 is a strict local minimum of E. Moreover, if $p_1 < p_2$ and $E(p_i) > E(0)$ then $S(p_1) \subset S(p_2)$. It follows that, as $p_i \to p^*$, the sets S_{p_i} shrink to the equilibrium 0.

Moreover, by the continuity of E, given $\epsilon > 0$ there exists $\delta > 0$ such that $S_p \subset B(\epsilon)$ provided $p \in B(\delta)$; see Fig. 10.8.

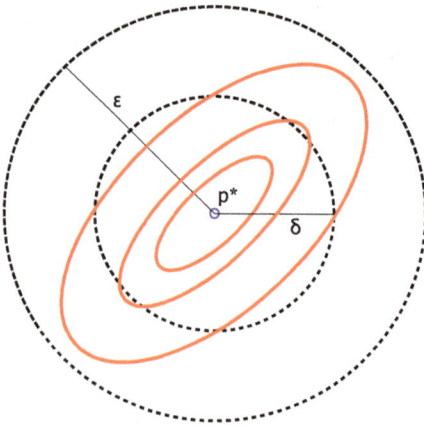

Figure 10.8: Plot of the sets $S_{p_1} \subset S(p_2) \subset S(p_3) \subset B(\epsilon)$.

Since E is constant on solutions, we find that $E(\bar{u}(t; p)) = E(\bar{u}(0; p)) = E(p)$ for all $t \in \mathbb{R}$. Thus, if $\|p\| < \delta$ it follows that $\bar{u}(t; p)$ remains in the ball $B(\epsilon)$ and this implies stability. Notice that 0 is not asymptotically stable since, if $p \neq 0$, $\bar{u}(t; p)$ does not tend to 0 since $E(\bar{u}(t; p)) = E(p)$. □

Example 10.3. The nontrivial equilibrium $x_0 = \frac{c}{d}, y_0 = \frac{a}{b}$ of the Lotka–Volterra system

$$\begin{cases} x' = x(a - by), \\ y' = y(dx - c), \end{cases}$$

is stable. Actually (x_0, y_0) is the proper minimum of the energy $h(x, y) = dx + by - c \ln x - a \ln y$. Notice that the trivial equilibrium $(0,0)$ is unstable. For any initial value $p = (p_0, 0)$, $p_0 > 0$, we find that the solution is given by $x(t) = p e^{at}$, $y(t) = 0$. Since $a > 0$ we have $x(t) \to +\infty$ as $t \to +\infty$ and thus we have instability. □

Remark 10.2. Theorem 10.5 could also be proved using more general arguments, known as the Lyapunov direct methods. We will give an outline of this topic in the next section.

From Theorem 10.5 we can infer the following stability criterion for gradient systems.

Theorem 10.6 (Dirichlet–Lagrange stability criterion). *Let $p^* = 0$ be an equilibrium of the gradient system*

$$\overline{x}'' + \nabla U(\overline{x}) = 0.$$

If 0 is a strict local minimum of the potential U, then 0 is a stable equilibrium.

Proof. If $p^* = 0$ is a strict local minimum of the potential U, then 0 is a strict local minimum of the energy $E = \frac{1}{2}\|\overline{x}'\|^2 + U(\overline{x})$. Then the result follows immediately from Theorem 10.5. □

Example 10.4. The trivial solution $x(t) \equiv 0$ of the pendulum equation $x'' + \sin x = 0$ is stable since $U(x) = -\cos x$ has a proper local minimum at $x = 0$.

Similarly, the equilibrium $x = 0$ of the nonlinear harmonic oscillator $x'' + \omega^2 x - x^3 = 0$, $\omega \neq 0$, is stable since its potential $V(x) = \frac{1}{2}\omega^2 x^2 - \frac{1}{4}x^4$ has a strict local minimum at $x = 0$. □

10.4 The Lyapunov direct method

Given the system $\overline{x}' = f(\overline{x})$, namely

$$x_i' = f_i(x_1, \ldots, x_n), \quad i = 1, \ldots, n, \tag{10.5}$$

the Lyapunov direct method is a general procedure to establish the stability of an equilibrium p^* of (10.5). For the sake of simplicity we will assume that $f = (f_1, \ldots, f_n)$ is defined on all of $\mathbb{R}^n$.

Let $p^* = 0$ be an equilibrium of $\overline{x}' = f(\overline{x})$, i. e., $f(0) = 0$ (we use 0 to denote both, $0 \in \mathbb{R}$ and $0 = (0, \ldots, 0) \in \mathbb{R}^n$: the context will make the intended meaning clear).

Definition 10.3. $V \in C^1(\mathbb{R}^n, \mathbb{R})$ is said to be a Lyapunov function for $\overline{x}' = f(\overline{x})$ if
(V1) $V(\overline{x}) > V(0)$, $\forall \overline{x} \in \mathbb{R}^n$, $\overline{x} \neq 0$.
(V2) $(\nabla V(\overline{x}) \mid f(\overline{x})) = \sum_1^n V_{x_i}(\overline{x}) f_i(\overline{x}) \leq 0$, $\forall \overline{x} \in \mathbb{R}^n$.

Example 10.5. $V(\bar{x}) = \|\bar{x}\|^2$ is a Lyapunov function for $\bar{u}' = f(\bar{u})$ provided $(\bar{x} \mid f(\bar{x})) \leq 0$, $\forall \bar{x} \in \mathbb{R}^n$. Actually, $(V1)$ clearly holds with $\Omega = \mathbb{R}^n$. As for $(V2)$, since $\nabla V(\bar{x}) = 2\bar{x}$, we find

$$(\nabla V(\bar{x}) \mid f(\bar{x})) = 2(\bar{x} \mid f(\bar{x})).$$

In particular, dealing with the linear system $f(\bar{x}) = A\bar{x}$ we see that $V(\bar{x}) = \|\bar{x}\|^2$ is a Lyapunov function provided the quadratic form $(A\bar{x} \mid \bar{x})$ is semidefinite negative. Or, we can also take $V(\bar{x}) = (A\bar{x} \mid \bar{x})$. $\qquad\square$

Remark 10.3. We have taken V defined on all of $\mathbb{R}^n$ to simplify notations and proofs. But we could as well assume V is defined on a neighborhood of $p^* = 0$. Since stability is a local property, this does not make any difference, whereas, in the case of asymptotic stability, this changes the "basin of stability" as we will point out later on.

In the sequel we will set

$$(\nabla V(\bar{x}) \mid f(\bar{x})) = \dot{V}(\bar{x}).$$

Theorem 10.7.
1. *If $\bar{x}' = f(\bar{x})$ has a Lyapunov function then the equilibrium $p^* = 0$ is stable.*
2. *If we strengthen $(V1 - 2)$ by requiring*
   $(V1')$ $(V1)$ holds and $V(\bar{x}) \to +\infty$ as $\|\bar{x}\| \to +\infty$;
   $(V2')$ $\dot{V}(\bar{u}) < 0$, $\forall \bar{u} \neq 0$;
   then p^* is asymptotically stable in the sense that $\lim_{t \to +\infty} \bar{u}(t; p) = 0$ for all $p \in \mathbb{R}^n$.

Proof. The proof of (1) is similar to the proof of Theorem 10.5 with E replaced by V and is left as an exercise.

Let us prove (2). First of all let us show that the solutions $\bar{u}(t; p)$ are bounded for all p. For this we notice that $(V2')$ yields

$$\frac{d}{dt} V(\bar{u}(t;p)) = \sum_{1}^{n} V_{x_i}(\bar{u}(t;p)) u_i'(t) = (\nabla V(\bar{u}(t;p)) \mid \bar{u}'(t;p)) = \dot{V}(\bar{u}(t;p)) < 0, \quad \forall t > 0.$$

Thus $t \mapsto V(\bar{u}(t;p))$ is a decreasing function and hence $V(\bar{u}(t;p)) < V(\bar{u}(0;p)) = V(p)$. From this and $(V1')$ it follows that $\exists C > 0$ such that $\|\bar{u}(t;p)\| \leq C$ for all $t \geq 0$, as claimed.

Now, we want to show that $\bar{u}(t;p) \to 0$, namely that $\|\bar{u}(t;p)\| \to 0$, as $t \to +\infty$. Arguing by contradiction, suppose that there exists $c > 0$ such that $\|\bar{u}(t;p)\| \geq c$ for all t. Thus one has $c \leq \|\bar{u}(t;p)\| \leq C$ for all t. Since $\bar{u}(t;p)$ ranges in the compact annulus $c \leq \|\bar{x}\| \leq C$ and $\dot{V} < 0$,

$$-L = \min\{\dot{V}(\bar{x}) : c \leq \|\bar{x}\| \leq C\} < 0.$$

This implies

$$V(\bar{u}(T;p)) = V(\bar{u}(0;p)) + \int_0^T \dot{V}(\bar{u}(t;p))\,dt \le V(p) - LT.$$

Taking $T > \frac{V(p)}{L}$ we find $V(\bar{u}(T;p)) < 0$, in contradiction with $(V1)$. $\qquad\square$

Roughly, referring to Fig. 10.9, we can give the following suggestive explanation. From the geometrical point of view, the condition $(V'(p) \mid f(p)) < 0$ means that the vector field $f(p)$ is pointing inward the closed surface $\{V(\bar{x}) = V(p)\}$, since $V'(p)$ is orthogonal to the surface. This implies that the trajectory of $\bar{x}' = f(\bar{x})$ with initial point p enters the region $\{V(\bar{x}) < V(p)\}$. Since this holds for every surface $\{V(\bar{x}) = V(p)\}$ and the surfaces shrink to 0, the trajectory is forced to converge to 0.

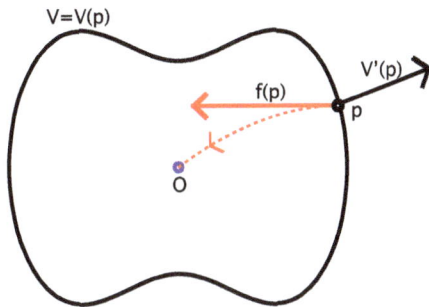

Figure 10.9: Asymptotic stability as in Theorem 10.7.

Remark 10.4. If $(V1'-2')$ hold in a neighborhood N of the equilibrium $p^* = 0$, then the preceding result has to be modified by saying that $\bar{u}(t;p) \to 0 \; \forall p \in N$. The set N is called the *basin of attraction* or *basin of stability* and $p^* = 0$ is referred to as a *locally asymptotically stable equilibrium*.

The counterpart of Theorem 10.7 is the following instability criterion.

Theorem 10.8. *Suppose that there exists $W \in C^1(\mathbb{R}^n, \mathbb{R})$ such that*
$(W1)$ $W(0) = 0$;
$(W2)$ $\dot{W}(\bar{x}) := (\nabla W(\bar{x}) \mid f(\bar{x}))$ *is either positive or negative, defined for all $\bar{x} \ne 0$;*
$(W3)$ $\exists \bar{x}_k \to 0$ *such that* $W(\bar{x}_k) \cdot \dot{W}(\bar{x}_k) > 0$.

Then the equilibrium $p^* = 0$ is unstable for $\bar{x}' = f(\bar{x})$.

For example, $W(\bar{x}) = -\|\bar{x}\|^2$ satisfies $(W1-2-3)$ provided $(f(\bar{x}) \mid \bar{x}) > 0$ for $\bar{x} \ne 0$. Clearly $(W1)$ is satisfied. Moreover, $(W2-3)$ hold since $\dot{W}(\bar{x}) = -2(f(\bar{x}) \mid \bar{x}) < 0$ for $\bar{x} \ne 0$ and $W(\bar{x}) \cdot \dot{W}(\bar{x}) = 2(f(\bar{x}) \mid \bar{x}) \cdot \|\bar{x}\|^2 > 0$ for $\bar{x} \ne 0$.

For a broader discussion of the Lyapunov direct method we refer the reader, e. g., to the book by J. LaSalle–S. Lefschetz, *Stability by Liapunov's Direct Method with Appli-*

cations, *VII + 134 S., New York/London, 1961, Academic Press* or the one by H. Amann, *Ordinary Differential Equations, De Gruyter Studies in Mathematics 13, 1990.*

10.5 Stability by linearization

Let us again consider the nonlinear autonomous system

$$x_i = f_i(x_1,\ldots,x_n), \quad i = 1,\ldots,n \tag{10.5}$$

where $f_i(0,\ldots,0) = 0$, $\forall i = 1,\ldots,n$. Developing $f = (f_1,\ldots,f_n)$ in Taylor expansion we can write

$$f(\overline{x}) = A\overline{x} + o(\|\overline{x}\|), {}^3$$

where $A = \nabla f(\overline{x})$.

Theorem 10.9. *Suppose that all the eigenvalues of A have negative real part. Then 0 is locally asymptotically stable.*

To prove the theorem, we need the following lemma

Lemma 10.1. *If all the eigenvalues of A have negative real parts, then $(A\overline{x} \mid \overline{x}) < 0$.*

Proof. Consider J, the Jordan normal form of A. Since $AX = XJ$ for some nonsingular matrix X, we find

$$(A\overline{x} \mid \overline{x}) = (XJX^{-1}\overline{x} \mid \overline{x}) = (J\underbrace{X^{-1}\overline{x}}_{\overline{y}} \mid \underbrace{X^{-1}\overline{x}}_{\overline{y}}) = (J\overline{y} \mid \overline{y}).$$

Hence it suffices to prove that $(J\overline{y} \mid \overline{y}) < 0$. For simplicity, we limit ourselves to deal with the case $n = 2$. There are three possible cases, depending on the eigenvalues $\lambda_{1,2}$ of A:

(1) If $\lambda_1 \neq \lambda_2$ then $J = \left(\begin{smallmatrix}\lambda_1 & 0\\0 & \lambda_2\end{smallmatrix}\right)$. Letting $y = \left(\begin{smallmatrix}y_1\\y_2\end{smallmatrix}\right)$ we infer

$$J\overline{y} = \begin{pmatrix}\lambda_1 y_1\\\lambda_2 y_2\end{pmatrix} \implies (J\overline{y} \mid \overline{y}) = \lambda_1 y_1^2 + \lambda_2 y_2^2 < 0,$$

since λ_1 and λ_2 are negative.

(2) If $\lambda_1 = \lambda_2 = \lambda < 0$ then either $J = \left(\begin{smallmatrix}\lambda & 0\\0 & \lambda\end{smallmatrix}\right)$ or $J = \left(\begin{smallmatrix}\lambda & 1\\0 & \lambda\end{smallmatrix}\right)$. In the former case one has $(J\overline{y} \mid \overline{y}) = \lambda\|\overline{y}\|^2 < 0$. In the latter case we find

$$J\overline{y} = \begin{pmatrix}\lambda y_1 + y_2\\\lambda y_2\end{pmatrix} \implies (J\overline{y} \mid \overline{y}) = \lambda y_1^2 + y_1 y_2 + \lambda y_2^2 < 0,$$

---

3 Recall that $o(\|\overline{x}\|)$ means that $\lim_{\|\overline{x}\|\to 0} \frac{\|o(\overline{x})\|}{\|\overline{x}\|} = 0$.

since the quadratic form $\lambda y_1^2 + y_1 y_2 + \lambda y_2^2$ is negative definite for $\lambda < 0$.

(3) If $\lambda = \alpha \pm i\beta$ then $J = \left(\begin{smallmatrix} \alpha & -\beta \\ \beta & \alpha \end{smallmatrix} \right)$ and we find

$$J\bar{y} = \left(\begin{array}{c} \alpha y_1 - \beta y_2 \\ \beta y_1 + \alpha y_2 \end{array} \right) \quad \Longrightarrow \quad (J\bar{y} \mid \bar{y}) = \alpha y_1^2 - \beta y_1 y_2 + \beta y_1 y_2 + \alpha y_2^2 = \alpha \|oy\|^2 < 0,$$

since $\alpha < 0$. $\qquad \square$

Proof of Theorem 10.9. Since all the eigenvalues of A have negative real parts, the preceding lemma yields $(A\bar{x} \mid \bar{x}) < 0$. Moreover, one has $(f(\bar{x}) \mid \bar{x}) = (A\bar{x} \mid \bar{x}) + o(\|\bar{x}\|^2)$ and hence $\exists \epsilon > 0$ such that $(f(\bar{x}) \mid \bar{x}) < 0, \forall \bar{x} \in B(\epsilon)$. Then, as remarked in Example 10.5, $V(\bar{x}) = (f(\bar{x}) \mid \bar{x})$ is a Lyapunov function for (10.5) in $B(\epsilon)$. Thus $\bar{x} = 0$ is locally asymptotically stable with basin of attraction $B(\epsilon)$. $\qquad \square$

Example 10.6. Given the system

$$\begin{cases} x' = -x + y^2, \\ y' = -y + x^2, \end{cases}$$

the origin $(0,0)$ is stable since the linearized system

$$\begin{cases} x' = -x, \\ y' = -y, \end{cases}$$

has a double negative eigenvalue $\lambda = -1$. $\qquad \square$

The counterpart of Theorem 10.9 is

Theorem 10.10. *Suppose that at least one eigenvalue of A has a positive real part. Then 0 is unstable.*

Example 10.7. Consider the Lotka–Volterra system

$$\begin{cases} x' = x(a - by), \\ y' = y(dx - c), \end{cases}$$

and show that the origin $(0,0)$ is unstable. It suffices to notice that the linearized system

$$\begin{cases} x' = ax, \\ y' = -cy, \end{cases}$$

has a positive eigenvalue, namely $\lambda = a$, and apply Theorem 10.10. $\qquad \square$

Theorem 10.9 rules out the case in which some eigenvalue of A is zero or has zero real part. The following example shows that in such a case we cannot deduce the stability of the equilibrium from the linearized system.

Example 10.8. Consider the system

$$\begin{cases} x' = y + \epsilon x(x^2 + y^2), \\ y' = -x + \epsilon y(x^2 + y^2). \end{cases}$$

Setting $\rho(t) = x^2(t) + y^2(t)$, we get $\rho'(t) = 2(xx' + yy')$. Multiplying the first equation by x and the second by y and summing up, we find $xx' + yy' = \epsilon(x^2 + y^2)^2$, yielding

$$\rho'(t) = 2\epsilon\rho^2(t).$$

Thus $\rho(t)$ is increasing if $\epsilon > 0$ and decreasing if $\epsilon < 0$. In the former case we have instability. In the latter case $\rho(t) \to L$ and $\rho'(t) \to 0$ as $t \to +\infty$. From the equation we infer that $L = 0$. Thus $x^2(t) + y^2(t) \to 0$ as $t \to +\infty$, which implies that $x(t) \to 0$ and $y(t) \to 0$, and hence we have asymptotic stability.

Here the linearization at the equilibrium $(0, 0)$ is

$$\begin{cases} x' = y, \\ y' = -x, \end{cases}$$

with matrix $A = \begin{pmatrix} 0 & 1 \\ -1 & 0 \end{pmatrix}$ whose eigenvalues are $\pm i$. $\quad\square$

Dealing with a nonlinear equation, we can either linearize the equivalent system at the equilibrium, or we can directly study the linearized equation. For example, consider the nonlinear second order equation

$$x'' + g(x)x' + f(x) = 0$$

with $f(0) = 0$. Since $g(x) = g(0) + g'(0)x + o(|x|)$ and $f(x) = f'(0)x + o(|x|)$ we find that the linearized equation at the equilibrium $x = 0$ is given by

$$y'' + g(0)y' + f'(0)y = 0$$

where we prefer to change the name of the dependent variable in the linearized equation. If the two solutions of the characteristic polynomial corresponding to this equation are negative (or if their real parts are negative), the trivial solution $x = 0$ is locally asymptotically stable. On the other hand, if at least one of them is positive (or it has positive real part), then $x = 0$ is unstable.

Example 10.9 (van der Pol equation). Consider the van der Pol equation

$$x'' - \mu(1 - x^2)x' + x = 0,$$

studied in Chapter 9. This equation is of the type discussed earlier, with $g(x) = -\mu(1 - x^2)$ and $f(x) = x$. Since $g(0) = -\mu$ and $f'(0) = 1$, then the linearization at $x = 0$ is given by $y'' - \mu y' + y = 0$. The roots of the characteristic equation corresponding to this linear equation are $\lambda = \frac{1}{2}[\mu \pm \sqrt{\mu^2 - 4}]$. Thus $x = 0$ is unstable for $\mu > 0$, whereas it is stable for $\mu < 0$.

Let us verify that we find the same result, as before, for the equivalent Lienard system

$$\begin{cases} x' = y - \mu(x - \frac{1}{3}x^3), \\ y' = -x. \end{cases}$$

The linearized problem at the origin is

$$\begin{cases} u' = v - \mu u, \\ v' = -u. \end{cases}$$

The matrix of this linear system is

$$A = \begin{pmatrix} -\mu & 1 \\ -1 & 0 \end{pmatrix},$$

whose eigenvalues are $\lambda = \frac{1}{2}[\mu \pm \sqrt{\mu^2 - 4}]$, as before. ☐

10.6 Further remarks

10.6.1 Stable and unstable limit cycles

In the previous chapter we introduced the notion of *limit cycle* of a planar system. If a limit cycle is such that all the nearby trajectories of the system approach the cycle from inside and outside we will say that the limit cycle is a *stable limit cycle*; see Fig. 10.10. Otherwise, it is unstable.

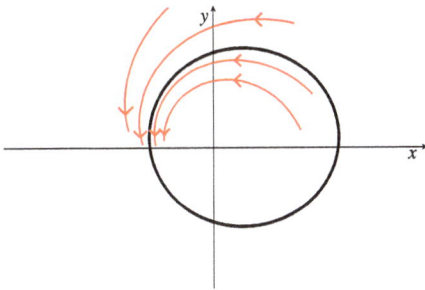

Figure 10.10: A stable limit cycle.

A system where a stable limit cycle arises is the Liénard planar system

$$\begin{cases} x' = y - \mu(x - \frac{1}{3}x^3), \\ y' = -x, \end{cases}$$

with $\mu > 0$.

10.6.2 Hyperbolic equilibria: stable and unstable manifolds

Theorem 10.10 can be made more precise when $A = \nabla f(0)$ has no eigenvalue with zero real part. An equilibrium with this property is called a *hyperbolic equilibrium*. If $n = 2$ a hyperbolic equilibrium is a saddle.

Theorem 10.11. *Suppose that the matrix A has k eigenvalues with negative real part and $n - k$ eigenvalues with positive real part. Then there exist a neighborhood $N \subseteq \mathbb{R}^n$ of 0, a surface $M^s \subset N$, called a stable manifold, and a surface $M^u \subset N$, called an unstable manifold, such that:*

1. *M^s, M^u are invariant, namely $p \in M^s \implies u(t;p) \in M^s, \forall t$, resp. $p \in M^u \implies u(t;p) \in M^u, \forall t$;*
2. *$\dim(M^s) = k$ and $p \in M^s \iff \lim_{t \to +\infty} u(t;p) = 0$;*
3. *$\dim(M^u) = k$ and $p \in M^s \iff \lim_{t \to -\infty} u(t;p) = 0$.*

Furthermore, if $N = \mathbb{R}^n$ then $\forall p \in \mathbb{R}^n \setminus M^u$ the distance from $\bar{u}(t;p)$ to M^u tends to 0 as $t \to +\infty$; see Fig. 10.11.

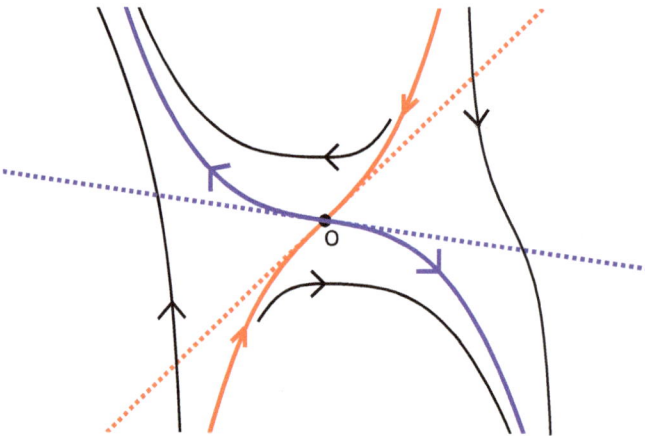

Figure 10.11: A hyperbolic equilibrium with the stable (in red) and the unstable (in blue) manifold.

If $f(\bar{x}) = A\bar{x}$ the surfaces M^s, M^u are linear manifolds (if $n = 2$ they are simply the linear stable and unstable manifolds discussed in Remark 10.1). Furthermore, if $f(\bar{x}) = A\bar{x} + o(\|\bar{x}\|)$, then one can show that M^s, M^u are close to the linear stable and unstable manifolds of the linearized system $\bar{x}' = A\bar{x}$.

10.6.3 Change of stability: bifurcation of equilibria

Bifurcation is a phenomenon that is concerned with equations depending on a parameter $\lambda \in \mathbb{R}$ that possess a trivial solution $x = 0$ for every λ. A bifurcation point is a value λ^* where there is, at least locally, a branching off of the nontrivial solutions.

Dealing with equilibria, this branching arises typically when the stability of $x = 0$ changes when λ crosses λ^*. Two examples will illustrate this phenomenon.

(1) Consider the logistic equation

$$x' = \lambda x - x^2,$$

discussed in Chapter 4, section 4.2. The linearization at the equilibrium $x = 0$ is $y' = \lambda y$. Therefore, according to Theorem 10.9 if $\lambda < 0$ the equilibrium $x = 0$ is asymptotically stable whereas if $\lambda > 0$ Theorem 10.10 implies that 0 is unstable. A second equilibrium of the equation is $x_\lambda = \lambda$. The linearization at x_λ is given by $y' = \lambda y - 2x_\lambda y$. For $x_\lambda = \lambda$ we find $y' = \lambda y - 2\lambda y = -\lambda y$. It follows that x_λ is stable for $\lambda > 0$ and unstable for $\lambda < 0$.

In Fig. 10.12 the asymptotic behavior of solutions with initial condition p is indicated by arrows.

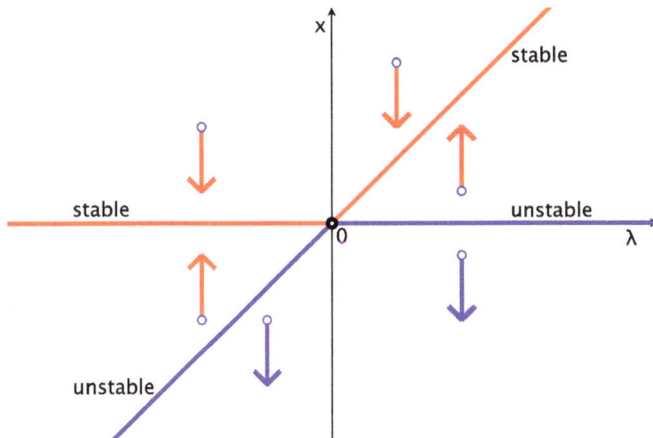

Figure 10.12: Bifurcation for $x' = \lambda x - x^2$.

For any fixed $\lambda < 0$:
1. the solution with initial condition $p > -1$ converges to the stable equilibrium $x = 0$;
2. the solution with initial condition $p < -1$ diverges;
3. if $p = -1$ the solution is simply the constant λ.

On the other hand, for any fixed $\lambda > 0$,
1. the solution with initial condition $p > 0$ converges to the stable equilibrium $x = \lambda$;
2. the solution with initial condition $p < 0$ diverges;
3. if $p = 0$ the solution is simply the constant $x = 0$.

(2) Consider the equation

$$x' = \lambda x - x^3$$

whose equilibria are $x = 0$ and $x_\lambda = \pm\sqrt{\lambda}, \lambda > 0$. According to Theorem 10.9 if $\lambda < 0$ the equilibrium $x = 0$ is asymptotically stable, whereas if $\lambda > 0$ Theorem 10.10 implies that 0 is unstable. From $\lambda = 0, x = 0$ in the plane (λ, x) two new branches of equilibria appear $x_\lambda = \pm\sqrt{\lambda}, \lambda > 0$. Both new equilibria x_λ turn out to be asymptotically stable, since the linearized problem at x_λ is $y' = (\lambda - 3x_\lambda^2)y = -2\lambda y$ and λ is positive; see Fig. 10.13.

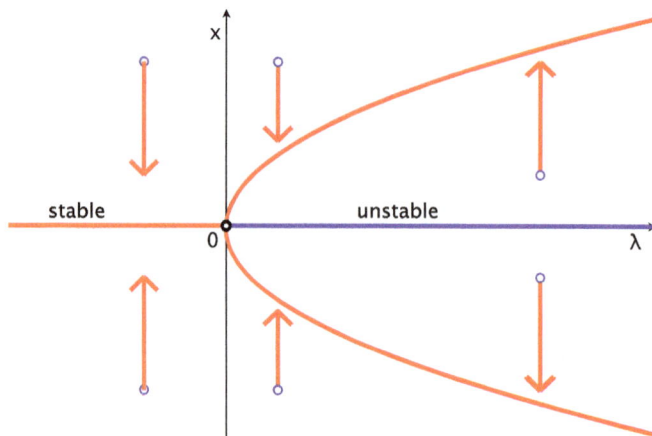

Figure 10.13: Supercritical pitchfork bifurcation for $x' = \lambda x - x^3$.

For any fixed $\lambda < 0$, the solution starting at any initial point p tends to the asymptotically stable equilibrium 0.

On the other hand, if $\lambda > 0$, the equilibrium $x = 0$ becomes unstable and one has:
1. the solution starting at any initial point $p > 0$ tends to the asymptotically stable equilibrium $\sqrt{\lambda}$;

2. the solution starting at any initial point $p < 0$ tends to the asymptotically stable equilibrium $-\sqrt{\lambda}$;
3. the solution starting at the initial point $p = 0$ is simply the constant $x = 0$.

This branching is called *pitchfork bifurcation*, the name being suggested by the form of the branches plotted in Fig. 10.13.

This pitchfork bifurcation is called *supercritical* to distinguish it from the *subcritical pitchfork bifurcation* that arises dealing with the equation

$$x' = \lambda x + x^3.$$

Notice that in this case the linearized equation at the nontrivial equilibria $x_\lambda = \pm\sqrt{-\lambda}$, $\lambda < 0$, is $y' = (\lambda + 3x_\lambda^2)y = -2\lambda y$. Since $\lambda < 0$, the branches $x = \pm\sqrt{-\lambda}$ are both unstable; see Fig. 10.14.

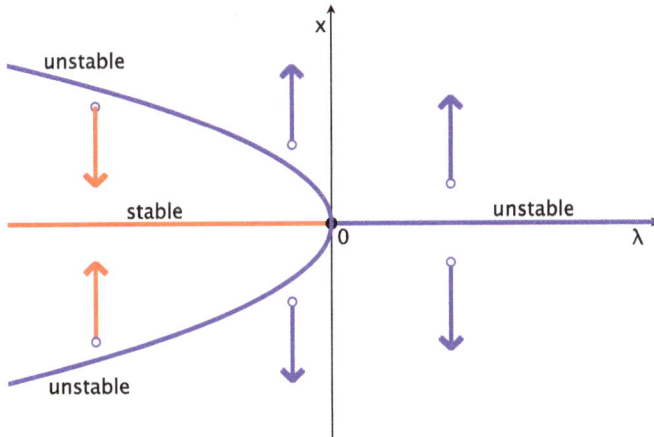

Figure 10.14: Subcritical pitchfork bifurcation for $x' = \lambda x + x^3$.

10.7 Exercises

1. Establish the stability of the equilibrium of
$$\begin{cases} x' = -x + y, \\ y' = -x + 2y. \end{cases}$$

2. Establish the stability of the equilibrium of
$$\begin{cases} x' = -y + ax, \\ y' = x + ay, \end{cases}$$

in terms of a.

3. Show that $\forall a \in \mathbb{R}$ the equilibrium of

$$\begin{cases} x' = 2x + y, \\ y' = -ax + 2y, \end{cases}$$

is unstable.

4. Given the 3×3 system

$$\begin{cases} x' = -x + y, \\ y' = -x - y, \\ z' = y - z, \end{cases}$$

show that $(0, 0, 0)$ is a stable equilibrium.

5. Establish the stability of the equilibrium of the 3×3 system

$$\begin{cases} x' = -x + y + z, \\ y' = -2y - z, \\ z' = az, \end{cases}$$

in terms of $a \neq 0$.

6. Establish the stability of the equilibrium $(0, 0)$ of the system

$$\begin{cases} x'' = -2x + y, \\ y'' = x. \end{cases}$$

7. Show that the origin is a saddle for the system

$$\begin{cases} x' = x - y, \\ y' = -y, \end{cases}$$

and find the stable manifolds.

8. Show that the origin is a center for the harmonic oscillator $x'' + \omega^2 x = 0$, $\omega \neq 0$.

9. Show that $x = 0$ is a stable solution of the equation $x''' + 3x'' + 3x' + x = 0$.

10. Using stability by linearization, show that the trivial solution of $x'' + 2x' + x + x^2 = 0$ is stable.

11. Show that if $a > 0$, the trivial solution $x = 0$ of $x''' - 2x' - ax = 0$ is unstable.

12. Show that $x = 0$ is an unstable solution of $x'''' - 4x'' + x = 0$.

13. Show that $x = 0$ is unstable for any equation of the form $x'''' + 2ax'' - b^2 x = 0$, $a, b \neq 0$.

14. Show that the solution $x = 0$ of $x'' + \omega^2 x \pm x^{2k} = 0$, $\omega \neq 0$, $k \in \mathbb{N}$, is stable provided $k \geq 1$.

15. Show that if $F'(0) = 0$ and $F''(0) > -1$ then $x = 0$ is stable relative to $x'' + x + F'(x) = 0$.

16. Study the stability of the equilibria $k\pi$ of the pendulum equation $x'' + \sin x = 0$.

17. Show that $(0, 0)$ is a stable equilibrium of

$$\begin{cases} x_1'' = -x_1 + 4x_1(x_1^2 - x_2^2), \\ x_2'' = -x_2 - 4x_2(x_1^2 - x_2^2). \end{cases}$$

    [Hint: Find $U(x_1, x_2)$ such that the system has the form $\overline{x}'' + \nabla U(\overline{x}) = 0$, $\overline{x} = (x_1, x_2)$.]

18. Using stability by linearization, show that if $g'(0) = 0$ then $x = 0$ is unstable for $x'' + g(x') - x = 0$.

19. Using the stability by linearization, show that the trivial solution of $x'' + xx' - x = 0$ is unstable.

20. Study the stability of the solution $x = 0$ of $x'' - \mu(1 - x^2)x' - x = 0$.

11 Series solutions for linear differential equations

In this chapter we discuss the method of solving linear differential equations by infinite power series. Although series solutions may be found for linear equations of higher order, we will focus, for simplicity, on one of the most important classes of equations, namely the second order linear homogeneous equations of the form

$$L[x] = a_2(t)x'' + a_1(t)x' + a_0(t)x = 0,$$

where a_0, a_1, a_2 are polynomials.

11.1 A brief overview of power series

Let us recall some basic properties of infinite series.

1. A power series

$$x(t) = \sum_{0}^{+\infty} A_k(t - t_0)^k$$

is convergent if $x(t) = \lim_{n \to \infty} S_n$, where $S_n = \sum_{0}^{n} A_k(t - t_0)^k$. It is clear that it always converges at $t = t_0$. It may or may not converge for other values of t.

2. The power series

$$x(t) = \sum_{0}^{+\infty} A_k(t - t_0)^k$$

is absolutely convergent at $t = t_1$ if

$$\sum_{0}^{+\infty} |A_k||t_1 - t_0|^k$$

is convergent. Absolute convergence implies convergence, but the converse is false, in general.

3. Every power series has a radius of convergence R, $R \geq 0$, such that if $R = 0$, it converges only at $t = t_0$. If $R > 0$, then it converges if $|t - t_0| < R$ and diverges if $|t - t_0| > R$; it can go either way at $|t - t_0| = R$, depending on the particular series. The ratio test is one of the most effective ways to determine R.

4. A convergent power series with radius of convergence R can be differentiated and integrated term by term, with the resulting series having the same radius of convergence R.

5. Sometimes it is convenient to shift indices of power series, which is easily done, as indicted by the example

$$\sum_{3}^{\infty} A_k(t - t_0)^k = \sum_{2}^{\infty} A_{k+1}(t - t_0)^{k+1} = \sum_{1}^{\infty} A_{k+2}(t - t_0)^{k+2} = \sum_{4}^{\infty} A_{k-1}(t - t_0)^{k-1}.$$

https://doi.org/10.1515/9783111185675-011

The points t_0 where $a_2(t_0) = 0$ are called *singular points* of the differential equation

$$a_2(t)x'' + a_1(t)x' + a_0(t)x = 0.$$

All other points are called *ordinary points.* We will treat the two cases separately.

11.2 Ordinary points

If the equation has no singular points, then we can divide by $a_2(t) \neq 0$ and obtain

$$x'' + \frac{a_1(t)}{a_2(t)}x' + \frac{a_0(t)}{a_2(t)}x = 0.$$

In order to find a power series solution of an equation $x'' + a_1(t)x' + a_0(t)x = 0$, we set

$$x(t) = \sum_0^{+\infty} A_k t^k$$

and determine the coefficients A_k as follows:
1. Differentiating the series term by term, formally, we have

$$x' = \sum_1^{+\infty} kA_k t^{k-1}, \quad x'' = \sum_2^{+\infty} k(k-1)A_k t^{k-2}.$$

2. Substituting these series into the equation yields

$$\sum_2^{+\infty} k(k-1)A_k t^{k-2} + a_1(t) \sum_1^{+\infty} kA_k t^{k-1} + a_0(t) \sum_0^{+\infty} A_k t^k = 0. \tag{11.1}$$

3. In order to find a recursive formula, first of all, the exponent of t should be the same in each sum, which can be accomplished by shifting indices. Then the lower limit on each sum should be the same, which may be accomplished by treating certain values of k separately.
4. We check the series for convergence, so that the preceding steps are not mere formality.

In our first example below we explain the method, giving complete details.

Example 11.1. As an introduction, we first explain, in detail, how to find the series solution to the first order linear equation

$$x' - \frac{1}{2}x = 0, \quad x(0) = 1.$$

Before we start, we want to point out that one can use any letter to describe the terms of series. In this example we use $c_p t^p$, in the examples below we use $A_k t^k$, or later we may use $a_n t^n$, etc. Let

$$x(t) = \sum_0^{+\infty} c_p t^p = c_0 + c_1 t + c_2 t^2 + c_3 t^3 + \cdots .$$

The initial value $x(0) = 1$ implies that $c_0 = 1$. Then

$$x'(t) - \frac{1}{2}x = \sum_1^{+\infty} pc_p t^{p-1} - \frac{1}{2}\sum_0^{+\infty} c_p t^p = 0.$$

Shifting indices in the first sum on the right side, we have

$$\sum_0^{+\infty} (p+1)c_{(p+1)} t^p - \sum_0^{+\infty} \frac{1}{2}c_p t^p = \sum_0^{+\infty} \left[(p+1)c_{(p+1)} - \frac{1}{2}c_p \right] t^p = 0.$$

Therefore, we must have $(p+1)c_{p+1} = \frac{1}{2}c_p$ from which we deduce the recursive formula

$$c_{p+1} = \frac{c_p}{2(p+1)}.$$

Calculating the first 4 terms, we have

$$c_1 = \frac{1}{2}c_0 = \frac{1}{2}, \quad c_2 = \frac{\frac{1}{2}}{4} = \frac{1}{2^3}, \quad c_3 = \frac{\frac{1}{4}}{6} = \frac{1}{2^4 \cdot 3}, \quad c_4 = \frac{c_3}{8} = \frac{1}{2^7 \cdot 3}, \quad c_5 = \frac{c_4}{10} = \frac{1}{2^8 \cdot 3.5}.$$

Therefore,

$$x(t) = 1 + \frac{1}{2}t + \frac{1}{2^3}t^2 + \frac{1}{2^4 \cdot 3}t^3 + \frac{1}{2^7 \cdot 3}t^4 + \frac{1}{2^8 \cdot 3.5}t^5 + \cdots .$$

Now, we can rewrite the series as

$$1 + \frac{1}{2}t + \frac{(\frac{1}{2}t)^2}{2!} + \frac{(\frac{1}{2}t)^3}{3!} + \frac{(\frac{1}{2}t)^4}{4!} + \frac{(\frac{1}{2}t)^5}{5!} + \cdots .$$

We conjecture that the series is the Taylor expansion of $e^{\frac{1}{2}t}$; one can use mathematical induction to verify it.

Example 11.2. Consider the differential equation

$$x'' + x = 0.$$

Of course, we can solve this equation by substituting $x(t) = e^{\lambda t}$ and solve the corresponding characteristic equation, as explained in Chapter 6. But we will solve it by using power series.

In order to find the general power series solution of $x'' + x = 0$ around $t = 0$, we substitute

$$x(t) = \sum_0^{+\infty} A_k t^k$$

into the equation and obtain

$$\sum_{2}^{+\infty} k(k-1)A_k t^{k-2} + \sum_{0}^{+\infty} A_k t^k = 0.$$

In order to change t^{k-2} in the first sum to t^k, we write

$$\sum_{0}^{+\infty} (k+1)(k+2)A_{k+2} t^k + \sum_{0}^{+\infty} A_k t^k = \sum_{0}^{+\infty} [(k+1)(k+2)A_{k+2} + A_k] t^k = 0.$$

Setting $(k+1)(k+2)A_{k+2} + A_k = 0$ yields the recursive formula

$$A_{k+2} = -\frac{1}{(k+1)(k+2)} A_k, \quad k \geq 0.$$

Evaluating A_{k+2} for $k = 0, 1, 2, 3, 4$, we obtain

$$A_2 = -\frac{1}{1 \cdot 2} A_0, \quad A_3 = -\frac{1}{1 \cdot 2 \cdot 3} A_1,$$

$$A_4 = -\frac{1}{3 \cdot 4} A_2 = \frac{1}{1 \cdot 2 \cdot 3 \cdot 4} A_0, \quad A_5 = -\frac{1}{4 \cdot 5} A_3 = \frac{1}{1 \cdot 2 \cdot 3 \cdot 4 \cdot 5} A_1.$$

It can be conjectured and verified inductively that

$$A_{2k} = \frac{(-1)^k}{(2k)!} A_0, \quad A_{2k+1} = \frac{(-1)^k}{(2k+1)!} A_1.$$

It is easy to see that the alternating series

$$\sum_{0}^{+\infty} A_k t^k$$

is convergent.

Recalling the Taylor series for $\sin t$ and $\cos t$, we have

$$\sum_{0}^{+\infty} A_{2k} t^{2k} = \sum_{0}^{+\infty} \frac{(-1)^k}{(2k)!} t^{2k} A_0 = A_0 \cos t$$

$$\sum_{0}^{+\infty} A_{2k+1} t^{2k+1} = \sum_{0}^{+\infty} \frac{(-1)^k}{(2k+1)!} A_1 = A_1 \sin t$$

Therefore, the general solution is given by

$$x(t) = \sum_{0}^{+\infty} A_k t^k = \sum_{0}^{+\infty} A_{2k} t^{2k} + \sum_{0}^{+\infty} A_{2k+1} t^{2k+1} = A_0 \cos t + A_1 \sin t.$$

Example 11.3. Find the series solution to the initial value problem

$$x'' - x = 0, \quad x(0) = 1, \ x'(0) = -1.$$

Let

$$x(t) = \sum_{0}^{+\infty} A_k t^k.$$

First of all, as we noted in the preceding example, the first two terms of the series play a key role in fulfilling the initial value requirement. More precisely, if the initial value requirements are $x(0) = x_0$ and $x'(0) = x'_0$, then $A_0 = x_0$ and $A_1 = x'_0$. This is because

$$x(t) = A_0 + A_1 t + A_2 t^2 + \cdots, \quad x'(t) = A_1 + 2A_2 t + \cdots$$

implies that $x(0) = A_0$ and $x'(0) = A_1$.

In order to obtain a recursive formula, we substitute the summation into the equation and obtain

$$\sum_{2}^{+\infty} k(k-1)A_k t^{k-2} - \sum_{0}^{+\infty} A_k t^k = \sum_{0}^{+\infty} [(k+1)(k+2)A_{k+2} - A_k] t^k = 0,$$

which yields the recursive formula

$$A_{k+2} = \frac{A_k}{(k+1)(k+2)}, \quad k \geq 0.$$

Letting $k = 0, 1, 2, \ldots$, we obtain

$$A_2 = \frac{A_0}{1 \cdot 2} = \frac{x(0)}{2!} = \frac{1}{2!}, \quad A_3 = \frac{A_1}{2 \cdot 3} = \frac{x'(0)}{2 \cdot 3} = \frac{-1}{3!}, \quad A_4 = \frac{1}{4!}, \quad A_5 = \frac{a_3}{4 \cdot 5} = \frac{-1}{5!},$$
$$A_6 = \frac{1}{6!}, \ldots.$$

Therefore

$$x(t) = \sum_{0}^{+\infty} \frac{(-1)^k t^k}{k!},$$

which we recognize to be the series expansion of e^{-t}.

Example 11.4. Find two linearly independent series solutions of

$$z'' + tz = 0.$$

Let us find two linearly independent solutions $x(t)$ and $y(t)$ such that

(A) $\quad x(t) = \sum_{0}^{+\infty} a_n t^n, \quad x(0) = 1, \quad x'(0) = 0,$

(B) $\quad y(t) = \sum_{0}^{+\infty} a_k t^k, \quad y(0) = 0, \quad y'(0) = 1.$

Then they will obviously be linearly independent since their Wronskian will be nonzero at $t = 0$.

(A). We have $x(0) = a_0 = 1$, $x'(0) = a_1 = 0$. In order to find find $x(t)$, we note that:

$$x(t) = \sum_0^{+\infty} a_n t^n \Rightarrow x'(t) = \sum_1^{+\infty} n a_n t^{n-1} \Rightarrow x''(t) = \sum_2^{+\infty} n(n-1) a_n t^{n-2}.$$

Substituting into the equation, we have

$$x'' + tx = \sum_2^{+\infty} n(n-1) a_n t^{n-2} + \sum_0^{+\infty} a_n t^{n+1} = \sum_0^{+\infty} (n+1)(n+2) a_{n+2} t^n + \sum_1^{+\infty} a_{n-1} t^n$$

$$= 2a_2 + \sum_1^{+\infty} [(n+1)(n+2) a_{n+2} + a_{n-1}] t^n = 0,$$

yielding $a_2 = 0$ and the recursive formula

$$a_{n+2} = \frac{-a_{n-1}}{(n+1)(n+2)} \qquad n \geq 1.$$

Now we have $x(0) = a_0 = 1$ and $x'(0) = a_1 = 0$, and also $a_2 = 0$. Let us compute a few terms and then write the general series representing $x(t)$.

$$n = 1 \Rightarrow a_3 = \frac{-a_0}{2 \cdot 3} = \frac{-1}{2 \cdot 3}, \quad n = 2 \Rightarrow a_4 = \frac{-a_1}{3 \cdot 4} = 0, \quad a_5 = \frac{-a_2}{4 \cdot 5} = 0,$$

$$a_6 = \frac{1}{2 \cdot 3 \cdot 5 \cdot 6}, \quad a_7 = \frac{-a_4}{6 \cdot 7} = 0, \quad a_8 = \frac{-a_5}{7 \cdot 8} = 0, \quad a_9 = \frac{-a_6}{8 \cdot 9} = \frac{-1}{2 \cdot 3 \cdot 5 \cdot 6 \cdot 8 \cdot 9},$$

$$a_{10} = a_{11} = 0.$$

Although one could use Mathematical Induction but the pattern is pretty clear, so we write

$$x(t) = 1 + \sum_1^{\infty} \frac{(-1)^k t^{3k}}{2 \cdot 3 \cdot 5 \cdot 6 \cdot 8 \cdot 9 \cdots (3k-1)3k}.$$

(B). In this case, we have $y(0) = a_0 = 0, y'(0) = a_1 = 1$ and we also have $a_2 = 0$, and

$$a_{n+2} = \frac{-a_{n-1}}{(n+1)(n+2)}, \qquad n \geq 1.$$

Hence,

$$a_3 = 0, \quad a_4 = \frac{-1}{3 \cdot 4}, \quad a_5 = a_6 = 0, \quad a_7 = \frac{1}{3 \cdot 4 \cdot 6 \cdot 7}, \quad a_8 = a_9 = 0,$$

$$a_{10} = \frac{1}{3 \cdot 4 \cdot 6 \cdot 7 \cdot 9 \cdot 10}, \quad a_{11} = a_{12} = 0, \quad a_{13} = \frac{1}{3 \cdot 4 \cdot 6 \cdot 7 \cdot 9 \cdot 10 \cdot 12 \cdot 13}, \cdots$$

We conclude that

$$y(t) = t + \sum_{1}^{\infty} \frac{(-1)^k t^{3k+1}}{3 \cdot 4 \cdot 6 \cdot 7 \cdots 3k(3k+1)}.$$

By the ratio test, both series are absolutely convergent.

We point out that the method of substituting infinite series does not always produce answers in infinite series; sometimes the answers can be finite series, i. e. a polynomial, see the next example.

Example 11.5. Use the power series method to solve the initial value problem

$$x'' - tx' + 3x = 0, \quad x(0) = 0, \quad x'(0) = 1.$$

We start by letting

$$x(t) = \sum_{0}^{+\infty} c_n t^n.$$

Then

$$x'' - tx' + 3x = \sum_{2}^{+\infty} n(n-1)c_n t^{n-2} - \sum_{1}^{+\infty} nc_n t^n + \sum_{0}^{+\infty} 3c_n t^n = 0,$$

$$\sum_{0}^{+\infty} (n+1)(n+2)c_{n+2} t^n - \sum_{1}^{+\infty} nc_n t^n + \sum_{0}^{+\infty} 3c_n t^n.$$

For $n = 0$ the first and last sums yield $2c_2 + 3c_0 = 0$, which implies that $c_2 = 0$, since the initial condition imply that $c_0 = 0$ and $c_1 = 1$. For $n \geq 1$ we have

$$\sum_{1}^{+\infty} (n+1)(n+2)c_{n+2} t^n - \sum_{1}^{+\infty} nc_n t^n + \sum_{1}^{+\infty} 3c_n t^n = \sum_{1}^{+\infty} [(n+1)(n+2)c_{n+2} - nc_n + 3c_n] t^n = 0$$

yielding the recursive formula

$$c_{n+2} = \frac{n-3}{(n+1)(n+2)} c_n, \quad n \geq 1.$$

Computing a few terms, we have

$$c_3 = \frac{-2c_1}{2.3} = \frac{-1}{3}, \quad c_4 = \frac{-c_2}{3.4} = 0, \quad c_5 = 0.$$

Now, we can see that all the remaining terms are zero. Consequently, the only nonzero terms are c_1 and c_3, and hence

$$x(t) = t - \frac{1}{3}t^3,$$

a third degree polynomial.

11.3 Bessel functions

Solutions of

$$t^2 x'' + tx' + (t^2 - m^2)x = 0 \qquad (B_m)$$

are referred to as *Bessel functions of order m*.

Although $t = 0$ is a singular point, we can still solve Bessel equations by using the solution series method discussed in the previous section.

We carry out the calculations for $m = 0$ and $m = 1$.

11.3.1 Bessel functions of first kind

Here we discuss solving Bessel equations of the first kind, $m = 0, 1$.

Case m = 0. For $m = 0$, the Bessel equation is given by

$$t^2 x'' + tx' + t^2 x = 0. \qquad (B_0)$$

Setting $x = \sum_0^\infty A_k t^k$, we have $x' = \sum_1^\infty k A_k t^{k-1}$, $x'' = \sum_2^\infty k(k-1)A_k t^{k-2}$. Hence

$$t^2 x'' + tx' + t^2 x = \sum_2^\infty k(k-1)A_k t^k + \sum_1^\infty k A_k t^k + \sum_0^\infty A_k t^{k+2} = 0.$$

After shifting indices in the last sum and combining, we obtain

$$A_1 t + \sum_2^\infty [k^2 A_k + A_{k-2}] t^k = 0.$$

This implies

$$A_1 = 0, \quad k^2 A_k + A_{k-2} = 0, \quad k = 2, 3, \ldots,$$

yielding the recursive formula

$$A_k = -\frac{A_{k-2}}{k^2}, \quad k \geq 2.$$

Therefore, for $k = 2n + 1$, $n = 1, 2, \ldots$, we find

$$A_3 = -\frac{A_1}{3^2} = 0, \quad A_5 = -\frac{A_3}{5^2} = 0, \ldots, A_{2n+1} = 0.$$

For $k = 2n$, $n = 1, 2, \ldots$, one has

$$A_2 = -\frac{A_0}{2^2}, \quad A_4 = -\frac{A_2}{4^2} = \frac{A_0}{2^2 \cdot 4^2}, \ldots, A_{2n} = (-1)^n \frac{A_0}{2^2 \cdot 4^2 \cdots (2n)^2}.$$

In conclusion, we find

$$x(t) = A_0\left(1 - \frac{t^2}{2^2} + \frac{t^4}{2^2 \cdot 4^2} + \cdots + \frac{(-1)^n t^{2n}}{2^2 \cdot 4^2 \cdots (2n)^2} + \cdots\right),$$

which is uniformly convergent on $\mathbb{R}$.

The solution is usually given as a power series in terms of $\frac{t}{2}$, which can be achieved by writing the general term as

$$\frac{(-1)^n}{2^2 \cdot 4^2 \cdots (2n)^2} t^{2n} = \frac{(-1)^n}{2^2 \cdot 4^2 \cdots (2n)^2} \frac{2^{2n}}{2^{2n}} t^{2n}$$

$$= \frac{(-1)^n 2^{2n}}{2^2 \cdot 4^2 \cdots (2n)^2}\left(\frac{t}{2}\right)^{2n} = \frac{(-1)^n}{(1 \cdot 2 \cdot 3 \cdots n)^2}\left(\frac{t}{2}\right)^{2n} = \frac{(-1)^n}{(n!)^2}\left(\frac{t}{2}\right)^{2n}.$$

Setting

$$J_0(t) = \sum_{0}^{\infty} \frac{(-1)^n}{(n!)^2}\left(\frac{t}{2}\right)^{2n} = 1 - \left(\frac{t}{2}\right)^2 + \frac{1}{(2!)^2}\left(\frac{t}{2}\right)^4 + \frac{1}{(3!)^2}\left(\frac{t}{2}\right)^6 + \cdots$$

and renaming A_0 as c, we have $x(t) = cJ_0(t)$.

The function J_0 is called *Bessel function of order $m = 0$ of the first kind.*

We see that J_0 is analytic and even; $J_0(0) = 1$, $J_0'(0) = 0$. Moreover, it is possible to show that J_0 has the following asymptotic behavior as $t \to +\infty$:

$$J_0(t) \simeq \frac{c}{\sqrt{t}} \cdot \cos\left(t - \frac{\pi}{4}\right).$$

This implies that J_0 has infinitely many zeros and decays to zero at infinity. One might also show that

$$\int_{0}^{+\infty} J_0(t)\, dt = 1.$$

Case $m = 1$. When $m = 1$ the Bessel equation is given by

$$t^2 x'' + tx' + (t^2 - 1)x = 0. \tag{B_1}$$

Repeating the previous calculations we now find

$$t^2 \sum_{2}^{\infty} k(k-1)A_k t^{k-2} + t \sum_{1}^{\infty} kA_k t^{k-1} + (t^2 - 1)\sum_{0}^{\infty} A_k t^k = 0.$$

It follows that

$$\sum_{2}^{\infty} k(k-1)A_k t^k + \sum_{1}^{\infty} kA_k t^k + \sum_{0}^{\infty} A_k t^{k+2} - \sum_{0}^{\infty} A_k t^k = 0$$

whereby

$$\sum_2^\infty k^2 A_k t^k + A_1 t + \sum_2^\infty A_{k-2} t^k - \sum_0^\infty A_k t^k$$

$$= \sum_2^\infty k^2 A_k t^k + A_1 t + \sum_2^\infty A_{k-2} t^k - A_0 - A_1 t - \sum_2^\infty A_k t^k$$

$$= \sum_2^\infty k^2 A_k t^k + \sum_2^\infty A_{k-2} t^k - A_0 - \sum_2^\infty A_k t^k = 0.$$

This yields

$$A_0 = 0, \quad k^2 A_k + A_{k-2} - A_k = 0,$$

which gives rise to the recursive formula

$$A_k = -\frac{A_{k-2}}{k^2 - 1}, \quad k = 2, 3, \ldots$$

For $k = 2n = 2, 4, 6, \ldots$, we find

$$A_2 = -\frac{A_0}{3} = 0, \quad A_4 = -\frac{A_2}{3 \cdot 5} = 0, \ldots, \quad A_{2n} = 0.$$

For $k = 2n + 1 = 3, 5, 7, \ldots, k^2 - 1 = 4n^2 + 4n = 4n(n+1) = 2^2 n(n+1)$, hence we find

$$A_3 = -\frac{A_1}{2^2 \cdot 1 \cdot 2} = \frac{A_1}{2^2 \cdot 2!},$$

$$A_5 = -\frac{A_3}{2^2 \cdot 2 \cdot 3} = \frac{A_1}{2^2 \cdot 2! \cdot 2^2 \cdot \underbrace{2 \cdot 3}_{=3!}} = \frac{A_1}{2^4 \cdot 2! \cdot 3!},$$

$$A_7 = -\frac{A_5}{2^2 \cdot 3 \cdot 4} = -\frac{A_1}{2^4 \cdot 2! \cdot 3! \cdot 2^2 \cdot 3 \cdot 4} = -\frac{A_1}{2^6 \cdot 2! \cdot 3 \cdot 3! \cdot 4} = -\frac{A_1}{2^6 \cdot 3! \cdot 4!} = \cdots$$

It is convenient to let $A_1 = \frac{c}{2}$. Then we have

$$A_{2n+1} = (-1)^n \frac{c}{2^{2n+1} \cdot n! \cdot (n+1)!}.$$

In conclusion we find

$$x(t) = c \cdot \left[\frac{t}{2} - \frac{t^3}{2!2^3} + \frac{t^5}{2!3!2^5} - \cdots \right] = c \cdot \left[\left(\frac{t}{2}\right) - \frac{1}{2!} \cdot \left(\frac{t}{2}\right)^3 + \frac{1}{2!3!} \cdot \left(\frac{t}{2}\right)^5 - \cdots \right]$$

$$= c \cdot \left(\frac{t}{2}\right) \cdot \left[1 - \frac{1}{2!} \cdot \left(\frac{t}{2}\right)^2 + \frac{1}{2!3!} \cdot \left(\frac{t}{2}\right)^4 - \cdots \right],$$

the series being uniformly convergent on $\mathbb{R}$. Setting

$$J_1(t) = \left(\frac{t}{2}\right) \cdot \sum_0^\infty \frac{(-1)^n}{n! \cdot (n+1)!} \left(\frac{t}{2}\right)^{2n}$$

we find that $x(t) = cJ_1(t)$ is a solution of (B_1).

The function J_1 is called a *Bessel function of order* $m = 1$ *of the first kind*. It is an analytic odd function and $J_1(0) = 0$. One can show that $J_1'(0) = \frac{1}{2}$.

Moreover, similar to J_0, we have

$$J_1(t) \simeq \frac{c}{\sqrt{t}} \cdot \cos\left(t - \frac{\pi}{4} - \frac{\pi}{2}\right), \quad \text{as } t \to +\infty,$$

and thus J_1 has infinitely many zeros and decays to zero at infinity.

Figure 11.1 shows the graphs of J_0 and J_1.

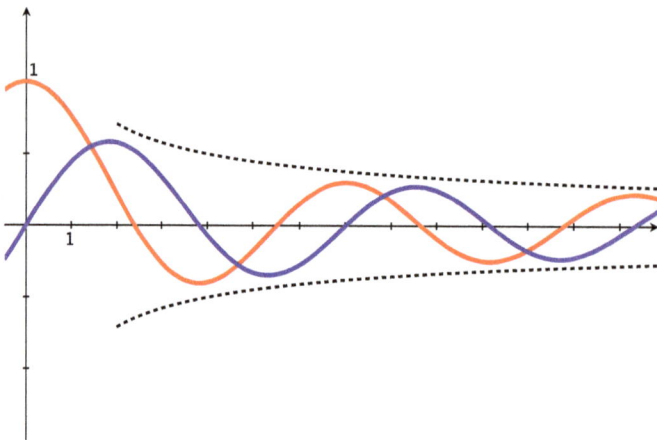

Figure 11.1: Plot of $J_0(t)$, in red, and $J_1(t)$, in blue. The dotted curves are $x(t) \simeq c/\sqrt{t}$.

For completeness, we write below the Bessel function of first kind of order m, $m \in \mathbb{N}$.

$$J_m(t) = \left(\frac{t}{2}\right)^m \cdot \sum_0^\infty \frac{(-1)^n}{n! \cdot (n+m)!} \left(\frac{t}{2}\right)^{2n}. \tag{11.2}$$

For $m < 0$, although the Bessel equation remains the same, one usually sets

$$J_m(t) = (-1)^m J_{-m}(t).$$

Finally, one could use (11.2) to define Bessel functions of any order v, giving a suitable meaning to $(n + v)!$ by means of the gamma function Γ.

Further properties of J_m are presented as exercises.

11.3.2 General solution of (B_m) and Bessel functions of second kind

Here we briefly discuss how to find the general solution of (B_m).

One can define, for $t > 0$, another solution Y_m of (B_m), called a *Bessel function of the second kind* of order m. One can show that

$$Y_m(t) = J_m(t) \cdot \left[k_1 \int_1^t \frac{ds}{sJ_m^2(s)} + k_2 \right], \quad t > 0,$$

for suitable constants $k_1, k_2 \neq 0$.

An important property of Y_m is that, unlike J_m, $Y_m(t) \to -\infty$ as $t \to 0$ so that Y_m has a singularity at $t = 0$. Actually, since $J_0(t) = 1 + o(1)$ and $J_m(t) \approx t^m$ for $m \in \mathbb{N}$, $m \neq 0$ and $t \approx 0+$ we infer

$$\int_1^t \frac{ds}{sJ_0^2(s)} \approx \ln t, \quad \int_1^t \frac{ds}{sJ_m^2(s)} \approx -t^{-2m} \quad \text{if } m \in \mathbb{N}, \, m \neq 0, \quad \Longrightarrow \quad \begin{cases} Y_0(t) \approx \ln t, \\ Y_m(t) \approx -t^{-m}. \end{cases}$$

A plot of $Y_0(t)$ and $Y_1(t)$ are reported in Fig. 11.2.

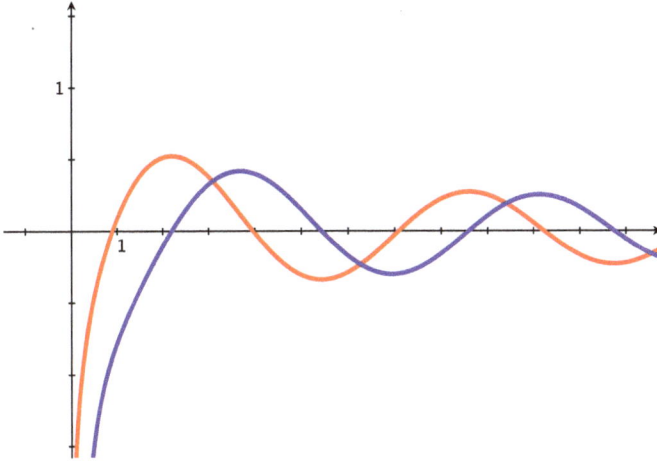

Figure 11.2: Plot of $Y_0(t)$, in red, and $Y_1(t)$, in blue.

As a consequence we infer that J_m and Y_m are linearly independent, otherwise we would have $Y_m(t) = \mu J_m(t)$ for some $\mu \in \mathbb{R}$ and this would imply that $Y_m(t)$ and $J_m(t)$ have the same behavior as $t \to 0+$.

We can conclude that the general solution of (B_m) is given by

$$x(t) = c_1 J_m(t) + c_2 Y_m(t), \quad t > 0.$$

The book by Watson, *A Treatise on the Theory of Bessel Functions*, Cambridge University Press 1966, contains further properties of Y_m including the precise definition of the Bessel functions of the second kind as well as their series expansions.

Remark 11.1. In some textbooks the Bessel functions of the second kind are called Neumann functions and denoted by N_m.

11.4 Singular points: the Frobenius method

Apart from some cases, like the Bessel equations, in general when there are singular points, solving second order linear equations is much more complicated, since near singular points there may not exist analytical solutions.

Suppose that $t = 0$ is a singular point of $L[x] = a_2(t)x'' + a_1(t)x' + a_0(t)x = 0$. We say that it is a *regular singular point* of this equation if

$$\frac{ta_1(t)}{a_2(t)}, \quad \frac{t^2 a_0(t)}{a_2(t)},$$

are analytical at $t = 0$. For example, the Euler equation $t^2 x'' + tx' + x = 0$ has a regular singular point at $t = 0$ since $ta_1 = t^2 a_0 = 1$. On the other hand the singular point $t = 0$ is not regular singular for the equation $t^2 x'' + x' = 0$.

To find a series solution near a regular singular point we can use a method due to Frobenius, which we are going to discuss in the specific case of

$$t^2 x'' + ptx' + (q + q_1 t)x = 0, \tag{11.3}$$

where p, q, q_1 are real numbers. Clearly, $t = 0$ is a regular singular point since

$$\frac{ta_1(t)}{a_2(t)} = \frac{t \cdot pt}{t^2} = p, \quad \frac{t^2 a_0(t)}{a_2(t)} = \frac{t^2 \cdot (q + q_1 t)}{t^2} = q + q_1 t,$$

are analytic.

We seek series solutions of (11.3) of the form

$$x(t) = \sum_0^\infty A_k t^{k+r}$$

where r is a number to be determined. Substituting

$$x' = \sum_0^\infty (k + r)A_k t^{k+r-1}$$

$$x'' = \sum_0^\infty (k + r)(k + r - 1)A_k t^{k+r-2}$$

into the equation we find

$$t^2 x'' + ptx' + (q + q_1 t)x = t^2 \sum_0^\infty (k + r)(k + r - 1)A_k t^{k+r-2}$$

$$+ pt \sum_0^\infty (k + r)A_k t^{k+r-1} + (q + q_1 t) \sum_0^\infty A_k t^{k+r}$$

$$= \sum_0^\infty (k + r)(k + r - 1)A_k t^{k+r}$$

$$+ p \sum_0^\infty (k + r)A_k t^{k+r} + (q + q_1 t) \sum_0^\infty A_k t^{k+r} = 0.$$

Rearranging

$$\sum_0^\infty [(k + r)(k + r - 1) + (k + r)p + (q + q_1 t)]A_k t^{k+r} = 0$$

we have

$$[r(r - 1) + rp + q]A_0 t^r + \sum_1^\infty [(k + r)(k + r - 1) + (k + r)p + (q + q_1 t)]A_k t^{k+r} = 0.$$

Introducing the *indicial polynomial* as the quadratic polynomial

$$\mathcal{I}(r) = r(r - 1) + rp + q,$$

we can rewrite the equation as

$$\mathcal{I}(r)A_0 t^r + \sum_1^\infty [(k + r)(k + r - 1) + (k + r)p + (q + q_1 t)]A_k t^{k+r} = 0,$$

namely

$$\mathcal{I}(r)A_0 t^r + q_1 A_0 t^{r+1} + \mathcal{I}(r + 1)A_1 t^{r+1} + q_1 A_1 t^{r+2} + \mathcal{I}(r + 2)A_2 t^{r+2} + \cdots = 0,$$

which yields

$$\begin{cases} \mathcal{I}(r)A_0 & = 0, \\ q_1 A_0 + \mathcal{I}(r + 1)A_1 & = 0, \\ q_1 A_1 + \mathcal{I}(r + 2)A_2 & = 0, \\ \cdots & = \cdots \\ q_1 A_{k-1} + \mathcal{I}(r + k)A_k & = 0. \end{cases} \qquad (11.4)$$

We limit ourselves to consider the following two cases:
1. $\mathcal{I}(r + k) \neq 0$ for any $k \in \mathbb{N}$;

2. $\mathcal{I}(r) = 0$ has two real roots $\rho_{1,2}$ (distinct or coincident) and $\mathcal{I}(\rho_{1,2} + k) \neq 0$ for any $k \in \mathbb{N}, k \neq 0$.

Case 1. Since $\mathcal{I}(r) \neq 0$ the first equation yields $A_0 = 0$. The second equation becomes $\mathcal{I}(r + 1)A_1 = 0$, which implies $A_1 = 0$, for $\mathcal{I}(r + 1) \neq 0$. Then the third equation becomes $\mathcal{I}(r + 2)A_2 = 0$, yielding $A_2 = 0$, for $\mathcal{I}(r + 2) \neq 0$. Iterating the procedure, we infer that $A_k = 0$ for $k \in \mathbb{N}$, yielding the trivial solution.

Case 2. We first consider a real root ρ_1. In the first equation of (11.4) with $r = \rho_1$, A_0 is undetermined whereas from the remaining equations we deduce

$$
\begin{cases}
A_1 = -\dfrac{q_1 A_0}{\mathcal{I}(\rho_1 + 1)}, \\[2mm]
A_2 = -\dfrac{q_1 A_1}{\mathcal{I}(\rho_1 + 2)} = +\dfrac{q_1^2 A_0}{\mathcal{I}(\rho_1 + 1)\mathcal{I}(\rho_1 + 2)}, \\[2mm]
A_3 = -\dfrac{q_1 A_2}{\mathcal{I}(\rho_1 + 3)} = -\dfrac{q_1^3 A_0}{\mathcal{I}(\rho_1 + 1)\mathcal{I}(\rho_1 + 2)\mathcal{I}(\rho_1 + 3)}, \\[2mm]
\dots
\end{cases}
$$

and, in general, we have the recursive formula

$$
A_k = (-1)^k \frac{q_1^k A_0}{\mathcal{I}(\rho_1 + 1)\mathcal{I}(\rho_1 + 2)\cdots\mathcal{I}(\rho_1 + k)}. \tag{11.5}
$$

Notice that

$$
\left|\frac{A_{k+1}}{A_k}\right| = \left|\frac{q_1 A_0}{\mathcal{I}(\rho_1 + k)}\right| \implies \lim_{k\to\infty}\left|\frac{A_{k+1}}{A_k}\right| = 0.
$$

Therefore the ratio test implies that the series $\sum_0^\infty A_k t^{k+\rho_1}$ is absolutely convergent on $\mathbb{R}$ and we can conclude that

$$
x_1(t) = \sum_0^\infty A_k t^{k+\rho_1}
$$

gives rise to a solution of (11.3).

Obviously, by the same arguments, a second solution of (11.3) is given by

$$
x_2(t) = \sum_0^\infty B_k t^{k+\rho_2}
$$

where

$$
B_k = (-1)^k \frac{q_1^k A_0}{\mathcal{I}(\rho_2 + 1)\mathcal{I}(\rho_2 + 2)\cdots\mathcal{I}(\rho_2 + k)}.
$$

If $\rho_1 \neq \rho_2$ and $\rho_2 - \rho_1$ is not an integer, then one can show that x_1 and x_2 are linearly independent and hence the general solution of (11.3) is a linear combination of x_1 and x_2.

On the other hand, if $\rho_1 = \rho_2$ or $\rho_1 \neq \rho_2$ and $\rho_2 - \rho_1$ is an integer, a solution linearly independent of x_1 is given by, respectively,

$$x_2(t) = x_1(t) \cdot \ln t + \sum_0^\infty C_k t^{k+\rho_1}, \quad x_2(t) = c \cdot \ln t + \sum_0^\infty C_k t^{k+\rho_2},$$

where c and the C_k have to be determined. Notice that in the second case, c might be zero.

Remark 11.2. If $q_1 = 0$ and the indicial equation has two distinct real roots $\rho_1 \neq \rho_2$ such that $\mathcal{I}(\rho_{1,2} + k) \neq 0$ for all $k \in \mathbb{N}$, $k \neq 0$, we infer from the recursive formulas that $A_k = B_k = 0$ for $k = 1, 2, \ldots$ Thus in such a case we simply find $x_1(t) = A_0 t^{\rho_1}$ and $x_2(t) = B_0 t^{\rho_2}$.

Example 11.6. Solve $t^2 x'' + \frac{1}{2} t x' = 0$.

In this case $q = q_1 = 0$ and $p = \frac{1}{2}$. The indicial polynomial is $\mathcal{I}(r) = r(r-1) + \frac{1}{2} r = r^2 - \frac{1}{2} r$, whose its roots are $\rho_1 = 0$ and $\rho_2 = \frac{1}{2}$. Since $\mathcal{I}(\rho_1 + k) = \mathcal{I}(k) = k^2 - \frac{1}{2} k \neq 0$ for $k = 1, 2, \ldots$, the recursive formula (11.5) implies (recall that here $q_1 = 0$) $A_k = 0$ for $k = 1, 2, \ldots$ Thus $x_1(t) = \sum_0^\infty A_k t^k = A_0$.

Next consider $\rho_2 = \frac{1}{2}$. Since $\mathcal{I}(\rho_2 + k) = (\frac{1}{2} + k)^2 - \frac{1}{2}(\frac{1}{2} + k) = \frac{1}{4} + k + k^2 - \frac{1}{4} - \frac{1}{2} k = k^2 + \frac{1}{2} k > 0$ for $k = 1, 2, \ldots$ we find $x_2(t) = \sum_0^\infty B_k t^{k+2}$, where the recursive formula yields $B_k = 0$ for $k = 1, 2, \ldots$ Thus a second solution is given by $x_2(t) = B_0 \cdot t^{1/2}$.

The two solutions x_1 and x_2 are clearly linearly independent and hence the general solution is a linear combination of x_1 and x_2, namely $x = c_1 + c_2 t^{1/2}$. The reader should have noticed that the given equation is an Euler equation, studied in Chapter 6, section 6.4. As an exercise, it could be shown that using the method carried out therein one finds the same result. □

The next example shows that in some cases a second linearly independent solution can be found in a more direct way.

Example 11.7. Using the Frobenius method, find the general solution of the Euler equation $t^2 x'' + 2t x' - 2x = 0$. The indicial equation is $\mathcal{I}(r) = r(r-1) + 2r - 2 = r^2 + r - 2 = 0$ with roots $\rho_1 = 1$, $\rho_2 = -2$. Moreover, $\mathcal{I}(k+1) = (k+1)^2 + (k+1) - 2 = k^2 + 3k > 0$ for $k = 1, 2, \ldots$, as well as $\mathcal{I}(k-2) = (k-2)^2 + (k-1) - 2 = k^2 - 3k + 1 \neq 0$ for $k = 1, 2, \ldots$ Since here $q_1 = 0$ we infer that $A_k = B_k = 0$ for $k = 1, 2, \ldots$ Thus $x_1(t) = A_0 t$ and $x_2(t) = B_0 t^{-2}$. It is clear that x_1, x_2 are linearly independent and thus the general solution is given by $x(t) = A_0 t + B_0 t^{-2}$. Notice that in this case $\rho_1 - \rho_2 = 3$ is an integer. □

Remark 11.3. Bessel equations have $t = 0$ as a regular singular point. As a useful exercise, the reader might wish to recover the results of the previous section by using the Frobenius method.

11.5 Exercises

1. Use the series method to solve $x' + 2x = 0$, $x(0) = 1$.

2. Find the recursive formula and the first five nonzero terms of the series solution of $x'' + tx' = 0$.

3. Use the infinite series method to find the function that solves $x'' - 3x = 0$, $x(0) = 1$, $x'(0) = 3$, and find the interval of convergence.

4. Use the infinite series method to find the function that solves $x'' - 4x = 0$, $x(0) = 1$, $x'(0) = 2$.

5. Find the recursive formula and the first five nonzero terms of the series solution for $x'' - tx' + 2x = 0$, $x(0) = 0$, $x'(0) = 1$.

6. Use the series solution method to find the function that solves the initial value problem $x' - 2tx = 0$, $x(0) = 1$.

7. Use the infinite series method to solve the initial value problem $x'' + tx' - 3x = 0$, $x(0) = 0$, $x'(0) = 1$.

8. Find the general series solution of $x'' - tx' = 0$.

9. Find two linearly independent solutions of $z'' - tz = 0$ (see Example 11.4).

10. Find the infinite series solution for $(1-t)x' - x = 0$, $x(0) = 1$ and identify the function represented by the series.

11. Find the infinite series solution for $(1+t)x' + x = 0$, $x(0) = 1$ and identify the function represented by the series.

12. Show that $t = 0$ is a strict maximum of J_0.

13. Describe the behavior of $J_2(t)$ as $t \to 0$.

14. Show that if $a \neq 0$ is such that $J_0(a) = 0$ then $J'(a) \neq 0$.

15. Show that $J_0'(t) = -J_1(t)$.

16. Show that $(tJ_1(t))' = tJ_0(t)$.

17. Using that $J_0'(t) = -J_1(t)$ (see Exercise n. 7), show that between two consecutive zeros of $J_0(t)$ there is one zero of $J_1(t)$.

18. Show that $x(t) = tJ_1(t)$ solves $tx'' - x' + tx = 0$.

19. Find the general solution of $tx'' + x' + \omega^2 tx = 0$, $\omega \neq 0$.

20. Find the equation satisfied by $J_0(\frac{1}{2}t^2)$.

21. Find $m \in \mathbb{N}$ such that the Bessel equation (B_m) $t^2 x'' + tx' + (t^2 - m^2)x = 0$ has a solution satisfying $x(0) = 0$, $x'(0) \neq 0$.

22. Solve the problem $t^2 x'' + tx' + t^2 x = 0$ such that $x(0) = b$ and $x(a) = 0$, where $a > 0$ is a zero of J_0.

23. Solve the ivp $t^2 x'' + tx' + (t^2 - 1)x = 0$, $x(0) = 0$, $x'(0) = 1$.

24. Solve the Euler equation $t^2 x'' - 2x = 0$ by means of the Frobenius method.

25. Solve the Euler equation $t^2 x'' - 3tx' + x = 0$ by means of the Frobenius method.

26. Find a solution of $t^2 x'' - tx' + (1 - t)x = 0$ such that $x(0) = x_0$.

12 Laplace transform

12.1 Definition

Let $\mathscr{C}$ denote the class of real valued continuous functions $f(t)$, $t \in \mathbb{R}$, such that there exist numbers a and C, $C > 0$, satisfying the inequality

$$|f(t)| \le C\,e^{at}, \quad \forall t \in \mathbb{R}. \tag{12.1}$$

Such functions are said to be of *exponential order*. For example, continuous bounded functions are of exponential order (with $a = 0$), while e^{t^2} is not.

Example 12.1. If $f \in \mathscr{C}$, then $t^n f \in \mathscr{C}$ for $t \ge 0$ and $n \in \mathbb{N}$.

To verify the assertion, let $c > 0$ be such that $t^n \le ce^{at}$. Then $|t^n f(t)| \le ce^{at} . Ce^{at} = C_1 e^{2at}$ ($C_1 = cC$, $t \ge 0$).

Let us note that if $f \in \mathscr{C}$, then there exist numbers a and C such that $|f(t)| \le Ce^{at}$. Hence the function $|e^{-st}f(t)|$ is integrable on $(0, +\infty)$ for $s > a$, since

$$|e^{-st}f(t)| \le Ce^{-st}e^{at} = Ce^{(a-s)t}$$

and $e^{(a-s)t}$ decays exponentially to 0 as $t \to +\infty$ provided $a - s < 0$. We know from Calculus that if $|g|$ is integrable, then so is g. Thus, if $s > a$, the integral

$$\int\limits_0^{+\infty} e^{-st}f(t)\,dt$$

exists and is finite.

Now we define the *Laplace transform* of f, $f \in \mathscr{C}$, to be the function

$$\mathscr{L}[f](s) = \int\limits_0^{+\infty} e^{-st}f(t)\,dt.$$

Notice that the value of f for $t < 0$ is irrelevant. The Laplace transform is a function of s defined for all $s > a$, for some real number a ($|f(t)| \le Ce^{at}$), called the *domain of convergence of* f. For convenience of notation, we let

$$\mathscr{L}[f](s) = F(s), \quad \mathscr{L}[g](s) = G(s), \quad \mathscr{L}[j](s) = J(s), \text{ etc.}$$

Remark 12.1. Although here we are mainly interested in real valued continuous functions, the definition of Laplace transform can be extended to include functions such as piecewise continuous functions and functions of complex numbers. Recall that $f(t)$ is piecewise continuous on $(0, \infty)$ if for any number $c > 0$, it has at most a finite number of jump discontinuities on the interval $[0, c]$.

https://doi.org/10.1515/9783111185675-012

As an example, let us consider the shifted Heaviside function ($b > 0$)

$$H_b(t) = \begin{cases} 1 & \text{if } t \geq b, \\ 0 & \text{if } 0 \leq t < b. \end{cases}$$

It is clear that H_b, being bounded, satisfies (12.1) for any $a > 0$. One has

$$\int_0^{+\infty} H_b(t)e^{-st}\,dt = \int_b^{+\infty} e^{-st}\,dt = -\frac{e^{-st}}{s}\Big|_b^{+\infty} = \frac{e^{-bs}}{s}, \quad s > 0$$

and thus $\mathscr{L}[H_b] = \frac{e^{-bs}}{s}, s > 0$.

12.2 Properties of $\mathscr{L}$

In the sequel we will assume, unless stated otherwise, that the functions we deal with belong to the class $\mathscr{C}$.

(A) Linearity property.

$$\mathscr{L}[a_1f_1 + a_2f_2 + \cdots + a_nf_n] = a_1\mathscr{L}[f_1] + a_2\mathscr{L}[f_2] + \cdots + a_n\mathscr{L}[f_n], \quad a_i \in \mathbb{R}, \quad f_i \in \mathscr{C}.$$

Proof. We give the proof for $n = 2$, which can be extended similarly to any number of functions.

$$\mathscr{L}[a_1f_1 + a_2f_2] = \int_0^{+\infty} e^{-st}(a_1f_1(t) + a_2f_2(t))\,dt$$

$$= a_1 \int_0^{+\infty} e^{-st}f_1(t)\,dt + a_2 \int_0^{+\infty} e^{-st}f_2(t)\,dt = a_1\mathscr{L}[f_1] + a_2\mathscr{L}[f_2]. \qquad \square$$

(B) If $f^{(k)}$, $0 \leq k \leq n$, belong to $\mathscr{C}$, then

$$\mathscr{L}[f^{(n)}] = s^n\mathscr{L}[f(t)] - s^{n-1}f(0) - s^{n-2}f'(0) - \cdots - f^{(n-1)}(0). \tag{12.2}$$

Proof. We will show the assertion for $n = 1, 2, 3$. It will then be clear that the general statement easily follows by mathematical induction.

Using integration by parts, we have

$$\mathscr{L}[f'] = \int_0^{+\infty} e^{-st}f'(t)\,dt = e^{-st}f(t)\Big|_0^{\infty} + s \int_0^{+\infty} e^{-st}f(t)\,dt = -f(0) + s\mathscr{L}[f],$$

$$\mathscr{L}[f''] = \mathscr{L}[(f')'] = s\mathscr{L}[f'] - f'(0) = s(s\mathscr{L}[f] - f(0)) - f'(0) = s^2\mathscr{L}[f] - sf(0) - f'(0).$$

Similarly, $\mathscr{L}[f''']$ can be treated as

$$\mathscr{L}[(f'')'] = s\mathscr{L}[f''] - f''(0) = s(s^2\mathscr{L}[f] - sf(0) - f'(0)) - f''(0)$$
$$= s^3\mathscr{L}[f] - s^2f(0) - sf'(0) - f''(0). \qquad \square$$

(C) If $\phi(t) = \int_0^t f(r)\,dr,\ f \in \mathscr{C}$, then $\phi \in \mathscr{C}$ and

$$\mathscr{L}[\phi] = \frac{1}{s} \cdot \mathscr{L}[f]. \qquad (12.3)$$

Proof. One has

$$|\phi(t)| \le \int_0^t |f(r)|\,dr \le C \int_0^t e^{ar}\,dr = \frac{C}{a}(e^{at} - 1) \le \left|\frac{C}{a}\right|e^{at}$$

whereby $\phi \in \mathscr{C}$. Furthermore, since $\phi' = f$ and $\phi(0) = 0$, an application of (B) yields

$$\mathscr{L}[f] = \mathscr{L}[\phi'] = s\mathscr{L}[\phi]$$

from which (12.3) follows. $\qquad \square$

(D) For $t \ge 0$ one has

$$\mathscr{L}[t^n f(t)] = (-1)^n \frac{d^n}{ds^n}\,\mathscr{L}[f]. \qquad (12.4)$$

Proof. Notice that for $t \ge 0$ one has $tf \in \mathscr{C}$ (see Example 12.1). Setting

$$F(s) = \mathscr{L}[f](s) = \int_0^{+\infty} e^{-st}f(t)\,dt,$$

it follows that

$$\frac{d}{ds}F(s) = \frac{d}{ds}\int_0^{+\infty} e^{-st}f(t)\,dt = \int_0^{+\infty} \frac{\partial}{\partial s}e^{-st}f(t)\,dt = -\int_0^{+\infty} e^{-st}tf(t)\,dt = -\mathscr{L}[tf(t)].$$

This shows that the statement is true for $n = 1$. $\qquad \square$

By using mathematical induction, one can easily show that for any $n \in N$,

$$\mathscr{L}[t^n f(t)] = (-1)^n \frac{d^n}{ds^n}\,\mathscr{L}[f].$$

As we will see below, properties (A)–(D) are useful in determining the Laplace transforms of some other elementary functions.

(L1) $\mathcal{L}[k] = \dfrac{k}{s}, s > 0.$

Proof.

$$\int_0^{+\infty} ke^{-st}\,dt = k\cdot -\dfrac{e^{-st}}{s}\Big|_0^{+\infty} = k\cdot\lim_{T\to\infty} -\dfrac{e^{-st}}{s}\Big|_0^T = k\cdot\left(0 + \dfrac{1}{s}\right) = \dfrac{k}{s}.$$

In particular, $\mathcal{L}[1] = \dfrac{1}{s}.$ □

(L2) $\mathcal{L}[e^{at}] = \dfrac{1}{s-a}, s > a.$

Proof. Notice that $e^{at} \in \mathscr{C}$, and

$$\int_0^{+\infty} e^{-st}e^{at}\,dt = \int_0^{+\infty} e^{(a-s)t}\,dt = \dfrac{1}{a-s}e^{(a-s)t}\Big|_0^{+\infty} = \lim_{T\to\infty}\dfrac{e^{(a-s)t}}{(a-s)}\Big|_0^T = \dfrac{1}{s-a}, s > a.\ \square$$

(L3) $\mathcal{L}[t^n] = \dfrac{n!}{s^{n+1}}, s > 0.$

Proof. Using (D), with $f(t) = 1$, we see that

$$\mathcal{L}[t^n] = (-1)^n\dfrac{d^n\mathcal{L}[1]}{ds^n}.$$

Since $\mathcal{L}[1] = \dfrac{1}{s}$, see (L1), it follows that

$$\mathcal{L}[t^n] = \dfrac{d^n}{ds^n}\left(\dfrac{1}{s}\right) = \dfrac{n!}{s^{n+1}}.$$ □

(L4) $\mathcal{L}[t^n e^t] = n!(s-1)^{-(n+1)}.$

Proof. By (D),

$$\mathcal{L}[t^n \cdot e^t] = (-1)^n\dfrac{d^n}{ds^n}\mathcal{L}[e^t] = (-1)^n\dfrac{d^n}{ds^n}\dfrac{1}{s-1} = n!(s-1)^{-(n+1)}.$$ □

(L5) $\mathcal{L}[\sin\omega t] = \dfrac{\omega}{s^2 + \omega^2}.$

(L6) $\mathcal{L}[\cos\omega t] = \dfrac{s}{s^2 + \omega^2}.$

Proofs of (L5) and (L6). We use (B) with $f(t) = \sin\omega t$ to infer

$$\mathcal{L}[\underbrace{\omega\cos\omega t}_{f'}] = s\mathcal{L}[\sin\omega t].$$

Similarly, using (B) with $f(t) = \cos\omega t$ we get

$$\mathcal{L}[\underbrace{-\omega\sin\omega t}_{f'}] = s\mathcal{L}[\cos\omega t] - 1.$$

Since $\mathcal{L}[\omega f] = \omega\mathcal{L}[f]$, see (A), then

$$\omega\mathcal{L}[\cos \omega t] = s\mathcal{L}[\sin \omega t], \quad -\omega\mathcal{L}[\sin \omega t] = s\mathcal{L}[\cos \omega t] - 1.$$

Form the above we find

$$\mathcal{L}[\cos \omega t] = \frac{s\mathcal{L}[\sin \omega t]}{\omega}. \tag{12.5}$$

Substituting into the second equation, it follows that

$$-\omega\mathcal{L}[\sin \omega t] = s\frac{s\mathcal{L}[\sin \omega t]}{\omega} - 1 \Rightarrow \omega\mathcal{L}[\sin \omega t] = -\frac{s^2\mathcal{L}[\sin \omega t]}{\omega} + 1.$$

Rearranging, $(\omega^2 + s^2)\mathcal{L}[\sin \omega t] = \omega$ and finally solving for $\mathcal{L}[\sin \omega t]$, we obtain (L5).
On the other hand, (12.5) yields

$$\mathcal{L}[\cos \omega t] = \frac{s\mathcal{L}[\sin \omega t]}{\omega} = \frac{s}{\omega} \cdot \frac{\omega}{s^2 + \omega^2} = \frac{s}{s^2 + \omega^2}. \qquad \square$$

(L7) $\mathcal{L}[\sinh \omega t] = \dfrac{\omega}{s^2 - \omega^2}, s > \omega.$

Proof. Since $\sinh \omega t = \frac{1}{2}(e^{\omega t} - e^{-\omega t})$ we use the linearity of $\mathcal{L}$ and (L2) to infer

$$\mathcal{L}\left[\frac{1}{2}(e^{\omega t} - e^{-\omega t})\right] = \frac{1}{2}\mathcal{L}[e^{\omega t}] - \frac{1}{2}\mathcal{L}[e^{-\omega t}] = \frac{1}{2}\left(\frac{1}{s - \omega}\right) - \frac{1}{2}\left(\frac{1}{s + \omega}\right) = \frac{\omega}{s^2 - \omega^2},$$

$s > \omega$. $\qquad \square$

(L8) $\mathcal{L}[\cosh \omega t] = \dfrac{s}{s^2 - \omega^2}, s > \omega.$

Proof.

$$\mathcal{L}\left[\frac{1}{2}(e^{\omega t} + e^{-\omega t})\right] = \frac{1}{2}\mathcal{L}[e^{\omega t}] + \frac{1}{2}\mathcal{L}[e^{-\omega t}] = \frac{1}{2}\frac{1}{s - \omega} + \frac{1}{2}\frac{1}{s + \omega} = \frac{s}{s^2 - \omega^2},$$

$s > \omega$. $\qquad \square$

(L9) $\mathcal{L}[e^{at}f(t)] = \mathcal{L}[f](s - a).$

Proof.

$$\mathcal{L}[e^{at}f(t)] = \int_0^\infty e^{-st} \cdot e^{at}f(t)\, dt = \int_0^\infty e^{-(s-a)t}f(t)\, dt = \mathcal{L}[f](s - a). \qquad \square$$

Example 12.2. (a) Evaluate $\mathscr{L}[t \sin t]$. Using (D) (12.4) with $n = 1$ and $f = \sin t$ we infer $\mathscr{L}[t \sin t] = -\frac{d}{ds}\mathscr{L}[\sin t]$. Then (L5) yields

$$\mathscr{L}[t \sin t] = -\frac{d}{ds}\frac{1}{1+s^2} = \frac{2s}{(1+s^2)^2}.$$

(b) Evaluate $\mathscr{L}[e^{2t} \sin t]$. Using (L9) and (L5) we find

$$\mathscr{L}[e^{2t} \sin t] = \mathscr{L}[\sin t](s-2) = \frac{1}{1+(s-2)^2}. \qquad \square$$

12.3 Inverse Laplace transform

The Laplace transform is unique in the sense that if f and g are piecewise continuous functions of exponential order then $\mathscr{L}[f] = \mathscr{L}[g]$ implies that $f(t) = g(t)$ on any interval $[0, T]$ except possibly for a finite number of points. So, we will assume that the Laplace transform has an inverse.

Under the assumption that the definition makes sense, that is $\mathscr{L}[f] = \mathscr{L}[g]$ $\Longleftrightarrow$ $f = g, \forall t \geq 0$, we define the *inverse Laplace transform* $\mathscr{L}^{-1}$ by

$$g(t) = \mathscr{L}^{-1}(F) \iff F = \mathscr{L}[g]$$

It is easy to see that the linearity of $\mathscr{L}$ implies that $\mathscr{L}^{-1}$ is also linear, that is

$$\mathscr{L}^{-1}[a_1F_1 + a_2F_2 + \cdots + a_nF_n] = a_1\mathscr{L}^{-1}(F_1) + a_2\mathscr{L}^{-1}(F_2) + \cdots + a_n\mathscr{L}^{-1}(F_n).$$

Example 12.3. Find

$$\mathscr{L}^{-1}\left[\frac{1}{(s-a)(s-b)}\right], \quad a \neq b.$$

Using partial fractions, we seek α, β such that

$$\frac{1}{(s-a)(s-b)} = \frac{\alpha}{s-a} + \frac{\beta}{s-b}.$$

We find $\alpha = \frac{1}{a-b}$ and $\beta = -\frac{1}{a-b}$. Thus

$$\mathscr{L}^{-1}\left[\frac{1}{(s-a)(s-b)}\right] = \frac{1}{a-b}\mathscr{L}^{-1}\left[\frac{1}{s-a}\right] - \frac{1}{a-b}\mathscr{L}^{-1}\left[\frac{1}{s-b}\right]$$

$$= \frac{1}{a-b}\cdot e^{at} - \frac{1}{a-b}\cdot e^{bt} = \frac{e^{at} - e^{bt}}{a-b}. \qquad \square$$

When $F(s)$ is of the type $\frac{1}{P(s)}$ where $P(s)$ is a quadratic polynomial with no real roots, it is not convenient to use the partial fraction method. In the next example we show how to overcome this difficulty by completing squares

Example 12.4. Find $\mathscr{L}^{-1}[\frac{1}{s^2+2s+2}]$.

Writing $s^2 + 2s + 2 = (s^2 + 2s + 1) + 1 = (s + 1)^2 + 1$, we find

$$\mathscr{L}^{-1}\left[\frac{1}{s^2 + 2s + 2}\right] = \mathscr{L}^{-1}\left[\frac{1}{(s + 1)^2 + 1}\right].$$

We can now use (L9) to infer

$$\mathscr{L}^{-1}\left[\frac{1}{s^2 + 2s + 2}\right] = e^{-t}\sin t. \qquad\qquad \square$$

Below is a brief list of few more common Laplace and Laplace inverse transforms:

f	$\mathscr{L}[f]$	$=$	F	$\mathscr{L}^{-1}[F]$
k	$\dfrac{k}{s}$	$=$	$\dfrac{k}{s}$	k
e^{at}	$\dfrac{1}{s-a}$	$=$	$\dfrac{1}{s-a}$	e^{at}
t^n	$\dfrac{n!}{s^{n+1}}$	$=$	$\dfrac{n!}{s^{n+1}}$	t^n
$\sin \omega t$	$\dfrac{\omega}{s^2 + \omega^2}$	$=$	$\dfrac{\omega}{s^2 + \omega^2}$	$\sin \omega t$
$\cos \omega t$	$\dfrac{s}{s^2 + \omega^2}$	$=$	$\dfrac{s}{s^2 + \omega^2}$	$\cos \omega t$
$\sinh \omega t$	$\dfrac{\omega}{s^2 - \omega^2}$	$=$	$\dfrac{\omega}{s^2 - \omega^2}$	$\sinh \omega t$
$\cosh \omega t$	$\dfrac{s}{s^2 - \omega^2}$	$=$	$\dfrac{s}{s^2 - \omega^2}$	$\cosh \omega t$

12.4 Solving differential equations by Laplace transform

One of the main features of the Laplace transform is that it is suitable for solving initial value problems of linear differential equations.

Here we give some examples in detail, which show how to use the Laplace transform in general. Below we proceed formally, first assuming $x \in \mathscr{C}$, and at the end the reader can easily verify that this is indeed the case. We prescribe the initial values at $t_0 = 0$, only for simplicity. When the initial value problem is prescribed at $t = t_0 \neq 0$ it suffices to perform the change of independent variable $t \mapsto t - t_0$.

(1) Solve $x' + x = 0$, $x(0) = k$.

Taking the Laplace transform of both sides of the equation, we find

$$\mathscr{L}[x' + x] = \mathscr{L}[x'] + \mathscr{L}[x] = 0 \implies s\mathscr{L}[x] - x(0) + \mathscr{L}[x] = 0.$$

Letting $X = \mathcal{L}[x]$, we infer

$$sX - k + X = 0 \implies (s+1)X = k \implies X = \frac{k}{s+1}.$$

Taking the inverse Laplace transform we get

$$x = \mathcal{L}^{-1}\left[\frac{k}{s+1}\right] = k\mathcal{L}^{-1}\left[\frac{1}{s+1}\right] = ke^{-t}.$$

It is clear that in this example we could have found the same result by using the integrating factor method. But then we would have to find the general solution first and then solve a pair of algebraic equations to get the solution satisfying the initial values. However, the solution obtained by using the Laplace Transform automatically satisfies the required initial values.

(2A) Solve the initial value problem

$$x'' - x' - 2x = 0, \quad x(0) = 1, \quad x'(0) = 2.$$

(2B) Find the general solution of $x'' - x' - 2x = 0$.

Solution (2A). Taking Laplace of both sides, we obtain

$$\mathcal{L}[x''] - \mathcal{L}[x'] - 2\mathcal{L}[x] = (s^2 X - s - 2) - (sX - 1) - 2X = 0 \implies (s^2 - s - 2)X = s + 1.$$

Therefore,

$$X = \frac{s+1}{(s^2 - s - 2)} = \frac{s+1}{(s-2)(s+1)} = \frac{1}{s-2}$$

which implies that $x(t) = \mathcal{L}^{-1}[X] = e^{2t}$ (see (L2)).

Solution (2B). In order to find the general solution, let us find a second solution linearly independent with the one in part (2A). There is more than one way to accomplish this. One convenient way may be to find the solution to the initial value problem

$$y'' - y' - 2y = 0, \quad y(0) = 0, \quad y'(0) = 1.$$

To find such a solution, we proceed as above and obtain

$$\mathcal{L}[y''] - \mathcal{L}[y'] - 2\mathcal{L}[y] = (s^2 Y - 1) - (sY) - 2Y = 0 \implies (s^2 - s - 2)Y = 1$$

yielding

$$Y = \frac{1}{(s-2)(s+1)}.$$

Using partial fractions, we write

$$Y = \frac{1}{3} \cdot \frac{1}{s-2} - \frac{1}{3} \cdot \frac{1}{s+1}$$

which yields the solution

$$y(t) = \mathscr{L}^{-1}[Y] = \frac{1}{3}e^{2t} - \frac{1}{3}e^{-t}.$$

It is clear that $x(t)$ and $y(t)$ are linearly independent, since $W(0) \neq 0$, where $W(t)$ represents their Wronskian. Consequently, the general solution can be expressed as

$$x(t) = c_1 e^{-t} + c_2 e^{2t}.$$

The next example deals with an equation with non constant coefficients.

For equations with variable coefficients, the Laplace Transform method is not, in general, a practical way to solve them. However, occasionally it can be used, as the following example shows.

(3) Solve the initial value problem

$$tx'' + tx' + x = 0, \quad x(0) = 0, \quad x'(0) = 1.$$

Proof. We note that

$$\mathscr{L}[tx''] = -\frac{d}{ds}\mathscr{L}[x''] = -\frac{d}{ds}(s^2 X(s) - sx(0) - x'(0)) = -s^2\frac{dX}{ds} + 2sX.$$

$$\mathscr{L}[tx'] = -\frac{d}{ds}\mathscr{L}[x'] = -\frac{d}{ds}(sX(s) - x(0)) = -sX'(s) - X(s).$$

Therefore, $\mathscr{L}[tx'' + tx' + x] = 0$ implies $-s^2 X' + 2sX - sX' - X(s) + X(s) = 0$. Solving for X', we obtain

$$X' = \frac{dX}{ds} = \frac{-2s}{-(s^2 + s)}X = \frac{2}{s+1}X.$$

Therefore,

$$\frac{dX}{X} = \frac{2ds}{s+1}$$

which yields $\ln X = \ln k(s+1)^{-2}$, or $X = k(s+1)^2$. Therefore, it follows from $(L4)$ that $x(t) = te^{-t}$ (the initial conditions imply that $k = 1$.) $\square$

The above example leads to some observations. First, concerning example (3), we notice that, unlike the constant coefficients case, we did not get an algebraic equation in $X = \mathscr{L}[x]$; instead we got a first order linear differential equation with constant co-efficients, which we had to solve for $X = \mathscr{L}[x]$ independently. Moreover, solving for $X = \mathscr{L}[x]$, we picked up a constant of integration. So, in order to get an answer to the initial value problem, we had to substitute the initial values and solve for the con-stant k.

When solving non-homogeneous linear equations by means of the Laplace trans-form, it is convenient to define the *convolution operator* $*$. The convolution $f * g$ is defined as

$$(f * g)(t) = \int_{-\infty}^{+\infty} f(t - r)g(r)\, dr, \qquad (12.6)$$

provided the integral on the right hand side makes sense.

Notice that the convolution is commutative namely $f * g = g * f$.

The main property of the convolution we will use here is the following one:

(†) If $f, g \in \mathscr{C}$ then $f * g \in \mathscr{C}$ and

$$\mathscr{L}[f * g] = \mathscr{L}[f] \cdot \mathscr{L}[g].$$

From (†) it follows immediately that:

(‡) If $f = \mathscr{L}^{-1}[F(s)]$ and $g = \mathscr{L}^{-1}[G(s)]$ then $(f * g$ makes sense and) one has

$$\mathscr{L}^{-1}[F(s) \cdot G(s)] = (f * g)(t).$$

Remark 12.2. Using (‡) we notice that the values of f and g for $t < 0$ are not relevant and can be set equal to zero. Therefore (12.6) becomes

$$(f * g)(t) = \int_{0}^{+\infty} f(t - r)g(r)\, dr.$$

For a broader discussion on convolutions we refer, e. g., to the books by W. Rudin, Real and Complex Analysis, McGraw-Hill, 1970 or H. Brezis, Functional Analysis, Sobolev Spaces and Partial Differential Equations, Springer V., 2010.

We will use convolutions in the next examples.

(4) Given $f \in \mathscr{C}$, solve the ivp for a general first order linear equation with constant coefficients

$$x' + kx = f, \quad x(0) = x_0, \quad k \in \mathbb{R}.$$

Setting $X = \mathscr{L}[x]$ and $F = \mathscr{L}[f]$, we find

$$sX - x_0 + kX = F \Rightarrow X = \frac{F + x_0}{s + k}$$

Thus

$$X = \mathcal{L}^{-1}\left[\frac{F + x_0}{s + k}\right] = \mathcal{L}^{-1}\left[\frac{F}{s + k}\right] + \mathcal{L}^{-1}\left[\frac{x_0}{s + k}\right] = \mathcal{L}^{-1}\left[\frac{F}{s + k}\right] + x_0\, e^{-kt}.$$

Setting $G = \frac{1}{s+k}$, then $\frac{F}{s+k} = F \cdot G$ and hence

$$\mathcal{L}^{-1}\left[\frac{F}{s + k}\right] = \mathcal{L}^{-1}[F \cdot G].$$

Using (‡) we find

$$\mathcal{L}^{-1}[F \cdot G] = f * g, \quad \text{where } g(t) = \mathcal{L}^{-1}[G]$$

Since

$$\mathcal{L}^{-1}[G] = \mathcal{L}^{-1}\left[\frac{1}{s + k}\right] = e^{-kt}$$

then $(f * g)(t) = \int_0^t f(t - r)e^{-kr}\, dr$ and finally

$$x(t) = x_0\, e^{-kt} + \int_0^t f(t - r)e^{-kr}\, dr, \tag{12.7}$$

which provides a solution for a general forcing term $f \in \mathcal{C}$.

Remark 12.3. The formula (12.7) can be seen as the general solution of $x' + kx = f$, depending on the parameter $x_0 = x(0) \in \mathbb{R}$.

Moreover, let us notice that in the preceding calculations it is not necessary to know what $\mathcal{L}[f]$ is.

It is also worth pointing out that in the last example we could have also handled a function $f \in \mathcal{C}$ with a finite number of discontinuities $t_1, \dots, t_n$. In such a case we would say that $x(t)$ is a *generalized solution*. By this we mean that:
1. $x(t)$ is continuous on $\mathbb{R}$;
2. $x(t)$ is differentiable on $\mathbb{R} \setminus \{t_1, \dots, t_n\}$;
3. $x(t)$ solves the differential equation on $\mathbb{R} \setminus \{t_1, \dots, t_n\}$.

Consider once again example (4) with $k = 1$, and $f = H_1$, where H_1 is the shifted Heaviside function introduced in remark 12.1.

(4') Solve the ivp $x' + x = H_1$, $x(0) = x_0$.
Here (12.7) becomes

$$x(t) = x_0\, e^{-t} + \int_0^t H_1(t - r)e^{-t}\, dr = x_0\, e^{-t} + \int_0^t e^{-(t-r)} H_1(r)\, dr.$$

From the definition of H_1 it follows that

$$\int_0^t e^{-(t-r)} H_1(r)\, dr = \begin{cases} 0 & \text{if } t < 1 \\ \int_1^t e^{r-t}\, dr = 1 - e^{1-t} & \text{if } t \geq 1 \end{cases}$$

and thus

$$x(t) = \begin{cases} x_0\, e^{-t} & \text{if } t < 1 \\ x_0\, e^{-t} + 1 - e^{1-t} & \text{if } t \geq 1 \end{cases}$$

We see that x is a generalized solution since x' has a simple discontinuity at $t = 1$, which is a consequence of the discontinuity of the forcing term H_1; see Fig. 12.1.

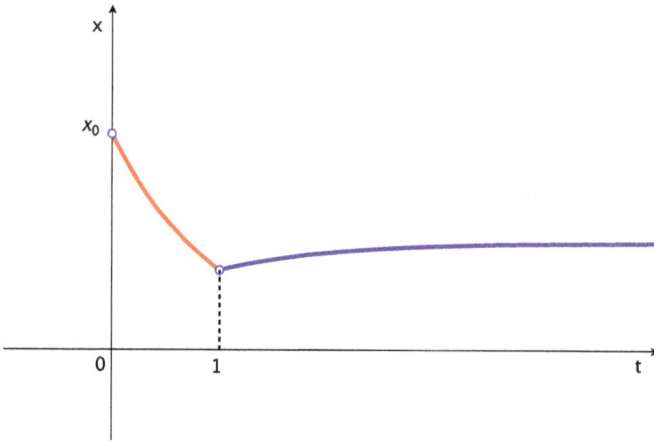

Figure 12.1: Solution of $x' + x = H_1$, $x(0) = x_0$.

(5) Solve $x'' + x = f(t)$, $x(0) = 0$, $x'(0) = 1$, with $f \in \mathscr{C}$.

Again, setting $X = \mathscr{L}[x]$, $F = \mathscr{L}[f]$, we find $s^2 X(s) - sx(0) - x'(0) + X(s) = F(s)$. Since $x(0) = 0$, $x'(0) = 1$ it follows that

$$s^2 X - 1 + X = F \Rightarrow X(s) = \frac{F(s) + 1}{1 + s^2} = \frac{F(s)}{1 + s^2} + \frac{1}{1 + s^2}.$$

We now take the inverse Laplace transform, yielding

$$x(t) = \mathscr{L}^{-1}[X(s)] = \mathscr{L}^{-1}\left[\frac{F(s)}{1 + s^2}\right] + \mathscr{L}^{-1}\left[\frac{1}{1 + s^2}\right] = \mathscr{L}^{-1}\left[\frac{F(s)}{1 + s^2}\right] + \sin t.$$

Since

$$\mathscr{L}^{-1}\left[\frac{F(s)}{1 + s^2}\right] = (f * g)(t), \quad \text{where } g(t) = \mathscr{L}^{-1}\left[\frac{1}{1 + s^2}\right] = \sin t$$

we find

$$x(t) = \sin t + (\sin *f)(t) = \sin t + \int_0^t \sin(t - r)f(r)\, dr.$$

Dealing with second order equations, if f has a finite number of discontinuities a generalized solution can have discontinuous first and/or second order derivatives.

(5′) In the previous example, if $f = H_1$ we get

$$x(t) = \sin t + (\sin *H_1)(t) = \sin t + \int_0^t \sin(t - r)H_1(r)\, dr.$$

Since

$$\int_0^t \sin(t - r)H_1(r)dr = \begin{cases} 0 & \text{if } 0 \le t \le 1, \\ \int_1^t \sin(t - r)\, dr = 1 - \cos(t - 1) & \text{if } t > 1, \end{cases}$$

we infer

$$x(t) = \begin{cases} \sin t & \text{if } 0 \le t \le 1, \\ \sin t + 1 - \cos(t - 1) & \text{if } t > 1. \end{cases}$$

In this case, although x is continuously differentiable, x is a generalized solution since x'' has a discontinuity at $t = 1$. Actually, $x''(1-) = -\sin 1$, whereas $x''(1+) = -\sin 1 + \cos 0 = 1 - \sin 1$; see Fig. 12.2.

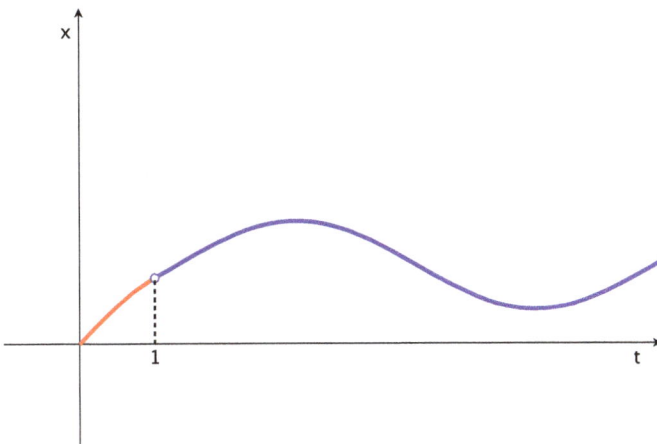

Figure 12.2: In red $x = \sin t$, $(0 < t < 1)$, in blue $x = \sin t + 1 - \cos(t - 1)$, $(t > 1)$.

12.5 Exercises

1. Find $\mathscr{L}[e^t - \sin t]$.
2. Find $\mathscr{L}[e^{-\frac{1}{2}t}]$.
3. Find $\mathscr{L}[e^{3t}]$ by using the definition of Laplace transform.
4. Find $\mathscr{L}[e^{2t} \sinh t]$.
5. Find $\mathscr{L}[te^{-t}]$.
6. Find $\mathscr{L}[e^{3t} \cos 2t]$.
7. Let $f \in \mathscr{C}$ be such that $tf' \in \mathscr{C}$. Show that $\mathscr{L}[tf'] = -F(s) - sF'(s)$, where $F(s) = \mathscr{L}[f]$.
8. If $\mathscr{L}[f(t)] = F(s) = \frac{4}{s^2-16}$, find $f(t) = \mathscr{L}^{-1}[F(s)]$.
9. Find $f(t) = \mathscr{L}^{-1}[F(s)]$ if $\mathscr{L}[f(t)] = F(s) = \frac{s+1}{s^2-s}$.
10. Find $\mathscr{L}[e^{-4t} \sin 3t]$.
11. Find $\mathscr{L}[t^2 e^t] =$ in two ways:
    (a) by using property (D)
    (b) by using property $(L9)$.
12. Find $f(t)$ that such that $\mathscr{L}[f(t)] = \frac{s+1}{s^2+s-2}$.
13. Find $\mathscr{L}[e^{-2t} \cos t]$.
14. * Show that $\mathscr{L}[J_0] = \frac{1}{\sqrt{1+s^2}}$, where J_0 is the Bessel function of order 0, such that $J_0(0) = 1, J_0'(0) = 0$.
15. Find $\mathscr{L}^{-1}[\frac{1}{s^2-2}]$.
16. Find $\mathscr{L}^{-1}[\frac{2s+3}{s^2}]$.
17. Find $\mathscr{L}^{-1}[\frac{s^2+s+1}{s^3+s}]$.
18. Find $\mathscr{L}^{-1}[\frac{1}{s^4+s^2}]$.
19. Find $\mathscr{L}^{-1}[\frac{1}{s^2-2s+5}]$.
20. Using the Laplace transform, solve the ivp $x' - x = 1, x(0) = 2$.
21. Using the Laplace transform, solve the ivp $x' + 3x = 0, x(1) = -1$.
22. Use the Laplace transform to solve $x'' - 2x' + x = 0, x(0) = 0, x'(0) = 1$.
23. Use the Laplace transform to solve $x'' + x = t, x(0) = 0, x'(0) = 1$.
24. Using the Laplace transform solve the ivp $x'' + x = 0, x(0) = x'(0) = 1$.
25. Using the Laplace transform solve the ivp $x'' + 3x' = 0, x(0) = -1, x'(0) = -2$.
26. Using the Laplace transform, solve the ivp $4x'' - 12x' + 5x = 0, x(0) = 1, x'(0) = k$, depending upon $k \in R$.
27. Solve $tx'' + 2tx' + 2x = 0, x(0) = 0, x'(0) = 1$ by the Laplace method.
28. Use the Laplace method to solve $x'' - x = f(t), x(0) = 1, x'(0) = 0$, where $f \in \mathscr{C}$.
29. Find the generalized solution of $x'' - 4x = H_1(t), x(0) = x'(0) = 0$.
30. Find the generalized solution of $x'' + x = f(t), x(0) = 0, x'(0) = 1$, where

$$f(t) = \begin{cases} 1 & \text{if } 0 \le t \le a, \\ 0 & \text{if } t > a, \end{cases}$$

$a > 0$.

13 A primer on equations of Sturm–Liouville type

Some of the topics we are going to present in this chapter are more advanced, but we make an effort to make the level of presentation simple. The topic is important in areas such as mechanics, the sciences, and the calculus of variations. Of course, we do not pretend to give a complete and exhaustive treatment of the subject.

13.1 Preliminaries

We are interested in studying the class of linear second order equations of the form

$$L[x] \stackrel{\text{def}}{=} (rx')' + px = 0. \tag{13.1}$$

Here and throughout in the sequel it is understood that $p, r \in C(\mathbb{R})$, $r(t) > 0$ and x and rx' are continuously differentiable on $\mathbb{R}$. For example, this is the case for any $r \in C^1(\mathbb{R})$ and $x \in C^2(\mathbb{R})$.

First, some preliminary lemmas are in order.

Lemma 13.1. *For every $a, b \in \mathbb{R}$ and every $x, y \in C^2([a, b])$ such that $x(a) = x(b) = 0$, $y(a) = y(b) = 0$ one has*

$$\int_a^b L[x]y \, dt = \int_a^b xL[y] \, dt. \tag{13.2}$$

Proof. Recalling that x, y vanish at $t = a, b$, integration by parts yields

$$\int_a^b L[x]y \, dt = \int_a^b (rx')'y \, dt + \int_a^b pxy \, dt = -\int_a^b rx'y' \, dt + \int_a^b pxy \, dt.$$

Similarly,

$$\int_a^b L[y]x \, dt = \int_a^b (ry')'x \, dt + \int_a^b pxy \, dt = -\int_a^b rx'y' \, dt + \int_a^b pxy \, dt,$$

yielding the conclusion of the lemma. □

In view of the symmetry property (13.2) the operator L is called *self-adjoint*, and the equation (13.1) a *self-adjoint equation*.

Lemma 13.2. *Any equation*

$$x'' + a_1(t)x' + a_0(t)x = 0, \quad a_1(t), a_0(t) \text{ continuous,}$$

can be transformed to self-adjoint form $L[x] = 0$.

https://doi.org/10.1515/9783111185675-013

Proof. If we multiply the equation by $e^{\int a_1(t)\,dt}$, then it can be written as

$$\left(e^{\int a_1(t)\,dt}x'\right)' + a_0(t)e^{\int a_1(t)\,dt}x = 0,$$

which is in self-adjoint form with $r(t) = e^{\int a_1(t)\,dt}$ and $p(t) = a_0(t)e^{\int a_1(t)\,dt}$. □

Example 13.1. In order to write the equation $x'' + tx' + t^2x = 0$ in self-adjoint form, we let

$$r = e^{\int t\,dt} = e^{\frac{1}{2}t^2}$$

obtaining $(r(t)x')' + t^2r(t)x = 0$. □

Lemma 13.3. *The self-adjoint equation* (13.1) *possesses the existence and uniqueness property.*

Proof. One way to see this is to note that

$$(rx')' + p(t)x = 0, \quad x(t_0) = x_0, \quad x'(t_0) = x_0',$$

is equivalent to the system

$$\bar{x}' = \begin{pmatrix} x_1' \\ x_2' \end{pmatrix} = \begin{pmatrix} 0 & \frac{1}{r(t)} \\ -p(t) & 0 \end{pmatrix}\begin{pmatrix} x_1 \\ x_2 \end{pmatrix}\bar{x}, \quad \bar{x}(t_0) = \begin{pmatrix} x_1(t_0) \\ x_2(t_0) \end{pmatrix} = \begin{pmatrix} x_0 \\ r(t_0)x_0' \end{pmatrix}$$

where $x_1 = x$ and $x_2 = rx_1'$. The coefficient matrix is continuous and the existence and uniqueness property holds. □

Lemma 13.4. *If* $x(t)$ *is a nontrivial solution of* (13.1), *then no finite interval* $[a,b] \subset \mathbb{R}$ *can contain more than a finite number of zeros of* $x(t)$; *in other words, the zeros of* $x(t)$ *are isolated.*

Proof. Let S be the set of zeros of $x(t)$. Suppose that $S \subset [a,b]$ for some bounded interval $[a,b]$. We will show that S cannot be an infinite set. Suppose that, on the contrary, it is infinite. Then, by the Bolzano–Weierstrass theorem, S has a limit point s in the interval $[a,b]$. Let $s_n \in S \cap [s - \frac{1}{n}, s + \frac{1}{n}]$, $n = 1,2,3,\ldots$ Clearly, $\lim_{n\to\infty} s_n = s$. Since, $x(t)$ is continuous, $\lim_{n\to\infty} x(s_n) = x(s)$. Since $x(s_n) = 0$ for each n, we conclude that $x(s) = 0$.

Next, we show that $x'(s) = 0$. But this follows from the fact that, by Rolle's lemma, for any natural number n, there is a number t_n between s_n and s_{n+1} such that $x'(t_n) = 0$. Since $s_n \to s$, we must also have $t_n \to s$. By continuity of $x'(t)$, we have $x'(t_n) \to x'(s)$, and hence $x'(s) = 0$.

We have shown that $x(s) = x'(s) = 0$ and hence, by uniqueness, $x(t) \equiv 0$, contradicting the assumption that $x(t)$ is a nontrivial solution. □

The above lemma justifies the assumption we will make in the sequel: that consecutive zeros of solutions of $L[x] = 0$ exist. This is important; otherwise discussing consecutive zeros is fruitless.

13.2 Oscillation for self-adjoint equations

A function $x(t)$ is said to be *oscillatory* on an interval (a, ∞), if $\forall \tau \ \exists t_0 > \tau$ such that such that $x(t_0) = 0$, or equivalently, if $\forall T > a$, $x(t)$ has infinitely many zeros in $(T, +\infty)$.

A function $x(t)$ is *nonoscillatory* if it is not oscillatory, namely if $\exists \tau > a$ such that $x(t) \neq 0$ for $t > \tau$. A second order linear equation is called *oscillatory* if it has one oscillatory solution.

In the sequel we will focus on the oscillatory properties of the self-adjoint equation

$$L[x] = (rx')' + px = 0. \tag{13.1}$$

Lemma 13.5 (Abel's identity for self-adjoint equations). *If $x(t)$ and $y(t)$ are solutions of* (13.1), *then*

$$r(t)[xy' - x'y] \equiv k, \quad k \ a \ constant.$$

Proof. Notice that corresponding to the scalar solutions x and y,

$$\bar{x}(t) = \begin{pmatrix} x \\ rx' \end{pmatrix} \quad \text{and} \quad \bar{y}(t) = \begin{pmatrix} y \\ ry' \end{pmatrix}$$

are solutions of the vector equation

$$\bar{z}' = A(t)\bar{z} = \begin{pmatrix} 0 & \frac{1}{r} \\ -p & 0 \end{pmatrix} \bar{z}.$$

Therefore, by Theorem 8.2 of Chapter 8, the Wronskian of $\bar{x}$ and $\bar{y}$ is given by

$$W(\bar{x}, \bar{y}) = W(t_0)e^{\int \text{trace } A(s) \, ds}.$$

Since the trace of A is 0, we have $W(\bar{x}, \bar{y}) = W(t_0)$. Renaming $W(t_0)$ as $W(t_0) = k$, we have $W(\bar{x}, \bar{y}) = k$. Since, by definition of the Wronskian of two vector $\bar{x}$ and $\bar{y}$ we have

$$W(\bar{x}, \bar{y}) = \begin{vmatrix} x & y \\ rx' & ry' \end{vmatrix} = r(xy' - yx'),$$

it follows that $r(xy' - yx') = k$. □

The following two simple and elegant results, due to the French mathematician J. C. F. Sturm, play an important role in the theory of oscillation.

13.2.1 Sturm separation and comparison theorems

Theorem 13.1 (Sturm separation theorem). *Let $x(t)$ be a nontrivial solution of* (13.1) *such that $x(t_1) = 0 = x(t_2)$ and assume that t_1 and t_2 are consecutive zeros ($t_1 < t_2$). If $y(t)$ is any solution such that x and y are linearly independent, then $y(t)$ has exactly one zero between t_1 and t_2; see Fig. 13.1.*

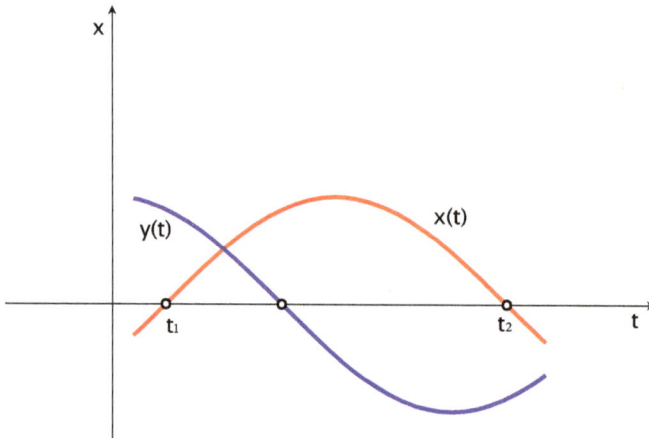

Figure 13.1: Sturm separation theorem.

Proof. Assume that $y(t)$ does not vanish in the interval (t_1, t_2). We will show that this leads to a contradiction. Since x and y are linearly independent, they cannot have a common zero; otherwise their wronskian would vanish at that point, which would imply that they are linearly dependent; this would be contrary to the hypothesis. Since t_1 and t_2 are consecutive zeros of $x(t)$, $x(t)$ does not change sign in (t_1, t_2), say $x(t) > 0, \forall t \in (t_1, t_2)$ (if $x(t) < 0$ we simply replace $x(t)$ by $-x(t)$). Similarly, we can assume that $y(t) > 0$ between t_1 and t_2.

Using Abel's identity (see Lemma 13.5) and the hypothesis $x(t_1) = 0$, we have $r(t_1)[x(t_1)y'(t_1) - x'(t_1)y(t_1)] = -r(t_1)x'(t_1)y(t_1) = k$. Since $x(t) > 0$ to the right of t_1 and $x(t_1) = 0$, we must have $x'(t_1) > 0$. This implies that $-r(t)x'(t_1)y(t_1) = k < 0$. Now, let us examine Abel's identity at $t = t_2$. We have $r(t_2)[x(t_2)y'(t_2) - x'(t_2)y(t_2)] = -r(t_2)x'(t_2)y(t_2) = k$. Since $x(t_2) = 0$ and $x(t) > 0$ to the immediate left of t_2, we must have $x'(t_2) < 0$. Consequently, $-r(t_2)x'(t_2)y(t_2) = k < 0$ implies that $y(t_2) < 0$. Since $y(t_1) > 0$ and $y(t_2) < 0$, it follows from the intermediate value theorem that $y(t)$ vanishes somewhere between t_1 and t_2.

To prove that $y(t)$ cannot have more than one zero between the two consecutive zeros of $x(t)$, we simply note that if $y(t)$ vanished at two points between t_1 and t_2, then, by the argument we just gave, switching the roles of x and y, we would have $x(t)$ vanishing at some point between t_1 and t_2, contradicting the assumption that t_1 and t_2 are consecutive zeros of $x(t)$. $\quad\square$

Corollary 13.1. *If* (13.1) *has one oscillatory solution, then all solutions are oscillatory.*

Theorem 13.2 (Sturm comparison theorem). *Let $x(t)$ and $y(t)$ be nontrivial solutions satisfying the equations*

$$\text{(I)} \quad (r(t)x')' + p(t)x = 0, \qquad \text{(II)} \quad (r(t)y')' + q(t)y = 0,$$

respectively.

Suppose that t_1 and t_2 in I are consecutive zeros of $x(t)$, $t_1 < t_2$, and $y(t_1) = 0$. Furthermore, assume that $p(t)$ and $q(t)$ are continuous in the interval $[t_1, t_2]$ with $q(t) \geq p(t)$, and strict inequality holding at some point in the interval $[t_1, t_2]$. Then $y(t)$ will vanish again somewhere between t_1 and t_2.

Proof. Suppose that $y(t) > 0$ for $t_1 < t < t_2$, which would imply $y'(t_1) > 0$. We will show that this leads to a contradiction. As in the proof of Theorem 13.1, we may assume, without loss of generality, that $x(t) > 0$ in the open interval (t_1, t_2) which would imply $x'(t_1) > 0, x'(t_2) < 0$.

Multiplying (I) by $-y$ and (II) by x and adding, we obtain

$$x(ry')' - y(rx')' = (p(t) - q(t))xy.$$

Integrating the above equation from t_1 to t_2 yields

$$rxy'\big|_{t_1}^{t_2} - \int_{t_1}^{t_2} rx'y' - yrx'\big|_{t_1}^{t_2} + \int_{t_1}^{t_2} y'rx' = \int_{t_1}^{t_2} (p(t) - q(t))xy\,dt.$$

Now, since $x(t_1) = x(t_2) = y(t_1) = 0$, simplifying the above equation we obtain

$$r(t_2)y(t_2)x'(t_2) = \int_{t_1}^{t_2} (q(t) - p(t))xy\,dt.$$

Since $q(t) - p(t) \geq 0$ with strict inequality at some point in the interval (t_1, t_2), on the right side we have

$$\int_{t_1}^{t_2} (q(t) - p(t))xy\,dt > 0,$$

while on the left side we have

$$r(t_2)x'(t_2)y(t_2) \leq 0,$$

a contradiction. $\quad\square$

Theorem 13.3. *Under the same assumptions as the Sturm comparison theorem above, any solution $z(t)$ of (II) vanishes between any two consecutive, and hence between any two, zeros of $x(t)$.*

Proof. We can use the existence theorem to obtain a nontrivial solution $y(t)$ of II such that $y(t_1) = 0, y'(t_1) = y_1$, for any $y_1 \neq 0$. Then it follows from the Sturm comparison theorem that $y(\bar{t}) = 0$, for some $\bar{t}$ between t_1 and t_2. If $z(t)$ and $y(t)$ are linearly dependent, then the assertion follows since they have the same zeros. If they are linearly independent, then, by the Sturm separation theorem, $z(t)$ vanishes somewhere between t_1 and $\bar{t}$ and hence between t_1 and t_2. □

Example 13.2. As an application of the preceding theorem, we can show a property of the zeros of Bessel functions (Bessel equations and Bessel functions are discussed in Chapter 11, section 11.3). Precisely, let us consider, e. g., a Bessel function of the first kind $J_m(t)$ (the arguments to deal with the Bessel functions of the second kind $Y_m(t)$ are similar). We will show that if $m_1 \neq m_2$ then between two zeros of $J_{m_1}(t)$ there is a zero of $J_{m_2}(t)$.

First of all, for $t > 0$ we can write the Bessel equation $t^2 x'' + t x' + (t^2 - m^2)x = 0$ as

$$tx'' + x' + \left(t - \frac{m^2}{t}\right)x = 0. \tag{B_m}$$

Since $tx'' + x' = (tx')'$, (B_m) is in the self-adjoint form (13.1) with $r(t) = t$. Let, e. g., $m_2 < m_1$. Taking $p(t) = t - \frac{m_1^2}{t}$ and $q(t) = t - \frac{m_2^2}{t}$ we get $q(t) > p(t)$ for $t > 0$. Hence we can apply Theorem 13.3 and the claim follows. □

Corollary 13.2. *If there exists a number $T > 0$ such that, for $t \geq T$, $p(t) > k > 0$, then the equation $x'' + p(t)x = 0$ is oscillatory. In particular, if $\lim_{t \to +\infty} p(t) = k > 0$, then $x'' + p(t)x = 0$ is oscillatory.*

Proof. A solution of $y'' + ky = 0$ is given by $y = \sin(\sqrt{k}t)$. Thus, an application of the Sturm comparison theorem shows that $x'' + p(t)x = 0$ is oscillatory. The last assertion follows since $\lim_{t \to +\infty} p(t) = k > 0$ implies that $p(t) > \frac{k}{2}$ for t larger than some number T. □

On the other hand we have the following nonoscillatory result.

Theorem 13.4. *If for $t \geq T$, $p(t) \leq 0$, $p(t) \neq 0$, then no nontrivial solution of (13.1) can have more than one zero in $[T, \infty)$.*

Proof. By way of contradiction, let t_1, t_2 be zeroes of $x(t)$ such that $t_2 > t_1 \geq T$. Then

$$\int_{t_1}^{t_2} (rx')' x + \int_{t_1}^{t_2} px^2 = 0.$$

$$z(t) + \int_c^t \frac{z^2}{r}\, ds = z(c) - \int_c^t p(s)\, ds.$$

We can choose a number $d > c$ such that, for $t \geq d$, the right side of the preceding equation is negative. Therefore, for $t \geq d$,

$$z(t) < -\int_c^t \frac{z^2}{r}\, ds,$$

which yields

$$z^2 > \left[\int_c^t \frac{z^2}{r}\, ds\right]^2.$$

Letting

$$y = \int_c^t \frac{z^2}{r}\, ds,$$

we find $y'(t) = \frac{z^2(t)}{r(t)}$ and thus $y'(t) > \frac{y^2(t)}{r(t)}$ for $t \geq d$. Notice that $y(t) > 0$ for $t > c$. Then $\frac{y'(t)}{y^2(t)} > \frac{1}{r(t)}$, whereby

$$\int_d^t \frac{y'}{y^2}\, ds > \int_d^t \frac{ds}{r(s)}.$$

Integrating the left side, we obtain

$$-\frac{1}{y(t)} + \frac{1}{y(d)} > \int_d^t \frac{ds}{r(s)},$$

which implies that

$$\frac{1}{y(d)} > \frac{1}{y(t)} + \int_d^t \frac{ds}{r(s)},$$

which is clearly a contradiction, since $y(t) > 0$ and $\int_d^\infty \frac{ds}{r(s)} = +\infty$. $\quad\square$

Example 13.3. Determine the oscillation status of the equation

$$\left(\frac{1}{3t-6}x'\right)' + \frac{5}{2t-7}x = 0.$$

Using Theorem 13.5, it is easy to see that it is oscillatory, since

$$\int_{10}^{\infty} \frac{1}{r}\, dt = \int_{10}^{\infty} (3t-6)\, dt = +\infty = \int_{10}^{\infty} \frac{5}{2t-7}\, dt = \int_{10}^{\infty} p\, dt. \qquad \square$$

13.2.2 Checking the oscillatory status of equations

Below we describe some useful transformations that may be used to change the form of an equation to one that is more convenient for using certain theorems and known results, in order to determine the oscillatory behavior of solutions.

1. Eliminating Second Highest Derivative. Consider

$$a(t)x'' + b(t)x' + c(t)x = 0 \qquad (13.3)$$

where $a(t)$, $b(t)$ and $c(t)$ are continuous, $a(t) > 0$. We substitute $x(t) = u(t)y(t)$ in the equation and determine u so that the x'-term gets eliminated.

Substituting $x' = uy' + u'y$, $x'' = uy'' + 2u'y' + u''y$, we obtain

$$a(t)[uy'' + 2u'y' + u''y] + b(t)[uy' + u'y] + c(t)uy = 0.$$

Rearranging terms, we have

$$auy'' + (2au' + bu)y' + (au'' + bu' + cu)y = 0.$$

Now if we set $2au' + bu = 0$ and solve for u (holding the constant of integration to be 1), we obtain

$$u = e^{-\frac{1}{2}\int \frac{b(t)}{a(t)}\, dt} \qquad (13.4)$$

thus transforming (13.3) into

$$auy'' + (au'' + bu' + cu)y = 0.$$

We point out that a similar transformation will also eliminate the second highest derivative in higher order equations. More precisely, substituting

$$x = y.e^{\frac{-1}{n}\int \frac{a_1(t)}{a_0(t)}\, dt}$$

will eliminate the $x^{(n-1)}$-term in

$$a_0(t)x^{(n)} + a_1 x^{(n-1)} + a_2 x^{(n-2)} + \cdots + a_n x = 0.$$

Example 13.4. (a) In order to determine the oscillatory status of the equation

$$tx'' + x' - tx = 0, \tag{13.5}$$

we first write it in selfadjoint form

$$(tx')' - tx = 0.$$

However, this does not seem to be helpful; $r(t) = t$ goes to $+\infty$ and $p(t) = -t$ goes to $-\infty$, so that Theorem 13.5 is not applicable. Now let us see if eliminating the x'-term can help. To this end, we substitute

$$x = t^{-\frac{1}{2}}y, \quad t > 1, \quad x' = t^{-\frac{1}{2}}y' - \frac{1}{2}t^{\frac{-3}{2}}y, \quad x'' = t^{-\frac{1}{2}}y'' - t^{-\frac{3}{2}}y' + \frac{3}{4}t^{-\frac{5}{2}}y$$

obtaining

$$u'' + \left[\frac{1}{4t^2} - 1\right]u = 0,$$

which is again nonoscillatory.

(b) Consider the equation

$$x'' - x' + tx = 0.$$

Writing it in self-adjoint form, we obtain

$$(e^{-t}x')' + te^{-t}x = 0,$$

which is not helpful since $p(t)$ and $r(t)$ are positive but approach zero as t approaches ∞. Now, let us eliminate the x'-term by letting $x(t) = e^{\frac{1}{2}t}u$. Then we obtain

$$u'' + \left[t - \frac{1}{4}\right]u = 0,$$

which is clearly oscillatory. □

2. A Useful Transformation of $x'' + p(t)x = 0$. As we will see later, in studying the oscillatory status of this seemingly simple equation, sometimes it is convenient to transform it to a self-adjoint form as follows. Let

$$x(t) = t^{\frac{1}{2}}y.$$

Then

$$x'' + p(t)x = t^{\frac{1}{2}}y'' + t^{-\frac{1}{2}}y' + \left(pt^{\frac{1}{2}} - \frac{1}{4}t^{\frac{-3}{2}}\right)y = 0.$$

Multiplying the last equation by $t^{\frac{1}{2}}$, we obtain

$$(ty')' + \left(pt - \frac{1}{4t}\right)y = 0.$$

The various transformations and methods discussed above sometimes allow us to use Theorem 13.5 in cases when, at first glance, it may not seem applicable. See the following example.

Example 13.5. Consider

$$x'' + t^{-\frac{3}{2}}x = 0. \tag{13.6}$$

We see that $p(t) = t^{-\frac{3}{2}}$ approaches 0 as t approaches infinity, and

$$\int_1^\infty t^{-\frac{3}{2}}\, dt$$

is finite. So, it is not obvious that any of the methods and discussions above can be applied directly.

However, using the transformation $x = t^{\frac{1}{2}}y$ changes the equation to the selfadjoint form

$$(ty')' + \left(\frac{1}{\sqrt{t}} - \frac{1}{4t}\right)y = 0. \tag{13.7}$$

We notice that for $t > 1$,

$$\frac{1}{\sqrt{t}} - \frac{1}{4t} > \frac{1}{t} - \frac{1}{4t} = \frac{3}{4t}.$$

Since $\int_1^\infty \frac{3}{4t} = \infty$, equation (13.7), and hence (13.6), is oscillatory by Theorem 13.5. □

In general, if the equation is given in the form

$$a(t)x'' + b(t)x' + c(t)x = 0$$

then we may either eliminate the x'-term or transform it into a self-adjoint form; and then either use Theorem 13.5 or try to find another way of resolving the issue. Of course, this is not always easy, even for some simple equations, as we will see in the following example.
Consider

$$x'' + p(t)x = 0$$

where $p(t)$ is some continuous function. We know that if $p(t) \geq k > 0$, k a constant, then the solutions are oscillatory, and if $p(t) \leq 0$, they are nonoscillatory. The difficult cases are among those where neither of these conditions holds. For example, $p(t)$ may be oscillatory, or $p(t)$ may be positive but such that it approaches 0 as t approaches ∞.

Example 13.6. Consider the equation

$$x'' + (8\cos t)x = 0. \tag{13.8}$$

In this simple looking example, none of the transformations seem to help. We will try to find a convenient equation so that we can use the Sturm comparison theorem directly. We note that on the interval $[-\frac{\pi}{4}, \frac{\pi}{4}]$,

$$8\cos t \geq 8\frac{\sqrt{2}}{2} > 4.$$

So, we consider the equation

$$y'' + 4y = 0. \tag{13.9}$$

We know that $y(t) = \cos 2t$ is a solution of (13.9) and

$$y\left(-\frac{\pi}{4}\right) = \cos\left(-\frac{\pi}{2}\right) = 0 = y\left(\frac{\pi}{4}\right) = \cos\left(\frac{\pi}{2}\right) = 0.$$

Now, since

$$8\cos t > 4$$

on the interval $[-\frac{\pi}{4}, \frac{\pi}{4}]$, it follows from the Sturm comparison theorem that $x(t)$ vanishes at some point $t = \tilde{t}$ between $-\frac{\pi}{4}$ and $\frac{\pi}{4}$. Therefore, $x(t)$ vanishes in the interval $[-\frac{\pi}{4} + 2n\pi, \frac{\pi}{4} + 2n\pi]$ for all natural numbers n. □

Unfortunately, there does not seem to be a general pattern for such problems. One has to cope with each such problem individually. In fact, the hardest problem of this type seems to be determining the oscillation status of

$$x'' + (\sin t)x = 0.$$

As one might expect, this equation is oscillatory. A proof can be given using a more advanced theorem, which is out of the scope of this book.

13.3 Sturm–Liouville eigenvalue problems

By a Sturm–Liouville *eigenvalue problem* we mean a boundary value problem as

$$\begin{cases} (rx')' + \lambda px = 0, \\ x(a) = x(b) = 0, \end{cases} \tag{EP}$$

where λ is a real parameter. Clearly one solution of (EP) is the trivial function $x = 0$ for all $\lambda \in \mathbb{R}$.

We say that $\bar{\lambda}$ is an eigenvalue for (EP) if for $\lambda = \bar{\lambda}$ problem (EP) has a nontrivial solution ϕ. Such a ϕ is called an eigenfunction corresponding to the eigenvalue $\bar{\lambda}$. Of course, if ϕ is an eigenfunction, so is $c\phi, \forall c \in \mathbb{R}, c \neq 0$.

Theorem 13.6.

(i) *If $p > 0$ then the eigenvalues of (EP) are strictly positive.*

(ii) *Let ϕ_i be eigenfunctions of (EP) corresponding to eigenvalues $\lambda_i, i = 1, 2$. If $\lambda_1 \neq \lambda_2$ then*

$$\int_a^b p(t)\phi_1(t)\phi_2(t)\, dt = 0.$$

Proof. (i) Let ϕ be an eigenfunction of (EP) corresponding to the eigenvalue $\bar{\lambda}$. Multiplying $(r\phi')' + \bar{\lambda}p\phi = 0$ by ϕ and integrating on (a, b) we find

$$\int_a^b (r\phi')'\phi + \bar{\lambda}\int_a^b p\phi^2 = 0.$$

Integrating by parts the integral on the left hand side (recall that $\phi(a) = \phi(b) = 0$) we infer

$$-\int_a^b r\phi'\phi' + \bar{\lambda}\int_a^b p\phi^2 = 0.$$

Thus

$$\bar{\lambda}\int_a^b p\phi^2 = \int_a^b r\phi'^2.$$

Since $r > 0, p > 0$ and $\phi \neq 0$, it follows that $\bar{\lambda} > 0$.

(ii) Multiplying $(r\phi_1')' + \lambda_1 p\phi_1 = 0$ by ϕ_2 and $(r\phi_2')' + \lambda_2 p\phi_2 = 0$ by ϕ_1 and integrating by parts as before we find

$$\int_a^b r\phi_1'\phi_2' = \lambda_1 \int_a^b p\phi_1\phi_2, \quad \int_a^b r\phi_2'\phi_1' = \lambda_2 \int_a^b p\phi_2\phi_1.$$

Then

$$\lambda_1 \int_a^b p\phi_1\phi_2 = \lambda_2 \int_a^b p\phi_1\phi_2$$

whereby

$$(\lambda_2 - \lambda_1) \int_a^b p\phi_1\phi_2 = 0.$$

Since $\lambda_2 \neq \lambda_1$ the result follows. □

Corollary 13.3. *If p does not change sign in (a,b) then ϕ_1, ϕ_2 are linearly independent.*

Proof. Otherwise, if $\phi_2 = c\phi_1$ we get $\int_a^b p\phi_1\phi_2 \, dt = c \int_a^b p\phi_1^2 \, dt$. Using (ii) we find $c \int_a^b p\phi_1^2 \, dt = 0$. Since p does not change sign in (a,b) this implies $c = 0$. □

Next, let us consider the case in which $r, p > 0$ are real numbers in more detail. Dividing by r and setting $m^2 = \frac{p}{r} > 0$, (EP) becomes

$$\begin{cases} x'' + \lambda m^2 x = 0, \\ x(a) = x(b) = 0. \end{cases} \tag{EP'}$$

It is convenient to perform the change of the independent variable

$$t \mapsto s = \frac{\pi(t-a)}{b-a},$$

which maps the interval $[a,b]$ into $[0, \pi]$.

Since $ds = \frac{\pi}{b-a} dt$, the equation $\frac{d^2x}{dt^2} + \lambda m^2 x = 0$ becomes

$$\frac{\pi^2}{(b-a)^2} \frac{d^2x}{ds^2} + \lambda m^2 x = 0;$$

namely

$$\frac{d^2x(s)}{ds^2} + \lambda \frac{(b-a)^2}{\pi^2} m^2 x(s) = 0.$$

Setting $\tilde{m}^2 = \frac{(b-a)^2}{\pi^2} m^2 > 0$, (EP') gets transformed into

$$\begin{cases} x''(s) + \lambda \tilde{m}^2 x(s) = 0, \\ x(0) = x(\pi) = 0. \end{cases} \tag{EP''}$$

If $x(s)$ solves (EP″) then

$$y(t) = x\left(\frac{\pi(t-a)}{b-a}\right)$$

solves (EP′).

Lemma 13.6. *The eigenvalues of (EP″) are given by*

$$\lambda_k = \frac{k^2}{\tilde{m}^2}, \quad k = 1, 2, \dots$$

with corresponding eigenfunctions $\phi_k(s) = \sin(ks)$.

Proof. The general solution of $x''(s) + \lambda \tilde{m}^2 x(s) = 0$ is given by $x = A\sin(\tilde{m}\sqrt{\lambda}s + \theta)$. Without loss of generality we can assume that $A \neq 0$ otherwise $x \equiv 0$. The boundary conditions $x(0) = x(\pi) = 0$ imply

$$\sin(\tilde{m}\sqrt{\lambda} \cdot 0 + \theta) = 0, \quad \sin(\tilde{m}\sqrt{\lambda} \cdot \pi + \theta) = 0.$$

The former yields $\theta = 0$, the latter $\tilde{m}\pi\sqrt{\lambda} \cdot \pi = k\pi$ whereby $\sqrt{\lambda} = \frac{k}{\tilde{m}}$, namely $\lambda_k = \frac{k^2}{\tilde{m}^2}$, $k = 1, 2, \dots$. To find the eigenfunctions ϕ_k we simply notice that $\sqrt{\lambda_k} = \frac{k}{\tilde{m}}$ and hence (taking $A = 1$ and $\theta = 0$) $\phi_k(s) = \sin(\tilde{m}\sqrt{\lambda_k}\, s) = \sin ks$ as claimed. □

Theorem 13.7. *The eigenvalues of (EP′) and the corresponding eigenfunctions are given by*

$$\lambda_k[m] = \frac{k^2\pi^2}{m^2(b-a)^2}, \quad \phi_k(t) = \sin\frac{\pi k(t-a)}{b-a}.$$

Proof. It suffices to use Lemma 13.6, taking into account that $\tilde{m}^2 = \frac{(b-a)^2}{\pi^2}m^2$ and that
$s = \frac{\pi(t-a)}{b-a}$. □

Example 13.7. The eigenvalues of $2x'' + 3\lambda x = 0$, $x(0) = x(2\pi) = 0$ are given by $\lambda_k = \frac{k^2\pi^2}{\frac{3}{2}(2\pi)^2} = \frac{k^2}{6}$. □

Corollary 13.4.
(i)   *Let $m > 0$. Then $\lambda_k[m] \to +\infty$ as $k \to +\infty$;*
(ii) *if $0 < m_1 \leq m_2$ then $\lambda_k[m_1] \geq \lambda_k[m_2]$;*
(iii) *$\phi_1(t) > 0$ in the open interval (a, b), whereas $\phi_k(t)$, $k = 2, 3, \dots$ has precisely $k - 1$ zeros in (a, b).*

Theorem 13.7 and Corollary 13.4 hold in general.

Theorem 13.8. *If $r, p > 0$ then (EP) has infinitely many positive eigenvalues λ_k such that $0 < \lambda_1 < \lambda_2 < \lambda_3 < \cdots$. Moreover,*

(i) $\lambda_k \to +\infty$ as $k \to +\infty$.

(ii) *For fixed $r > 0$, the eigenvalues $\lambda_k[p]$ satisfy the following comparison property: if $p_1 \leq p_2$ on $[a, b]$, then $\lambda_k[p_1] \geq \lambda_k[p_2]$ for any $k = 1, 2, \ldots$*

(iii) *Any eigenfunction ϕ_k associated to λ_k has exactly $k - 1$ zeros in the open interval (a, b).*

An outline of the proof of these results is based on several steps; it is given below for completeness.

Proof. For simplicity, we let $r = 1$. The general case requires only some more cumbersome notations.

Step 1. Problem (EP) is equivalent to the first order system

$$\begin{cases} x' = y, \\ y' = -\lambda px. \end{cases} \tag{13.10}$$

Let us introduce new coordinates $r = r(t)$ and $\theta = \theta(t)$ by setting

$$\begin{cases} x = r \sin \theta, \\ y = r \cos \theta, \end{cases}$$

where r, θ verify

$$r^2 = x^2 + y^2, \quad \tan \theta = \frac{x}{y}.$$

Differentiating the former we find $2rr' = 2xx' + 2yy'$. Using the equation it follows that $rr' = xy - \lambda pxy = (1 - \lambda p)xy$. Taking into account that $xy = r^2 \sin \theta \cos \theta = r^2 \sin 2\theta$ and dividing by r^2 if $r \neq 0$ (if $r = 0$ we get $x = y = 0$, which gives the trivial solution) we get

$$r' = (1 - \lambda p) \cdot r \sin 2\theta.$$

Similarly, from $\tan \theta = \frac{x}{y}$ we infer

$$\frac{1}{\cos^2 \theta} \cdot \theta' = \frac{x'y - y'x}{y^2} = \frac{x'y - y'x}{r^2 \cos^2 \theta}.$$

Since $x'y - y'x = y^2 + \lambda px^2 = r^2 \cos^2 \theta + \lambda pr^2 \sin^2 \theta$ we find, after simplifications, $\theta' = \lambda p \sin^2 \theta + \cos^2 \theta$. In conclusion we have shown that the equation $x'' + \lambda px = 0$ has been transformed into the system

$$\begin{cases} r' = (1 - \lambda p) \cdot r \sin 2\theta, \\ \theta' = \lambda p \sin^2 \theta + \cos^2 \theta. \end{cases} \tag{13.11}$$

The advantage of this system is that the second equation is independent of r and hence it can be integrated separately, as we are going to do in the next step.

Step 2. Consider the initial value problem

$$
\begin{cases}
\theta' = \lambda p \sin^2 \theta + \cos^2 \theta, \\
\theta(a) = 0,
\end{cases}
\tag{13.12}
$$

and let $\theta(\lambda; t)$ denote its solution. Let us remark that if θ solves (13.12) then the corresponding nontrivial solution of (EP) $x(t) = r(t) \sin \theta(t)$ is such that $x'(a) = r(a) \cos \theta(a) = r(a) > 0$.

We claim:

(†) *Any solution λ_k of $\theta(\lambda; b) = k\pi$, $k = 1, 2, \ldots$, is an eigenvalue of (EP).*

Actually, from $x = r \sin \theta$ it follows that $x(a) = r(a) \sin \theta(\lambda_k; a) = 0$ as well as $x(b) = r(b) \sin \theta(\lambda_k; b) = 0$.

Notice that $\theta(0, t)$ satisfies $\theta' = \cos^2 \theta$, $\theta(a) = 0$. Integrating, we get $\tan \theta(0, t) = t - a$ whereby $\theta(0, t) = \arctan(t - a)$. In particular one has $\theta(0, b) = \arctan(b - a)$ and hence

$$
0 < \theta(0, b) < \frac{\pi}{2}.
\tag{13.13}
$$

Moreover, using the equation $\theta' = \lambda p \sin^2 \theta + \cos^2 \theta$, it is possible to show that $\theta(\lambda, b)$ is an increasing function of λ and that $\lim_{\lambda \to +\infty} \theta(\lambda, b) = +\infty$. See Fig. 13.2.

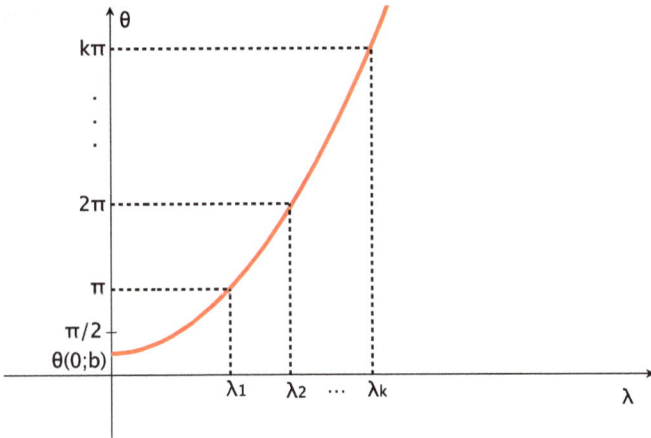

Figure 13.2: Plot of $\theta = \theta(\lambda, b)$.

From these properties and (13.13) it follows that the equation $\theta(\lambda; b) = k\pi$ has a solution $\lambda = \lambda_k$ for any $k = 1, 2, \ldots$ and $\lambda_k \to +\infty$ as $k \to +\infty$. According to (†) this suffices to prove the existence of infinitely many eigenvalues as well as (i).

As for the comparison property (ii), we denote by $\theta_i(\lambda; t)$, $i = 1, 2$, the solution of $\theta' = \lambda p_i \sin^2 \theta + \cos^2 \theta$. Using the equation, one shows that $p_1 \leq p_2$ implies $\theta_1(\lambda; b) \leq \theta_2(\lambda; b)$. Solving $\theta_1(\lambda; b) = k\pi$ and $\theta_2(\lambda; b)k\pi$ we infer that $\lambda_k[p_1] \geq \lambda_k[p_2]$; see Fig. 13.3.

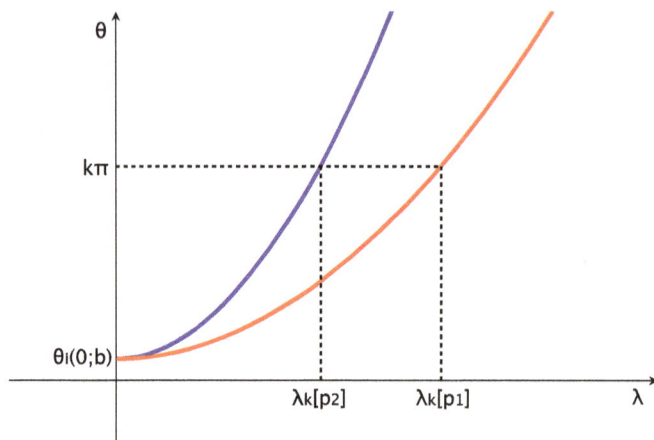

Figure 13.3: Plot of $\theta_1(\lambda, b) \leq \theta_2(\lambda, b)$.

Finally, to prove (iii) we recall that $\phi_k(t) = r_k(t) \sin \theta_k(t)$, where $r_k(t)$, $\lambda_k(t)$ solve (13.11) with $\lambda = \lambda_k$. Thus the zeros of $\phi_k(t)$ are found by solving $\sin \theta_k(t) = 0$ for t, which yields $\theta_k(t) = h\pi$, $h \in \mathbb{N}$. Notice that $\theta'_k = \lambda_k p \sin^2 \theta + \cos^2 \theta > 0$ implies that $\theta_k(t)$ is increasing as a function of t. Moreover, $\theta_k(a) = \theta(\lambda_k; a) = 0$ and $\theta_k(b) = \theta(\lambda_k; b) = k\pi$; see Fig. 13.4.

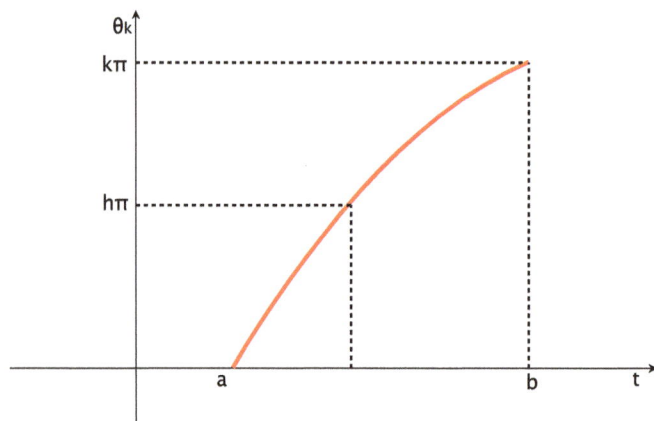

Figure 13.4: Plot of $\theta_k(t)$.

From these facts we infer the following information:

(1) if $k = 1$, then $0 < \theta_1(t) < \pi$, $\forall t \in (a, b)$ and hence $\phi_1(t) = r_1(t) \sin \theta_1(t) > 0$ in (a, b);

(2) if $k > 1$, then for each $h = 1, 2, \ldots, k - 1$ we can find $t_h \in (a, b)$ solving $\theta_k(t) = h\pi$. These t_h, $h = 1, 2, \ldots, k - 1$, are exactly the $k - 1$ zeros of ϕ_k we were looking for. □

13.4 Exercises

1. Write the following in selfadjoint form:

$$2x'' - 4x' + x = 0.$$

2. Write the following in selfadjoint form:

$$t^2 x'' + tx' - x = 0, \quad t > 0.$$

3. Write the following in selfadjoint form:

$$tx'' - x' - tx = 0, \quad t > 0.$$

4. Write the following in selfadjoint form:

$$t^2 x'' - tx' + x = 0.$$

5. Explain whether or not the following is oscillatory:

$$(t \ln t) x'' + x' + (t^{\frac{3}{2}} + 1)x = 0.$$

6. Show that for any continuous function $p(t)$,

$$x'' - x' + p(t)x = 0$$

is nonoscillatory if $p(t) \le 0$.

7. Show that the equation $x'' + \frac{t+1}{t-1}x = 0$ is oscillatory.

8. Let $P(t) = a_p t^p + a_{p-1} t^{p-1} + \cdots + a_1 t + a_0$ and $Q(t) = b_p t^p + b_{p-1} t^{p-1} + \cdots + b_1 t + b_0$. Show that the equation $x'' + \frac{P(t)}{Q(t)}x = 0$ is oscillatory provided $\frac{a_p}{b_p} > 0$.

9. Show that the nontrivial solutions of $x'' - e^t x = 0$, $x(0) = 0$, never vanishes elsewhere.

10. Write the following in selfadjoint form and determine its oscillation status: $\sqrt{t}x'' + (t + 1)x = 0$.

11. Determine the oscillation status of $x'' + x' + tx = 0$.

12. Write it in selfadjoint form in order to determine a condition on $c(t)$ so that

$$x'' - \frac{1}{t}x' + c(t)x = 0$$

is nonoscillatory.

13. Eliminate the x' term in order to determine a condition on $c(t)$ so that $x'' - \frac{1}{t}x' + c(t)x = 0$ is oscillatory.

14. Determine the oscillation status of

$$\left(\frac{2}{3t+2}x'\right)' + \frac{t}{t^2 + \sin t}x = 0, \quad t > 1. \tag{$*$}$$

15. Determine the oscillation status of

$$x'' - 2tx' + 4x = 0.$$

16. Determine the values of the constant a for which

$$tx'' - x' + atx = 0, \quad t > 1,$$

is oscillatory. Are there values of a that will make it nonoscillatory?

17. * Prove that

$$x'' + 20(\sin t)x = 0$$

is oscillatory.

18. Determine the oscillation status of

$$x'' - 2tx' + 2t^2x = 0.$$

19. * Consider the two equations

$$(I) \quad \left(\frac{1}{t^2}x'\right)' + x = 0, \quad \text{and} \quad (II) \quad \left(\frac{1}{t}x'\right)' + t^2x = 0, \quad t > 0.$$

It is easy to see that both are oscillatory, by Theorem 13.5. Which one oscillates faster?

[Hint: Expand each equation and write it in the form $z'' + p(t)z' + q(t)z = 0$, eliminate the z'-term and then compare the two equations of the form $Z'' + P(t)Z = 0$, using the Sturm comparison theorem.]

20. Prove (i) and (ii) of Corollary 13.4.

21. Find $b > 0$ such that the eigenvalues of $x'' + 8\lambda x = 0$, $x(0) = x(b) = 0$, are $\lambda_k = 2k^2$.

22. Show that the eigenvalues λ_k of $(1 + t^2)x'' + \lambda x = 0$, $x(0) = x(\pi) = 0$, satisfy $\lambda_k \geq k^2$.

23. Show that the eigenvalues λ_k of $x'' + \lambda(2 + \cos t)x = 0$, $x(0) = x(1) = 0$, satisfy $\frac{1}{3}k^2\pi^2 \leq \lambda_k \leq k^2\pi^2$.

24. Let $\lambda_k[r_i]$, $i = 1, 2$, denote the eigenvalues of $r_ix'' + \lambda px = 0$, $x(a) = x(b) = 0$. If $r_1 \geq r_2 > 0$, show that $\lambda_k[r_1] \geq \lambda_k[r_2]$.

25. Make the appropriate transformation to eliminate the x''-term in the equation

$$x''' - 3x'' + x = 0.$$

14 A primer on linear PDE in 2D
I: first order equations

14.1 First order linear equations in 2D

A general first order PDE in two dimensions is an equation such as

$$F(x,y,u,p,q) = 0, \quad p = u_x = \frac{\partial u}{\partial x}, \quad q = u_y = \frac{\partial u}{\partial y}, \tag{14.1}$$

where $(x,y) \in \mathbb{R}^2$ and F is a function defined on a set $S \subseteq \mathbb{R}^5$.

Notation: we are using $(x,y) \in \mathbb{R}^2$ as independent variables, but they can be labeled differently, depending on their physical meaning.

Equation (14.1) is a partial differential equation since it involves the partial derivatives u_x, u_y of the unknown function $u = u(x,y)$; it is of first order since there are only first order partial derivatives.

A solution of (14.1) is a function $u = u(x,y)$ defined on a set $D \subseteq \mathbb{R}^2$ such that

$$F(x,y,u(x,y),u_x(x,y),u_y(x,y)) = 0, \quad \forall(x,y) \in D.$$

If F is linear with respect to u,p,q we say that the equation is *linear*. If $F(x,y,0,0,0) = 0$, the equation is said to be *homogeneous*. So, a general first order linear PDE is

$$a_1(x,y)u_x + a_2(x,y)u_y + a_3(x,y)u + a_4(x,y) = 0,$$

and a general first order linear homogeneous PDE is

$$a_1(x,y)u_x + a_2(x,y)u_y + a_3(x,y)u = 0.$$

The main feature of a linear homogeneous equation is that if $u(x,y), v(x,y)$ are two solutions, then, as in the ODE case, $w(x,y) = au(x,y) + \beta v(x,y)$ is also a solution, for any $\alpha, \beta \in \mathbb{R}$. Actually (the dependence on (x,y) is understood)

$$
\begin{aligned}
a_1 w_x + a_2 w_y + a_3 w &= a_1 \cdot (\alpha u + \beta v)_x + a_2 \cdot (\alpha u + \beta v)_y + a_3 \cdot (\alpha u + \beta v) \\
&= a_1 \alpha u_x + a_1 \beta v_x + a_2 \alpha u_y + a_2 \beta v_y + a_3 \alpha u + a_3 \beta v \\
&= \alpha \underbrace{(a_1 u_x + a_2 u_y + a_3 u)}_{=0} + \beta \underbrace{(a_1 v_x + a_2 v_y + a_3 v)}_{=0} = 0.
\end{aligned}
$$

14.1.1 Linear transport equations

Given continuous functions $a(x,y), b(x,y), f(x,y)$, a general transport equation has the form

https://doi.org/10.1515/9783111185675-014

$$a(x,y)u_x + b(x,y)u_y = f(x,y).$$

The simplest linear transport equation is the case in which $a(x,y)$, $b(x,y)$ are constant and $f = 0$, namely

$$a_1 u_x + a_2 u_y = 0, \quad a_1, a_2 \in \mathbb{R}.$$

If $a_1 = 0$ and $a_2 \neq 0$, the equation becomes $u_y = 0$ whose solutions are $u = \phi(x)$ for any C^1 function ϕ. If $a_1 \neq 0$, dividing by a_1 and setting $b = \frac{a_2}{a_1}$, we obtain the equation

$$u_x + bu_y = 0, \quad b \in \mathbb{R}. \tag{14.2}$$

In the sequel it is always understood that $b \neq 0$, otherwise we find $u_x = 0$ which is trivially solved by $u = \phi(y)$, for any C^1 function ϕ.

The transport equation (14.2) arises in many physical applications. For example, consider a fluid, moving with constant speed b, (see Fig. 14.1) in a thin cylindrical channel parallel to x axis, which can be assimilated to a one dimensional line.

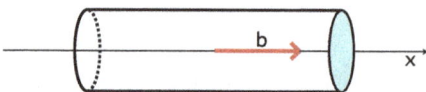

Figure 14.1: A part of the channel and the speed b.

If $u(t,x)$ stands for the density at time t and position x of the fluid, the *principle of mass conservation* implies that u satisfies $u_t + bu_x = 0$, which is nothing but (14.2), with independent variables (t,x).

One solution of (14.2) is the function identically equal to zero. To find all the solutions, let us consider the linear change of variable

$$\begin{cases} \xi = y - bx, \\ \eta = y + bx, \end{cases} \quad \text{i.e.,} \quad \begin{cases} x = \frac{1}{2b}(\eta - \xi), \\ y = \frac{1}{2}(\eta + \xi), \end{cases}$$

and set

$$v(\xi, \eta) = u(x,y) = u\left(\frac{1}{2b}(\eta - \xi), \frac{1}{2}(\eta + \xi)\right).$$

One has

$$\frac{\partial v}{\partial \eta} = \frac{\partial u}{\partial x}\frac{\partial x}{\partial \eta} + \frac{\partial u}{\partial y}\frac{\partial y}{\partial \eta} = \frac{1}{2b}u_x + \frac{1}{2}u_y = \frac{1}{2b}(u_x + bu_y) = 0,$$

because u solves (14.2). This means that v is independent of η, namely that $v = \phi(\xi)$ for some C^1 function $\phi : \mathbb{R} \to \mathbb{R}$. Since $\xi = y - bx$ we find that $v(\xi, \eta) = \phi(y - bx)$ and hence

$$u(x,y) = \phi(y - bx).$$

On the other hand, for any C^1 function $\phi(s)$, letting $u(x,y) = \phi(y - bx)$ one has

$$u_x = -b\phi'(y - bx), \quad u_y = \phi'(y - bx)$$

whereby

$$u_x + bu_y = -b\phi'(y - bx) + b\phi'(y - bx) = 0.$$

In conclusion we have shown the following.

Theorem 14.1. *The function $u(x,y)$ is a solution of (14.2) if and only if $u(x,y) = \phi(y - bx)$, where $\phi : \mathbb{R} \to \mathbb{R}$ is some C^1 function.*

We will call $u = \phi(y - bx)$ the *general solution* of (14.2).

Notice that from the preceding result it follows that along any straight line $y - bx = c$, $c \in \mathbb{R}$, the solution $u(x,y)$ is constant: actually, one has $u(x,y) = \phi(y - bx) = \phi(c)$. The straight lines $y - bx = c$ are called the *characteristics* of (14.2). They have the specific property that solutions are constant along them.

14.1.1.1 Homogeneous transport equations

Let us first deal with the homogeneous transport equation

$$a(x,y)u_x + b(x,y)u_y = 0, \tag{14.3}$$

where we assume $a(x,y), b(x,y)$ are defined on $\mathbb{R}^2$. The case when a, b are defined on a subset $D \subset \mathbb{R}^2$ requires minor changes.

Let us define the characteristics of (14.3).

Definition 14.1. *The characteristics of (14.3) are the curves $\xi = \xi(s), \eta = \eta(s)$ in the plane (x,y) on which the solutions $u(x,y)$ of (14.3) are constant.*

Let $\xi(s), \eta(s)$ satisfy the (autonomous) system of ordinary differential equations

$$\begin{cases} \xi'(s) = a(\xi(s), \eta(s)), \\ \eta'(s) = b(\xi(s), \eta(s)). \end{cases} \tag{14.4}$$

If $u(x,y)$ is any solution of (14.4), then one has

$$\frac{d}{ds} u(\xi(s), \eta(s)) = u_x \xi' + u_y \eta' = a(x,y)u_x + b(x,y)u_y = 0$$

and hence $u(\xi(s), \eta(s))$ is constant. We have shown the following.

Lemma 14.1. *The characteristics of (14.3) are solutions of (14.4).*

If $a(x,y) \equiv 1$ and $b(x,y) \equiv b \in \mathbb{R}$, namely if (14.3) is the equation with constant coefficients (14.2), system (14.4) becomes

$$\begin{cases} \xi'(s) = 1, \\ \eta'(s) = b. \end{cases} \tag{14.5}$$

Solving, we find $\xi = s + c_1$, $\eta = bs + c_2$. Eliminating the parameter s we get

$$\eta = b(\xi - c_1) + c_2 = b\xi + c, \quad c = c_2 - bc_1.$$

We recognize that $x = \xi(s)$, $y = \eta(s)$ are nothing but the parametric equations of $y - bx = c$. This justifies having given the name *characteristics* to $y - bx = c$ earlier.

According to Theorem 14.1, the transport equation $u_x + bu_y = 0$ has infinitely many solutions. How can we single out a unique solution?

To have an idea of a possible answer, it is convenient to recall what happens in the case of the ODE $y' = f(x,y)$, which also has infinitely many solutions, depending on an arbitrary real constant. As is well known, to single out a unique solution of $y' = f(x,y)$ we can require that $y(x_0) = y_0$, for some (x_0, y_0) in the domain of f. In other words, we look for a differentiable function $y = y(x)$ satisfying the Cauchy problem

$$\begin{cases} y' = f(x,y), \\ y(x_0) = y_0. \end{cases}$$

Dealing with a partial differential equation as (14.2), the initial value problem, or Cauchy problem, amounts to requiring that a solution u of (14.2) equals a given function h on a given curve y in the plane. We shall see that, in analogy of the Cauchy problem for ODE, this requirement allows one to single out a unique solution of the transport equation.

Though in general we can prescribe an initial condition on any curve but the characteristics, in the next theorem we deal, for brevity, with the specific case in which y is defined by $y = g(x)$. The corresponding initial value problem is equivalent to finding a differentiable function $u(x,y)$ such that

$$\begin{cases} a(x,y)u_x + b(x,y)u_y = 0, \\ u(x,y) = h(x,y), \quad \text{for } y = g(x). \end{cases} \tag{14.6}$$

The case in which y is defined by $x = \tilde{g}(y)$ is briefly discussed later.

We will use the characteristics to solve (14.6). Given $P = (x,y) \in \mathbb{R}^2$, we consider the characteristic passing through P, namely we consider the Cauchy problem

$$\begin{cases} \xi'(s) = a(\xi(s), \eta(s)), \quad \xi(0) = x, \\ \eta'(s) = b(\xi(s), \eta(s)), \quad \eta(0) = y. \end{cases} \tag{14.7}$$

It is worth pointing out that, even though the transport equation (14.3) is linear, system (14.7) might possibly be nonlinear. However, if we assume:

(H1) a and b are C^1 functions;

then (14.7) has a unique solution $\xi(s; x, y), \eta(s; x, y)$. Furthermore, for the sake of simplicity, we will also suppose that $\xi(s; x, y), \eta(s; x, y)$ are defined for all s.

Recall that the solution of (14.7) depends on the initial data (x, y) and form a family of curves filling up the plane. It is also known that, if (H1) holds then the map $(x, y) \mapsto (\xi(s; x, y), \eta(s; x, y))$ is continuously differentiable.

Next, we seek a point Q where this characteristic meets γ; see Fig. 14.2.

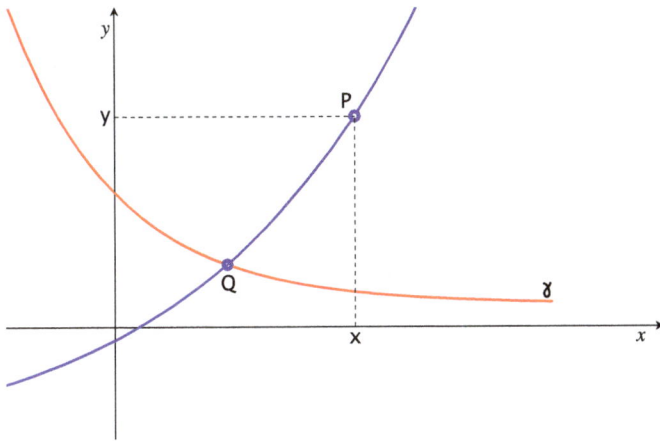

Figure 14.2: The characteristic through P (the blue line) that meets y at Q.

This amounts to considering the system

$$\begin{cases} y = g(x), \\ x = \xi(s; x, y), \\ y = \eta(s; x, y), \end{cases}$$

which yields the equation

$$\eta(s; x, y) = g(\xi(s; x, y)). \tag{14.8}$$

Let us suppose:

(H2) For all (x, y) equation (14.8) has a unique solution $s = s_0(x, y)$. Moreover, the map $(x, y) \mapsto s_0(x, y)$ is continuously differentiable.

Notice that assumption (H2) rules out the case in which $y = g(x)$ is itself a characteristic.

After these preliminaries, we are in a position to find the solution of the initial value problem (14.6).

Theorem 14.2. *Suppose that (H1)–(H2) hold and let $h = h(x,y)$ be a given C^1 function. Then there exists one and only one solution of the initial value problem (14.6) which is given by*

$$u(x,y) = h(\xi(s_0(x,y);x,y), \eta(s_0(x,y);x,y)).$$

Remark on notation: To simplify notation, hereafter we will often write $\xi(s)$, $\eta(s)$ instead of $\xi(s;x,y)$, $\eta(s;x,y)$.

Proof. Since $u(\xi(s),\eta(s))$ is constant, then taking $s = 0$ and $s = s_0$ we find

$$u(x,y) = u(\xi(0),\eta(0)) = u(\xi(s_0),\eta(s_0)).$$

Now $(\xi(s_0),\eta(s_0)) = Q \in \gamma$. So, using the initial condition $u(x,y) = h(x,y)$ for $y = g(x)$, we find $u(\xi(s_0),\eta(s_0)) = h(\xi(s_0),\eta(s_0))$. Finally, recall that, by assumption, h and the function $(x,y) \mapsto (\xi(s_0(x,y);x,y), \eta(s_0(x,y);x,y))$ are C^1 and hence so is

$$u(x,y) = h(\xi(s_0(x,y);x,y), \eta(s_0(x,y);x,y)).$$

In conclusion, it follows that such a u is the solution we were looking for. □

If the curve γ is defined by $x = \tilde{g}(y)$, (H2) will be substituted by:
(H̃2) for all (x,y) the equation $\xi(s;x,y) = \tilde{g}(\eta(s;x,y))$ a unique solution $s = \tilde{s}_0(x,y)$. Moreover, the map $(x,y) \mapsto \tilde{s}_0(x,y)$ is continuously differentiable.

Repeating the previous calculations, we find that the solution is given by

$$u(x,y) = h(\xi(\tilde{s}_0),\eta(\tilde{s}_0)). \tag{14.9}$$

Summarizing, to find the solution $u(x,y)$ to the ivp

$$a(x,y)u_x + b(x,y)u_y = 0, \quad u(x,y) = h(x,y) \text{ on } \gamma,$$

the following steps are in order.

Step 1. One finds the characteristic ξ,η by solving the ODE system

$$\begin{cases} \xi' = a(\xi,\eta), & \xi(0) = x, \\ \eta' = b(\xi,\eta), & \eta(0) = y. \end{cases}$$

Step 2. One finds s_0, resp. $\tilde{s}_0$ (depending on x,y) by solving $\eta(s) = g(\xi(s))$, resp. $\xi(s) = \tilde{g}(\eta(s))$.

Step 3. Since u is constant along the characteristic, we have $u(x, y) = h(\xi(s_0), \eta(s_0))$, resp. $u(x, y) = h(\xi(\bar{s}_0), \eta(\bar{s}_0))$.

It might be helpful to consider the constant coefficients case in more detail, where the equation becomes $u_x + bu_y = 0$ with $b \in \mathbb{R}$. Since $a(x, y) \equiv 1$ and $b(x, y) \equiv b \in \mathbb{R}$, it is clear that (H1) holds. As for (H2), this assumption can be formulated here in an analytical way. For brevity, we focus on the case where the initial condition is given by $u(x, g(x)) = h(x)$.

Recall that here the characteristics $\xi(s), \eta(s)$ satisfy (14.5) and thus are straight lines; see Fig. 14.3. In particular, the characteristic passing through $P = (x, y)$, marked in blue in Fig. 14.3, is given by

$$\xi(s) = s + x, \quad \eta(s) = bs + y.$$

This line meets $y = g(x)$ at a unique point Q provided $\eta(s) = g(\xi(s))$, namely $bs + y = g(s + x)$ has a unique solution. If we set $r = s + x$, the equation $bs + y = g(s + x)$ becomes $b \cdot (r - x) + y = g(r)$ or $g(r) - br = y - bx$. Let us assume that:

(*) *the map* $r \mapsto g(r) - br$ *has a* C^1 *inverse* ψ.

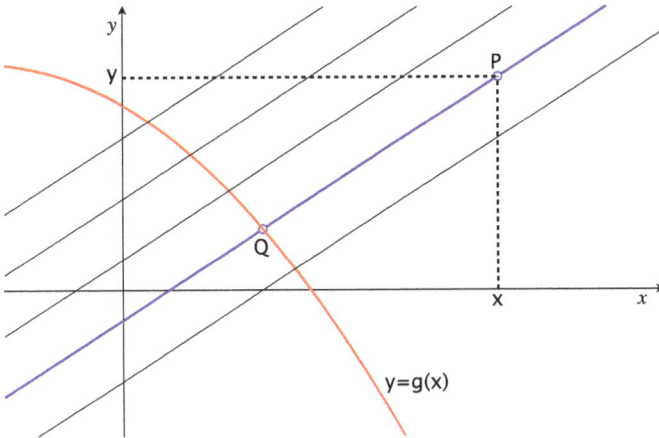

Figure 14.3: The characteristics of $u_x + bu_y = 0$.

If (*) holds, then for all x, y the equation $g(r) - br = y - bx$ has a unique solution $r_0(x, y)$ and the map $(x, y) \mapsto r_0(x, y)$ is C^1. It follows that, in such a case, the equation $\eta(s) = g(\xi(s))$ also has a unique solution given by $s_0 = s_0(x, y) = r_0(x, y) - x$, and the map $(x, y) \mapsto s_0(x, y)$ is C^1. This proves that (H2) holds. Since the converse is also true, we can conclude that (H2) is equivalent to (*).

Next, let us show a second method to find the solution of the ivp $u_x + bu_y = 0$, $u(x, g(x)) = h(x)$, by using Theorem 14.1 directly. Substituting $y = g(x)$ in the general solution $u = \phi(y - bx)$ of $u_x + bu_y = 0$, and taking into account the initial condition $u(x, g(x)) = h(x)$, we find $\phi(g(x) - bx) = h(x)$. Since $g(x) - bx = r$ if and only if $x = \psi(r)$, we infer $\phi(r) = h(\psi(r))$, which means that $\phi = h \circ \psi$. In this way we have singled out a unique ϕ such that $u(x,y) = \phi(y - bx) = h(\psi(y - b))$ is the solution of our ivp.

Of course, we could also have used the general method, which in this case yields

$$u(x,y) = h(\xi(s_0)) = h(s_0 + x),$$

and the two solutions clearly coincide.

When we deal with equations with constant coefficients, it can be convenient to use the general or the second method indicated before, depending on the data g and h. However, in the following examples, (a) and (b), we will carry out the calculations using both the methods, leaving it to the reader to choose the one that seems more convenient.

Example 14.1.
(a) Solve the ivp $u_x - u_y = 0$, $u(x, 0) = \sin x$. Here $g(x) = 0$ and $h(x) = \sin x$.
   1. The characteristics through (x, y) are $\xi = s + x$, $\eta = -s + y$.
   2. Solving $\eta(s) = g(\xi(s)) = 0$ yields $-s + y = 0$ namely $s_0 = y$.
   3. One has

$$u(x,y) = h(\xi(s_0)) = h(s_0 + x) = \sin(y + x).$$

Alternatively we can use the second method: from $u = \phi(y + x)$ and using the initial condition $u = \sin x$ for $y = 0$, we find $\phi(x) = \sin x$. Thus $u = \sin(y + x)$ as before.

(b) Solve the ivp $u_x + u_y = 0$, $u(x, 3x) = e^x$. Here $b = 1$, $g(x) = 3x$ and $h(x) = e^x$.
   1. The characteristics through (x, y) are $\xi = s + x$, $\eta = s + y$.
   2. Solving $\eta(s) = g(\xi(s))$ yields $s + y = 3(s + x)$ and we find $s_0 = \frac{1}{2}(y - 3x)$.
   3. One has

$$u(x,y) = h(\xi(s_0)) = h(s_0 + x) = e^{s_0 + x} = e^{\frac{1}{2}(y - 3x) + x} = e^{\frac{1}{2}(y - x)}.$$

If we use the second method then from $u = \phi(y - x)$ and the initial condition, we infer $\phi(3x - x) = e^x$, i.e., $\phi(2x) = e^x$. Thus $\phi(r) = e^{\frac{1}{2}r}$ and we find $u(x,y) = \phi(y - x) = e^{\frac{1}{2}(y-x)}$ as before. □

In the following examples we deal with the general ivp (14.6).

Example 14.2.
(a) Solve the ivp $xu_x - u_y = 0$, $u(x, 0) = h(x) = 2x$.
   1. The characteristics are the solutions of the system $\xi' = \xi$, $\eta' = -1$, namely $\xi = c_1 e^s$ and $\eta = -s + c_2$. The characteristics such that $\xi(0) = x$, $\eta(0) = y$ are found by setting $\xi(0) = c_1 e^0 = x$, $\eta(0) = c_2 = y$. Then we find $c_1 = x$, $c_2 = y$ and hence

$$\xi(s; x, y) = xe^s, \quad \eta(s; x, y) = -s + y.$$

2. The value s_0 is found by solving $\eta(s; x, y) = 0$ for s, namely $-s + y = 0$, yielding $s_0 = y$.
3. The solution is given by

$$u(x, y) = h(\xi(s_0)) = 2\xi(s_0) = 2xe^{s_0} = 2xe^y.$$

(b) Solve $u_x - yu_y = 0$, $u(x, e^x) = h(x) = x^2$.

1. The characteristics are the solutions of the system $\xi' = 1$, $\eta' = -\eta$, namely $\xi = s + c_1$ and $\eta = c_2 e^{-s}$. The characteristics such that $\xi(0) = x$, $\eta(0) = y$ are given by

$$\xi(s; x, y) = s + x, \quad \eta(s; x, y) = ye^{-s}.$$

2. To find s_0 we solve $\eta(s) = h(\xi(s)) = e^{\xi(s)}$ for s, namely

$$ye^{-s} = e^{s+x} \Rightarrow y = e^{x+2s} \Rightarrow x + 2s = \ln y, \quad y > 0,$$

and thus $s_0 = \frac{1}{2}[\ln y - x]$, $(y > 0)$. Notice that $(H2)$ holds. Let us also remark that for $y \leq 0$ the preceding equation $y = e^{x+2s}$ has no solution at all and hence the characteristics do not meet y.

3. The solution $u(x, y)$ is given by

$$u(x, y) = h(\xi(s_0)) = (s_0 + x)^2 = \left(x + \frac{1}{2}[\ln y - x]\right)^2.$$

(c) Solve $u_x + yu_y = 0$, $u(0, y) = h(y) = 2 - y$.

1. The system (14.4) becomes $\xi'(s) = 1$, $\eta'(s) = \eta(s)$, whose general solution is $\xi(s) = s + c_1$, $\eta(s) = c_2 e^s$. The characteristic through (x, y) is given by

$$\xi(s; x, y) = s + x, \quad \eta(s; x, y) = ye^s.$$

2. The value $\tilde{s}_0$ is found by solving $\xi(s) = 0$, yielding $\tilde{s}_0 = -x$ and $\eta(\tilde{s}_0) = ye^{\tilde{s}_0} = ye^{-x}$. Notice that $(\overline{H2})$ is satisfied.
3. The solution is given by

$$u(x, y) = h(\eta(\tilde{s}_0)) = 2 - ye^{-x}. \qquad \Box$$

14.1.1.2 Transport equations with forcing term

Given a continuous function $f(x, y)$, we want to solve a class of non-homogeneous transport equations such as

$$u_x + bu_y = f(x, y), \quad b \in \mathbb{R}. \tag{14.10}$$

As for every linear differential equation, the general solution of (14.10) is the sum of the general solution of the corresponding homogeneous equation $u_x + bu_y = 0$, and a particular solution u^* of (14.10).

Repeating the arguments discussed in the previous section we set

$$\begin{cases} \xi = y - bx, \\ \eta = y + bx, \end{cases} \quad \text{i.e.,} \quad \begin{cases} x = \frac{1}{2b}(\eta - \xi), \\ y = \frac{1}{2}(\eta + \xi), \end{cases}$$

and

$$v(\xi, \eta) = u(x, y) = u\left(\frac{1}{2b}(\eta - \xi), \frac{1}{2}(\eta + \xi)\right).$$

One has

$$\frac{\partial v}{\partial \eta} = \frac{\partial u}{\partial x}\frac{\partial x}{\partial \eta} + \frac{\partial u}{\partial y}\frac{\partial y}{\partial \eta} = \frac{1}{2b}u_x + \frac{1}{2}u_y = \frac{1}{2b}(u_x + bu_y) = \frac{1}{2b}f(\xi, \eta),$$

because u solves (14.10). One particular solution of this equation is

$$v^*(\xi, \eta) = \frac{1}{2b}\int_0^\eta f(\xi, \eta)\, d\eta.$$

Integrating and going back to the variables x, y, we find $u^*(x, y)$.

Example 14.3. Solve $u_x + u_y = 4xy$.

To use the preceding formula, we first notice that

$$f(\xi, \eta) = 8x(\xi, \eta)y(\xi, \eta) = 2(\eta - \xi)(\eta + \xi) = 2(\eta^2 - \xi^2).$$

Then we find

$$v^*(\xi, \eta) = \frac{1}{2}\int_0^\eta f(\xi, \eta)\, d\eta = \int_0^\eta (\eta^2 - \xi^2)\, d\eta = \frac{1}{3}\eta^3 - \eta\xi^2.$$

Going back to the variables x, y we find

$$u^*(x, y) = \frac{1}{3}(x + y)^3 - (x + y)(y - x)^2.$$

Thus the general solution is given by $u = \phi(y - x) + u^*$, for some C^1 function ϕ.

In some cases, finding a particular solution is easier and can be done in a more direct way. For example, let us solve $u_x + u_y = f(x)$. Looking for a solution independent of y, $u^* = u^*(x)$, we find $u_x^* = f(x)$ whereby $u^*(x) = \int f(x)\, dx$. Similarly, to find a particular solution of $u_x + u_y = f(y)$ we can search for a solution independent of x, $u^* = u^*(y)$, yielding $u_y^* = f(y)$ whereby $u^*(y) = \int f(y)\, dy$.

Example 14.4. Solve $u_x + u_y = 2x$. One finds $u^*(x) = x^2$. Then the general solution is given by $u = \phi(y - x) + x^2$.

14.2 Appendix: an introduction to inviscid Burgers' equation

So far we have studied linear transport equations using the method of the characteristics. This method can be extended to solve some non-linear equations as well. A general study of this topic is out of the scope of this book. We will only briefly discuss the inviscid Burger's equation

$$u_x + buu_y = 0, \quad b \in \mathbb{R}, \quad b \neq 0, \tag{14.11}$$

which is important in fluid dynamics as an approximation of the classical Euler equation of an incompressible, inviscid fluid.

We first consider the initial value problem of finding a function $u = u(x,y)$ which solves Burger's equation (14.11) and satisfies the initial condition

$$u(0,y) = y. \tag{14.12}$$

In the case of the linear transport equation we have seen that the solutions of $u_x + bu_y = 0$ are constant along the characteristics $y - bx = c$ or $y = \eta(x) \stackrel{\text{def}}{=} bx + c$. Notice that $\eta(x)$ solves the trivial ordinary differential equation $\eta'(x) = b$.

In Burger's equation, where the coefficient b is replaced by bu, it is natural to guess that the characteristics are the solution $\eta(x)$ satisfying

$$\eta'(x) = bu(x, \eta(x)).$$

Notice that this is a *nonlinear* ordinary differential equation, in contrast to the case of $u_x + bu_y = 0$ when the characteristics solved a linear ODE.

Recall that, for all given z, the above equation has a unique solution $\eta(x)$ such that $\eta(0) = z$. This follows from the theory of ODE, because u, being a solution of (14.11), is of class C^1. For the sake of simplicity, we assume that $\eta(x)$ is defined for all x.

To see that the solution $\eta(x;z)$ of the Cauchy problem

$$\begin{cases} \eta' = bu(x,\eta), \\ \eta(0) = z, \end{cases} \tag{14.13}$$

is a characteristic, we have to check that

$$u(x, \eta(x)) = \text{const.}$$

For this purpose, it suffices to evaluate

$$\frac{d}{dx} u(x, \eta(x)) = u_x + u_y \eta'(x) = u_x + buu_y = 0,$$

which implies that $u(x, \eta(x))$ is constant, as we wanted. Next, we can use this feature to solve (14.13). Since $u(x, \eta(x))$ is constant w. r. t. x,

$$u(x, \eta(x)) = u(0, \eta(0)) = u(0, z),$$

namely, using (14.12),

$$u(x, \eta(x)) = z. \tag{14.14}$$

Hence, the equation $\eta'(x) = bu(x, \eta(x))$ becomes simply $\eta'(x) = bz$, whose general solution is

$$\eta(x) = bxz + c, \quad c \in \mathbb{R}.$$

Taking into account the initial condition $\eta(0) = z$, it follows that $c = z$ and hence the solution of the Cauchy problem (14.13), namely the characteristics of the Burger equation (14.11), are given by

$$\eta(x) = bxz + z.$$

They are straight lines as in the linear case, but now their slope is variable. In particular they are not parallel, but have a common intersection point at $(-\frac{1}{b}, 0)$. We will see a consequence of this new feature.

Coming back to the problem of solving (14.11)–(14.12), we substitute $\eta(x) = bzx + z$ into (14.14), yielding

$$u(x, bxz + z) = z. \tag{14.15}$$

Changing the variable $y = bxz + z$, one finds $z = \frac{y}{bx+1}$, $x \neq -\frac{1}{b}$, and hence

$$u(x, y) = z = \frac{y}{bx + 1}, \quad x \neq -\frac{1}{b}.$$

A direct computation shows that, actually, $u(x, y)$ solves (14.11)–(14.12).

Remark 14.1. Notice that the solution found above is not defined for $x = -1/b$. This is closely related to the fact pointed out before, that the characteristics of (14.11) meet at the point $(-1/b, 0)$.

Example 14.5. The solution of $u_x - \frac{1}{2}uu_y = 0$, $u(0, y) = y$ is given by

$$u(x, y) = z = \frac{y}{1 - \frac{1}{2}x} = \frac{2y}{2 - x}, \quad x \neq 2.$$

□

The case in which $h(y) = \alpha + \beta y$ can be handled in a quite similar way, as shown in the following example.

Example 14.6. (a) Solve $u_x + uu_y = 0$, $u(0, y) = h(y) = 2 + 3y$.

Here (14.14) becomes $u(x, \eta(x)) = u(0, z) = 2 + 3z$ and hence $\eta' = u = 2 + 3z$. Integrating and taking into account that $\eta(0) = z$, we find $\eta(x) = 2x + (3x + 1)z$. Now (14.15) is substituted by $u(2x + (3x + 1)z) = 2 + 3z$.

Setting $y = 2x + (3x + 1)z$ we get $z = \frac{y - 2x}{3x + 1}$, yielding

$$u(x, y) = 2 + 3 \cdot \frac{y - 2x}{3x + 1} = \frac{3y + 2}{3x + 1}, \quad x \neq -\frac{1}{3}.$$

□

Next, let us consider the more general initial condition

$$u(0, y) = h(y), \tag{14.16}$$

where h is a C^1 function of one variable. For brevity we will be sketchy.

As in the preceding case, the characteristics are the solutions of $\eta'(x) = bu(x, \eta(x))$, $\eta(0) = z$. Repeating the previous calculations one finds that $u(x, \eta(x))$ is constant and hence

$$u(x, \eta(x)) = u(0, \eta(0)) = u(0, z) = h(z),$$

which is the counterpart of (14.14). It follows that the equation of characteristics is $\eta'(x) = bh(z)$, $\eta(0) = z$. Therefore

$$\eta(x) - \eta(0) = \int_0^x \eta'(x)\, dx = \int_0^x bh(z)\, dx = bxh(z) \quad \implies \quad \eta(x) = bxh(z) + z.$$

Inserting this into $u(x, \eta(x)) = h(z)$ we find

$$u(x, bxh(z) + z) = h(z).$$

Repeating the preceding procedure, we set

$$y = bxh(z) + z$$

and solve the *nonlinear* equation $bxh(z) + z = y$ for z. For this, we can use the implicit function theorem (see Theorem 1.2 in Chapter 1) which, in this specific case, states the following.

Theorem. *Given* $(\bar{x}, \bar{y}, \bar{z})$ *and the* C^1 *function* $H(x, z)$, *suppose that* $H(\bar{x}, \bar{z}) = \bar{y}$ *and that* $H_z(\bar{x}, \bar{z}) \neq 0$. *Then there exists a neighborhood* U *of* $(\bar{x}, \bar{y})$ *and a unique* C^1 *function* $z = z(x, y)$, *defined on* U, *such that* $z(\bar{x}, \bar{y}) = \bar{z}$ *and*

$$H(x, z(x, y)) = y, \quad \forall (x, y) \in U. \tag{14.17}$$

In the present case we have $H(x, z) = z + bxh(z)$. Hence $H_z(x, z) = 1 + bxh'(z)$. So, if

$$1 + bxh'(z) \neq 0,$$

then we can solve $z + bxh(z) = y$ for z, yielding $z = z(x, y)$ such that

$$z(x, y) + bxh(z(x, y)) = y.$$

Notice that, in particular, $z(0, y) = y$. Inserting $z(x, y)$ into $u(x, y) = u(x, z + xh(z)) = h(z)$ we find

$$u(x, y) = h(z(x, y)).$$

Let us check that this is a solution of (14.11)–(14.16).

As for (14.11) we evaluate $u_x + buu_y$. One has $u_x = h'(z)z_x$ and $u_y = h'(z)z_y$. The partial derivatives z_x, z_y can be evaluated using the identity (14.17). Actually, differentiating (14.17) we find

$$H_x(x, z) + H_z(x, z)z_x = 0, \quad H_z(x, z)z_y = 1$$

whereby (notice that, by continuity, $H_z(x, z) = 1 + bxh'(z) \neq 0$ in a neighborhood of $(\bar{x}, \bar{z})$)

$$z_x(x, y) = -\frac{H_x(x, z)}{H_z(x, z)} = -\frac{bh(z)}{1 + bxh'(z)},$$

$$z_y(x, y) = \frac{1}{H_z(x, z)} = \frac{b}{1 + bxh'(z)}.$$

Thus

$$u_x + buu_y = -h'(z) \cdot \frac{bh(z)}{1 + bxh'(z)} + bh(z) \cdot \frac{h'(z)}{1 + bxh'(z)} = 0.$$

Moreover, $u(x, 0) = h(z(0, y)) = h(y)$. Then $u = h(z(x, y))$ is the solution of the initial value problem (14.11)–(14.16).

In conclusion, the key point in solving (14.11)–(14.16) is to find $z = z(x, y)$ such that $z + bxh(z) = y$. Once we have z, the solution is given by

$$u(x, y) = h(z(x, y)).$$

Let us check that in the case of Example 14.6 we recover the result therein. We have to solve $z + xh(z) = y$ for z, namely $z + x(2 + 3z) = y$. Then one finds

$$z = z(x, y) = \frac{y - 2x}{1 + 3x}, \quad x \ne -\frac{1}{3},$$

and thus, as before, the solution is given by

$$u(x, y) = h(z(x, y)) = 2 + 3z(x, y) = 2 + 3\frac{y - 2x}{1 + 3x} = \frac{3y + 2}{3x + 1}, \quad x \ne -\frac{1}{3}.$$

Notice that $x \ne -\frac{1}{3}$ is nothing but the condition $1 + xh'(z) \ne 0$ in this specific case.

14.3 Exercises

1. Solve $3u_x + 2u_y = 0$.
2. Let u be a solution of $u_x - u_y = 0$. Knowing that $u(0, 1) = 2$, find $u(1, 0)$.
3. Given $r, s \in \mathbb{R}$, show that there are infinitely many u such that $u_x + u_y = 0$ and $u(0, r) = s$.
4. Solve the ivp $u_x + 2u_y = 0$, $u(0, y) = y$.
5. Solve the ivp $u_x - 3u_y = 0$, $u(0, y) = 2y$.
   Note to the reader: in exercises 6–10 we give two different solutions.
6. Solve the ivp $2u_x - 3u_y = 0$, $u(x, 0) = x^2$.
7. Solve the ivp $u_x + 2u_y = 0$, $u(x, 0) = e^{-x}$.
8. Solve the ivp $2u_x - u_y = 0$, $u(x, x) = x - 1$.
9. Solve the ivp $u_x - u_y = 0$, $u = e^x$, on the curve $y = e^x - x$.
10. Solve $u_x + u_y = 0$ such that $u = x$ on $y = \ln x + x$, $x > 0$.
11. Solve the ivp $u_x - u_y = 0$, $u(0, y) = y^2$.
12. Solve the ivp $3u_x + 2u_y = 0$, $u(y, y) = \frac{1}{3}y$.
13. Show that $u(x, y) \equiv k$, $k \in \mathbb{R}$, is the only solution of $a(x, y)u_x + b(x, y)u_y = 0$ such that $u = k$ on any curve y transversal to the characteristics.
14. Solve the ivp $2yu_x - u_y = 0$, $u(x, 0) = 2x$.
15. Solve the ivp $yu_x + xu_y = 0$, $u(x, x + 1) = x$.
16. Solve the ivp $xu_x + 2u_y = 0$, $u(x, 1) = -x$.
17. Solve the ivp $2u_x + yu_y = 0$, $u(0, y) = -y$.
18. Solve the ivp $u_x - yu_y = 0$, $u(0, y) = y^2$.
19. Solve the ivp $u_x + (1 + y)u_y = 0$, $u(x, 0) = 2x$.
20. Solve $u_x - u_y = y$.
21. Solve the ivp $u_x + u_y = x$, $u(x, 0) = 3x$.
22. Solve the ivp $u_x - 2u_y = y^2$, $u(0, y) = y^3$.
23. Solve the ivp $u_x + u_y = 2x + 2y$, $u(0, y) = y$.
24. Find the solution of Burger's equation $u_x - 3uu_y = 0$ such that $u(0, y) = 1 - \frac{1}{3}y$.

15 A primer on linear PDE in 2D
II: second order equations

A *second order* partial differential equation is an equation in which the higher partial derivatives are of second order. We will deal with linear second order PDE in two dimensions, whose general form is

$$au_{xx} + bu_{xy} + cu_{yy} + du_x + eu_y + fu = h(x,y), \quad (x,y) \in \mathbb{R}^2$$

with a, b, c not all zero. As usual the independent variables x, y can be labeled differently, such as t, x or t, y. The features of this equation essentially depend on the coefficients a, b, c of the second order partial derivatives. It is convenient to distinguish among three cases:
1. $b^2 - 4ac > 0$, elliptic equations;
2. $b^2 - 4ac = 0$, parabolic equations;
3. $b^2 - 4ac < 0$, hyperbolic equations.

We focus on very specific classes of elliptic, parabolic and hyperbolic equations:
1. the Laplace equation $u_{xx} + u_{yy} = 0$, which is the simplest *elliptic* linear second order equation;
2. the heat equation $u_t - u_{xx} = 0$, which is the simplest *parabolic* linear second order equation;
3. the vibrating string equations $u_{tt} - c^2 u_{xx} = 0$, which is the simplest *hyperbolic* linear second order equation.

15.1 The Laplace equation

The Laplace equation,

$$\Delta u = u_{xx} + u_{yy} = 0, \tag{L}$$

is the prototype of elliptic equations. Its solutions are called *harmonic functions*. This equation arises in very many situations. For example, a solution of (L) can be the stationary solution of an evolutionary equation, or it can be the electrostatic potential in a region with no charges. As another example, consider a holomorphic function $f(x + iy)$ in complex analysis. If $f(x + iy) = u(x,y) + iv(x,y)$ is holomorphic then u, v (are C^∞ and) satisfy the Cauchy–Riemann equations,

$$u_x = v_y, \quad u_y = -v_x.$$

https://doi.org/10.1515/9783111185675-015

It follows that $u_{xx} = v_{yx}$, $u_{yy} = -v_{xy}$ and hence $\Delta u = v_{yx} - v_{xy}$. Using the Schwarz theorem about mixed partial derivatives we infer that $v_{yx} = v_{xy}$ and hence $\Delta u = 0$. Similarly, $\Delta v = 0$. Thus the components of a holomorphic function are harmonic functions.

Given an open bounded set $\Omega \subset \mathbb{R}^2$ with boundary $\partial \Omega$ and a function $f(x,y)$ on $\partial \Omega$, the *Dirichlet problem on the set* Ω associated to (L) consists of searching a function $u = u(x,y)$ such that

$$\begin{cases} \Delta u(x,y) = 0, & \forall (x,y) \in \Omega, \\ u(x,y) = f(x,y) & \forall (x,y) \in \partial \Omega. \end{cases} \qquad \text{(D)}$$

In the sequel we will prove some existence and uniqueness results for the Dirichlet problem.

We could also seek solutions satisfying other *boundary conditions*, such as the Neumann boundary conditions in which we prescribe the normal derivative of u on the boundary $\partial \Omega$. We will briefly address this problem later.

15.1.1 Dirichlet problem on a ball and the method of separation of variables

We are now going to solve the Dirichlet problem on a ball. To start with, let us consider the unit ball $B = \{(x,y) \in \mathbb{R}^2 : x^2 + y^2 < 1\}$ and look for u satisfying

$$\Delta u = 0, \text{ on } B \quad u = f \text{ on } \partial B. \qquad (15.1)$$

It is convenient to introduce polar coordinates in $\mathbb{R}^2$, $x = r\cos\theta$, $y = r\sin\theta$, that is, $r = \sqrt{x^2 + y^2}$, $\theta = \arctan(\frac{y}{x})$. Given $u(x,y)$ we set $u(r,\theta) = u(r\cos\theta, r\sin\theta)$. Below we assume that u is twice differentiable.

Lemma 15.1. *The Laplacian in polar coordinates is given by*

$$\Delta u = u_{xx} + u_{yy} = u_{rr} + \frac{1}{r}u_r + \frac{1}{r^2}u_{\theta\theta} \quad (r > 0).$$

Proof. For $r > 0$ one has

$$r_x = \frac{x}{\sqrt{x^2+y^2}} = \cos\theta, \quad r_y = \frac{y}{\sqrt{x^2+y^2}} = \sin\theta,$$

$$\theta_x = -\frac{y}{x^2+y^2} = -\frac{\sin\theta}{r}, \quad \theta_y = \frac{x}{x^2+y^2} = \frac{\cos\theta}{r}.$$

Then

$$u_x = u_r r_x + u_\theta \theta_x = u_r \cos\theta - u_\theta \frac{\sin\theta}{r}, \quad u_y = u_r r_y + u_\theta \theta_y = u_r \sin\theta + u_\theta \frac{\cos\theta}{r}.$$

Moreover, we have

$$u_{xx} = \left(u_r \cos\theta - u_\theta \frac{\sin\theta}{r}\right)_r \cos\theta - \left(u_r \cos\theta - u_\theta \frac{\sin\theta}{r}\right)_\theta \frac{\sin\theta}{r},$$

$$= u_{rr} \cos^2\theta + u_\theta \frac{\sin\theta\cos\theta}{r^2} - u_{\theta r} \frac{\sin\theta\cos\theta}{r}$$

$$- u_{r\theta} \frac{\sin\theta\cos\theta}{r} + u_r \frac{\sin^2\theta}{r} + u_{\theta\theta} \frac{\sin^2\theta}{r^2} + u_\theta \frac{\sin\theta\cos\theta}{r^2}$$

$$= u_{rr} \cos^2\theta + 2u_\theta \frac{\sin\theta\cos\theta}{r^2} - 2u_{r\theta} \frac{\sin\theta\cos\theta}{r} + u_r \frac{\sin^2\theta}{r} + u_{\theta\theta} \frac{\sin^2\theta}{r^2}$$

as well as

$$u_{yy} = \left(u_r \sin\theta + \frac{\sin\theta}{r} u_\theta\right)_r \sin\theta + \left(u_r \sin\theta + \frac{\sin\theta}{r} u_\theta\right)_\theta \frac{\cos\theta}{r}$$

$$= u_{rr} \sin^2\theta - 2u_\theta \frac{\sin\theta\cos\theta}{r^2} + 2u_{r\theta} \frac{\sin\theta\cos\theta}{r} + u_r \frac{\cos^2\theta}{r} + u_{\theta\theta} \frac{\cos^2\theta}{r^2}.$$

Summing up and simplifying, we find $\Delta u = u_{xx} + u_{yy} = u_{rr} + \frac{1}{r}u_r + \frac{1}{r^2}u_{\theta\theta}$. □

From the lemma it follows that if $\Delta u = 0$ on the ball B then $u(r,\theta)$ satisfies

$$u_{rr} + \frac{1}{r}u_r + \frac{1}{r^2}u_{\theta\theta} = 0, \quad 0 < r < 1, \quad 0 \le \theta \le 2\pi. \tag{15.2}$$

We solve this equation by separation of variables by setting $u(r,\theta) = R(r)S(\theta)$. Then (15.2) becomes

$$R''(r)S(\theta) + \frac{1}{r}R'(r)S(\theta) + \frac{1}{r^2}R(r)S''(\theta) = 0.$$

Rearranging, we find $r^2 R'' S + rR'S = -RS''$, that is

$$\frac{r^2 R''(r) + rR'(r)}{R(r)} = -\frac{S''(\theta)}{S(\theta)}.$$

The left hand side is independent of θ whereas the right hand side is independent of R. Thus the two quantities are constant, say $= \lambda$, and hence we find the two independent eigenvalue problems

$$r^2 R''(r) + rR'(r) = \lambda R(r), \tag{15.3}$$

$$S''(\theta) + \lambda S(\theta) = 0. \tag{15.4}$$

We first solve (15.4) together the natural periodicity boundary condition $S(\theta + 2\pi) = S(\theta)$. Repeating the arguments carried out in the case of Sturm-Lioville problems (see Chapter 13, Section 13.3) it is easy to check that the eigenvalues of this problem are $\lambda_n = n^2$, $n = 0, 1, 2, \ldots$, with eigenfunctions $S_n(\theta) = A_n \sin n\theta + B_n \cos n\theta$. Notice that for $n = 0$ we simply find $S_0 = B_0$.

For $\lambda = \lambda_n = n^2$, equation (15.3) becomes $r^2 R''(r) + rR'(r) - n^2 R(r) = 0$. This is an Euler equation (see Chapter 6, Section 6.4) whose solutions are

$$R_0(r) = c_0 + c_0' \ln r, \quad R_n(r) = c_n r^n + c_n' r^{-n}, \quad n = 1, 2, \ldots \tag{15.5}$$

As a consequence, each

$$u_n(r, \theta) = (A_n \sin n\theta + B_n \cos n\theta) R_n(r), \quad n = 0, 1, 2, \ldots,$$

is harmonic in $\{(x, y) \in \mathbb{R}^2 : 0 < x^2 + y^2 < 1\}$. Until now we have put $r > 0$. If we want that u_n is bounded at $r = 0$ we shall take $c_0' = c_n' = 0$ in (15.5). So we find $R_n(r) = c_n r^n$, $n = 0, 1, 2, \ldots$ If we replace A_n by $c_n A_n$ and B_n by $c_n B_n$, we find

$$u_0 = B_0, \quad u_n = (A_n \sin n\theta + B_n \cos n\theta) \cdot r^n, \quad n = 1, 2, \ldots$$

We set

$$u(r, \theta) = \sum_0^\infty u_n(r, \theta) = B_0 + \sum_1^\infty (A_n \sin n\theta + B_n \cos n\theta) \cdot r^n.$$

Since $A_n \sin n\theta + B_n \cos n\theta \leq C$, and the geometric series $\sum C r^n$ is uniformly convergent for $0 < r < 1$, it follows that so is $\sum u_n$ on $0 \leq r < 1$ (notice that $u_n(0, \theta) = B_0$ for any $n = 0, 1, 2, \ldots$) Similarly, one shows that series obtained by differentiating $\sum u_n$ term by term twice, with respect to r and θ, are uniformly convergent on $0 \leq r < 1$. Since the u_n are harmonic in B, we infer that so is u.

Next, let a_n, b_n denote the Fourier coefficients of f, namely

$$a_n = \frac{1}{\pi} \int_{-\pi}^{\pi} f(\vartheta) \sin n\vartheta \, d\vartheta, \quad b_n = \frac{1}{\pi} \int_{-\pi}^{\pi} f(\vartheta) \cos n\vartheta \, d\vartheta \quad (n = 0, 1, 2, \ldots) \tag{15.6}$$

Choosing

$$A_n = a_n, \quad B_n = b_n,$$

we find

$$u(r, \theta) = \sum_0^\infty u_n(r, \theta) = \sum_0^\infty (a_n \sin n\theta + b_n \cos n\theta) \cdot r^n.$$

For $r = 1$ this series is $\sum_0^\infty (a_n \sin n\theta + b_n \cos n\theta)$ which is just $f(\theta)$. In particular, such a series converges and therefore, by Abel's theorem on series, we infer that

$$\lim_{r \to 1^-} u(r, \theta) = u(1, \theta) = f(\theta). \tag{15.7}$$

This shows that $u(r, \theta)$ satisfies the boundary condition. In conclusion, we have proved that $u(r, \theta)$ gives a solution of (15.1). Rescaling $r \mapsto \frac{r}{R}$, we find that $u(\frac{r}{R}, \theta)$ satisfies

$$\Delta u = 0, \text{ on } B_R = \{x^2 + y^2 < R\}, \quad u = f \text{ on } \partial B_R, \tag{15.8}$$

which is just the Dirichlet problem for (L) on the ball of radius R. We anticipate that (15.8) has a unique solution; see Theorem 15.5 below.

Theorem 15.1. *Let f be any continuous 2π-periodic function with Fourier coefficients a_n, b_n. Then the solution of (15.8) is given, in polar coordinates, by*

$$u\left(\frac{r}{R}, \theta\right) = \sum_0^\infty (a_n \sin n\theta + b_n \cos n\theta) \cdot \left(\frac{r}{R}\right)^n.$$

Example 15.1. Solve

$$\Delta u = 0 \text{ in } B_2, \quad u = 8 \sin 2\theta \text{ on } \partial B_2.$$

Since $f(\theta) = 8 \sin 2\theta$, we infer that $f_2 = 8$ and $f_n = 0$ for all $n \in \mathbb{N}, n \neq 2$. Thus $u(r, \theta) = 8(\frac{r}{2})^2 \sin 2\theta = 2r^2 \sin 2\theta$. Since $r^2 \sin 2\theta = 2r \sin \theta \cdot r \cos \theta = 2xy$, we find $u(x, y) = 4xy$. □

15.1.1.1 The Poisson integral formula

From Theorem 15.1, we can derive an integral representation of $u(r, \theta)$.

Theorem 15.2. *The solution u of the Dirichlet problem in the ball B_R, $\Delta u = 0$ in B_R, $u = f$ on ∂B_R, is given, in polar coordinates, by the Poisson integral formula*

$$u(r, \theta) = \frac{1}{2\pi} \int_{-\pi}^{\pi} \frac{R^2 - r^2}{R^2 - 2Rr \cos(\theta - \vartheta) + r^2} f(\vartheta) \, d\vartheta, \quad 0 \leq r < R. \tag{15.9}$$

Proof (Sketch). From (15.6) it follows for $n = 0, 1, 2, \ldots$

$$a_n \cdot (\sin n\theta r^n) = \frac{1}{2\pi} \int_{-\pi}^{\pi} f(\vartheta) \sin n\vartheta \, d\vartheta \cdot (\sin n\theta \cdot r^n),$$

$$b_n \cdot (\cos n\theta r^n) = \frac{1}{2\pi} \int_{-\pi}^{\pi} f(\vartheta) \cos n\vartheta \, d\vartheta \cdot (\cos n\theta \cdot r^n).$$

Since $\sin n\theta r^n$ and $\cos n\theta r^n$ are independent of ϑ we infer

$$a_n \cdot (\sin n\theta r^n) = \frac{1}{2\pi} \int_{-\pi}^{\pi} f(\vartheta) \sin n\vartheta \sin n\theta r^n \, d\vartheta,$$

$$b_n \cdot (\cos n\theta r^n) = \frac{1}{2\pi} \int_{-\pi}^{\pi} f(\vartheta) \cos n\vartheta \cos n\theta r^n \, d\vartheta.$$

Summing up,

$$a_n \sin n\theta r^n + b_n \cos n\theta r^n = \frac{1}{2\pi} \int_{-\pi}^{\pi} f(\vartheta) r^n (\sin n\vartheta \sin n\theta + \cos n\vartheta \cos n\theta) \, d\vartheta$$

$$= \frac{1}{2\pi} \int_{-\pi}^{\pi} f(\vartheta) r^n \cos n(\theta - \vartheta) \, d\vartheta.$$

Therefore

$$u(r,\theta) = \sum_{0}^{\infty} \frac{1}{2\pi} \int_{-\pi}^{\pi} f(\vartheta) r^n \cos n(\theta - \vartheta) \, d\vartheta.$$

Integrating term by term (recall that the series $\sum r^n \cos n(\theta - \vartheta)$ is uniformly convergent for $r < 1$), we find

$$u(r,\theta) = \frac{1}{2\pi} \int_{-\pi}^{\pi} f(\vartheta) \cdot \sum_{0}^{\infty} r^n \cos n(\theta - \vartheta) \, d\vartheta, \quad 0 \le r < 1.$$

It is possible to show that for $0 \le r < 1$ one has

$$\sum_{0}^{\infty} r^n \cos n(\theta - \vartheta) = \frac{1 - r^2}{1 - 2r \cos(\theta - \vartheta) + r^2}.$$

From this we deduce

$$u(r,\theta) = \frac{1}{2\pi} \int_{-\pi}^{\pi} \frac{1 - r^2}{1 - 2r \cos(\theta - \vartheta) + r^2} f(\vartheta) \, d\vartheta, \quad 0 \le r < 1.$$

After the rescaling $r \mapsto r/R$ we find

$$u(r,\theta) = \frac{1}{2\pi} \int_{-\pi}^{\pi} \frac{R^2 - r^2}{R^2 - 2Rr \cos(\theta - \vartheta) + r^2} f(\vartheta) \, d\vartheta, \quad 0 \le r < R. \qquad \square$$

The function

$$K(R,r,\theta,\vartheta) = \frac{R^2 - r^2}{R^2 - 2Rr \cos(\theta - \vartheta) + r^2}$$

is called the *Poisson kernel of the Dirichlet problem* $\Delta u = 0$ in B_R. With this notation (15.9) can be written for short

$$u(r, \theta) = \frac{1}{2\pi} \int\limits_{-\pi}^{\pi} K(R, r, \theta, \vartheta) f(\vartheta) \, d\vartheta.$$

Remark 15.1. The denominator of the Poisson kernel has a suggestive geometrical meaning. Consider the triangle OAP, where O is the origin and, in polar coordinates, $A = (R, \theta)$, $P = (r, \vartheta)$; see Fig. 15.1.

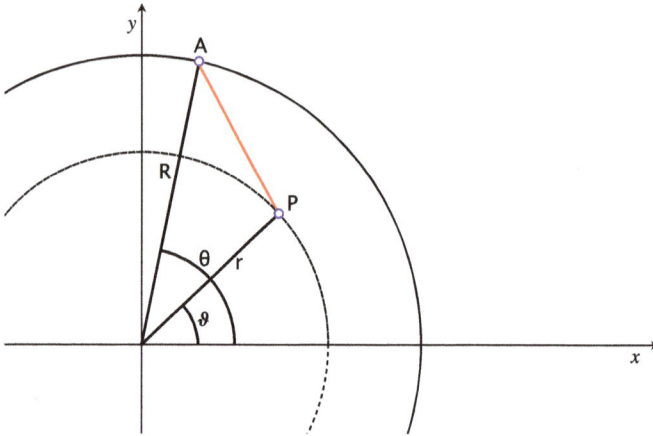

Figure 15.1: Plot of the triangle *OAP*.

Applying to OAP the law of cosines (also known as Carnot's theorem), we find

$$\overline{AP}^2 = \overline{OA}^2 - 2\overline{OA} \cdot \overline{OP} \cos(\theta - \vartheta) + \overline{OP}^2 = R^2 - 2Rr \cos(\theta - \vartheta) + r^2.$$

In particular, $K(R, r, \theta, \vartheta) = \frac{R^2 - r^2}{\overline{AP}^2}$.

For $r = 0$ one has $K(R, 0, \theta, \vartheta) = \frac{R^2}{R^2} = 1$. Then, using the Poisson integral formula with $r = 0$, we find

$$u(0, \theta) = \frac{1}{2\pi} \int\limits_{-\pi}^{\pi} f(\vartheta) \, d\vartheta.$$

Since $u(0, \theta)$ equals, in cartesian coordinates, the value of u at the origin $(0, 0)$, and $f(\vartheta) = u(R, \vartheta)$, we get

$$u(0, 0) = \frac{1}{2\pi} \int\limits_{-\pi}^{\pi} u(R, \vartheta) \, d\vartheta.$$

Let us show that the right hand side is the average of u on ∂B_R, defined by setting

$$\fint_{\partial B_R} u\, ds = \frac{1}{|\partial B_R|} \int_{\partial B_R} u\, ds = \frac{1}{2\pi R} \int_{\partial B_R} u\, ds.$$

Consider the line (or curvilinear) integral of u along the curve ∂B_R, $\int_{\partial B_R} u\, ds$ where ds is the elementary arc length of ∂B_R, see Chapter 1, Section 1.2. Using the parametrization $x = R\cos\vartheta$, $y = R\sin\vartheta$, $\vartheta \in [0, 2\pi]$, one has $ds = R\,d\vartheta$ whereby $\int_{\partial B_R} u\, ds = R\int_{-\pi}^{\pi} u(R, \vartheta)\, d\vartheta$. Thus

$$\fint_{\partial B_R} u\, ds = \frac{1}{2\pi R} \int_{\partial B_R} u\, ds = \frac{1}{2\pi} \int_{-\pi}^{\pi} u(R, \vartheta)\, d\vartheta.$$

We can summarize the previous arguments by stating the following result, known as the *mean value theorem for harmonic functions*.

Theorem 15.3. *If $u = u(x,y)$ is a harmonic function on the ball B_R and $u = f$ on the boundary ∂B_R, then the value of u at the center $(0,0)$ of B_R is equal to the average of u on the boundary of the ball:*

$$u(0,0) = \frac{1}{2\pi} \int_{-\pi}^{\pi} f(\vartheta)\, d\vartheta = \fint_{B_R} u\, ds.$$

15.1.1.2 The Neumann problem on a ball

We look for solutions of a Neumann problem on the unit ball $B = \{0 \le r < 1\}$. In polar coordinates, this amounts to finding a harmonic function $u(r, \theta)$ on B such that $u_r(1, \theta) = f(\theta)$, where f is a given periodic function.

Let a_n, b_n be the Fourier coefficients of f. We know that $u(r, \theta) = \sum_0^\infty (A_n \sin n\theta + B_n \cos n\theta)r^n = B_0 + \sum_1^\infty (A_n \sin n\theta + B_n \cos n\theta)r^n$ is harmonic in the ball B. Differentiating term by term with respect to r we find

$$u_r(r, \theta) = \sum_1^\infty (nA_n \sin n\theta + nB_n \cos n\theta)r^{n-1} \quad (r < 1).$$

If $nA_n = a_n$, $nB_n = b_n$ $(n = 1, 2, \ldots)$ and $r = 1$ the series on the right hand side becomes simply $f(\theta)$. Then, using Abel's theorem, we find

$$\lim_{r \to 1^-} u_r(r, \theta) = u_r(1, \theta) = f(\theta).$$

Therefore $u(r, \theta) = B_0 + \sum_1^\infty (\frac{a_n}{n} \sin n\theta + \frac{b_n}{n} \cos n\theta)r^{n-1}$ solves the Neumann problem, for any $B_0 \in \mathbb{R}$.

The solution depends upon the arbitrary constant B_0, This is not surprising since, if u is a solution of the Neumann problem $\Delta u = 0$ on Ω, $\frac{\partial u}{\partial n} = f$ on $\partial\Omega$ ($\frac{\partial}{\partial n}$ denotes the normal derivative), then $c + u$ is also a solution $\forall c \in \mathbb{R}$, since $\Delta(c + u) = \Delta u = 0$ and $\frac{\partial}{\partial n}(c + u) = \frac{\partial u}{\partial n}$.

Example 15.2. Solve $\Delta u = 0$ for $r < 1$, $u_r(1, \theta) = \sin 2\theta$.

Since all the Fourier coefficients of f are zero but $a_2 = 1$, we infer that $u(r, \theta) = B_0 + \frac{r^2}{2} \sin 2\theta$, $B_0 \in \mathbb{R}$. $\qquad\square$

15.1.2 The maximum principle

From the mean value theorem we can deduce the (weak) maximum principle for harmonic functions.

Theorem 15.4. *If u is harmonic on a bounded open set $\Omega \subset \mathbb{R}^2$, then*

$$\min_{\partial\Omega} u = \min_{\overline{\Omega}} u \leq \max_{\partial\Omega} u = \max_{\overline{\Omega}} u.$$

Proof. Notice that, since $\overline{\Omega} := \Omega \cup \partial\Omega$ is compact, i. e., closed and bounded, the previous maxima and minima are achieved. In addition, $\partial\Omega \subset \overline{\Omega}$ implies that $\max_{\partial\Omega} u \leq \max_{\overline{\Omega}} u$. If by contradiction $\max_{\partial\Omega} u < \max_{\overline{\Omega}} u$, then $\max_{\overline{\Omega}} u$ is achieved on $\overline{\Omega} \setminus \partial\Omega = \Omega$. Let $P = (\overline{x}, \overline{y}) \in \Omega$ be such that $u(\overline{x}, \overline{y}) = m = \max\{u(x, y) : (x, y) \in \overline{\Omega}\}$. Up to a change of coordinates, we can assume that $P = O = (0, 0)$. We can also suppose that u is not constant, otherwise we are done. Then, since $(0, 0) \in \Omega$ and Ω is open, $\exists R, \epsilon > 0$ such that $B_R \subset \Omega$ (see Fig. 15.2) and

$$\max_{\partial B_R} u \leq m - \epsilon. \tag{15.10}$$

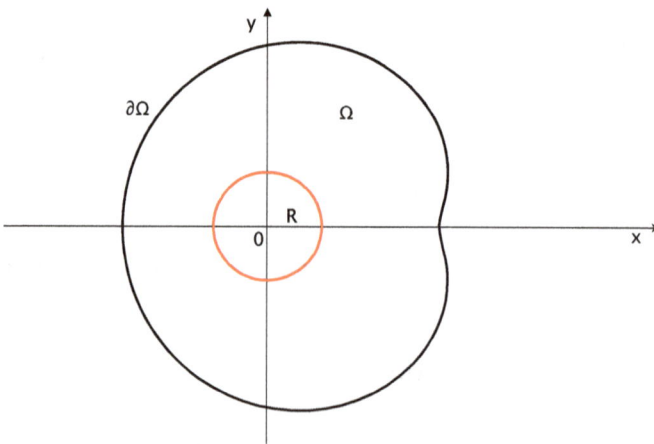

Figure 15.2: Plot of the ball $B_R \subset \Omega$.

Then, using the mean value Theorem 15.3, we infer

$$u(0,0) = \fint_{\partial B_R} u \, ds.$$

From $\max_{\partial B_R} u \le m - \epsilon$ it follows that $u(0,0) \le m - \epsilon$. Since $u(0,0) = m$ we find $m \le m - \epsilon$, a contradiction. This proves that u achieves its maximum on the boundary $\partial \Omega$ and hence $\max_{\overline{\Omega}} u = \max_{\partial \Omega} u$. The case of the minimum is handled in a similar way. □

15.1.3 Uniqueness of the Dirichlet problem

Lemma 15.2. *If u satisfies*

$$\begin{cases} \Delta u(x,y) = 0, & (x,y) \in \Omega, \\ u(x,y) = 0, & (x,y) \in \partial\Omega, \end{cases} \tag{15.11}$$

then $u \equiv 0$.

Proof. Using the maximum principle we infer that

$$\min_{\partial\Omega} u \le u(x,y) \le \max_{\partial\Omega} u.$$

Since $u = 0$ on $\partial\Omega$, we find that $u(x,y) = 0$ for all $(x,y) \in \Omega$. □

Theorem 15.5. *The Dirichlet problem*

$$\begin{cases} \Delta u(x,y) = h(x,y), & (x,y) \in \Omega, \\ u(x,y) = f(x,y), & (x,y) \in \partial\Omega, \end{cases} \tag{D}$$

has at most one solution.

Proof. If v, w are two solutions of (D), then $u = v - w$ satisfies

$$\begin{cases} \Delta u(x,y) = 0, & (x,y) \in \Omega, \\ u(x,y) = 0, & (x,y) \in \partial\Omega. \end{cases}$$

Then Lemma 15.2 implies that $u \equiv 0$, namely $v = w$. □

Proving the existence of the solution of (D) on a general domain Ω is out of the scope of this book. In the next appendix we give a suggestion for how to handle this problem when $h = 0$.

15.1.4 Appendix: Dirichlet's integral

The following proposition is an interesting property of harmonic functions.

Proposition 15.1. *Let u be any solution of* $\Delta u(x,y) = 0$ *for* $(x,y) \in \Omega$, *and suppose that* Ω *is a bounded open set such that its boundary* $\partial\Omega$ *is a smooth curve. Then for any smooth function v such that* $v = u$ *on* $\partial\Omega$, *one has*

$$\int_\Omega |\nabla u|^2 \, dx \le \int_\Omega |\nabla v|^2 \, dx.$$

Proof. Since $\partial\Omega$ is a smooth curve we can use the *Divergence theorem* (see Chapter 1, Section 1.4) which states that

$$\int_\Omega \mathrm{div}\, F = \int_{\partial\Omega} F \cdot \bar{n},$$

for any C^1 vector field F. Taking $F = (v - u)\nabla u$ one has $F = 0$ on $\partial\Omega$, since $v = u$ therein. Then it follows that

$$\int_\Omega \mathrm{div}\, F = \int_\Omega \mathrm{div}[(v - u)\nabla u] = 0. \tag{15.12}$$

On the other hand, setting $w = v - u$, we find

$$\mathrm{div}[w\nabla u] = (w \cdot u_x)_x + (w \cdot u_y)_y = w_x u_x + w u_{xx} + w_y u_y + w u_{yy}$$
$$= w(u_{xx} + u_{yy}) + w_x u_x + w_y u_y = w\Delta u + \nabla w \cdot \nabla u.$$

Then (15.12) yields

$$\int_\Omega (w\Delta u + \nabla w \cdot \nabla u) = 0. \tag{15.13}$$

Since $\Delta u = 0$ in Ω, we infer $0 = \int_\Omega \nabla w \cdot \nabla u = \int_\Omega \nabla v \cdot \nabla u - \int_\Omega |\nabla u|^2$. Using Schwarz's inequality we find

$$\int_\Omega |\nabla u|^2 = \int_\Omega \nabla v \cdot \nabla u \le \left(\int_\Omega |\nabla v|^2 \right)^{\frac{1}{2}} \cdot \left(\int_\Omega |\nabla u|^2 \right)^{\frac{1}{2}},$$

from which the result follows at once. □

The integral $\int_\Omega |\nabla u|^2$ is called the *Dirichlet integral*.

Remark 15.2. It is worth noting that (15.13), namely

$$\int_\Omega w\Delta u = -\int_\Omega \nabla w \cdot \nabla u,$$

valid for any w such that $w = 0$ on $\partial\Omega$, can be seen as the integration by parts for functions of two variables.

According to Proposition 15.1 one can expect that the solution of Dirichlet's problem

$$\Delta u = 0 \text{ in } \Omega, \quad u = f \text{ on } \partial\Omega$$

might be found as the minimum of the Dirichlet integral among the class of all smooth functions v such that $v = f$ on $\partial\Omega$. Unfortunately, the minimum could not exist in such a class. To overcome this difficulty one has to consider a larger class of "generalized" functions, but this involves topics of functional analysis.

15.2 The heat equation

The *heat equation* in its simplest form is

$$u_t - u_{xx} = 0.$$

A solution is a function $u(x,t)$ such that $u(x,t)$ is C^2 w. r. t. x, C^1 w. r. t. t and verifies $u_t(x,t) - u_{xx}(x,t) = 0$ for all (x,t).

In applications, the heat equation models, in suitable units, the propagation of the temperature u in a cylindrical bar with axis parallel to x, under the assumption that there are no external heat sources and that the lateral surface of the bar is insulated. In this model it is assumed that the thermal conductivity of the material, the mass density and the specific heat of the bar are constant and the temperature u depends only on the position x in the bar and on time t.

More precisely, we want to solve the following problem:

$$\begin{cases} u_t - u_{xx} = 0, & \forall x \in [0,\pi], \ \forall t \geq 0, \\ u(x,0) = f(x), & \forall x \in [0,\pi], \\ u(0,t) = u(\pi,t) = 0, & \forall t \geq 0. \end{cases} \qquad \text{(H)}$$

The differential equation is simply the heat equation for a bar of length π and is complemented with the conditions:
1. $u(x,0) = f(x)$, $\forall x \in [0,1]$, which prescribes the temperature on the bar at initial time $t = 0$, and
2. $u(0,t) = u(L,t) = 0$, $\forall t \geq 0$, which is a so-called *Dirichlet* boundary condition that says that the edges of the bar are kept at zero temperature for all time $t \geq 0$.

As for the Laplace equation, we can solve the heat equation (H) by means of the separation of variables. Since the arguments are quite similar, we will be brief. Setting $u(x, t) = X(x)T(t)$, $u_t - u_{xx} = 0$ yields

$$X(x)T'(t) = X''(x)T(t), \quad \text{that is} \quad \frac{X''(x)}{X(x)} = \frac{T'(t)}{T(t)}.$$

Then

$$\frac{X''(x)}{X(x)} = \frac{T'(t)}{T(t)} = -\lambda.$$

Using the Dirichlet boundary condition $u(0, t) = u(\pi, t)$, we are led to the eigenvalue problem

$$X''(x) + \lambda X(x) = 0, \quad X(0) = X(\pi) = 0,$$

which gives (see Chapter 13, Section 13.3) $\lambda = n^2$, $n = 1, 2, \ldots$, and

$$X_n(x) = A_n \sin nx, \quad n = 1, 2, \ldots$$

Now we solve $T'(t) + n^2 T(t) = 0$ yielding

$$T_n(t) = B_n e^{-n^2 t}, \quad n = 1, 2, \ldots$$

Calling $c_n = A_n B_n$, we are led to seek solutions in the form

$$u(x, t) = \sum_1^\infty c_n e^{-n^2 t} \sin nx. \tag{15.14}$$

Let us suppose that $f(x)$ is continuously differentiable and that $f(0) = f(\pi) = 0$. Under these conditions f can be expanded in a sine Fourier series (see, e. g., S. Suslov, *An Introduction to Basic Fourier Series*, Springer US, 2003)

$$f(x) = \sum_1^\infty a_n \sin nx, \quad a_n = \frac{2}{\pi} \int_0^\pi f(\vartheta) \sin n\vartheta \, d\vartheta. \tag{15.15}$$

The initial condition $u(x, 0) = f(x)$ yields

$$\sum_1^\infty c_n \sin nx = \sum_1^\infty a_n \sin nx$$

whereby $c_n = a_n$ so that (15.14) becomes

$$u(x, t) = \sum_1^\infty a_n e^{-n^2 t} \sin nx. \tag{15.16}$$

Theorem 15.6. *Suppose that $f(x)$ is C^1 on $[0, \pi]$ (actually f being Hölder continuous suffices) and such that $f(0) = f(\pi) = 0$. Then $u(x,t) = \sum_1^\infty a_n e^{-n^2 t} \sin nx$ solves the heat equation (H).*

Proof (Outline). Performing a term-by-term differentiation of the series $\sum_1^\infty a_n e^{-n^2 t} \sin nx$ with respect to t, we obtain the series

$$\sum_1^\infty (-n^2 f_n) e^{-n^2 t} \sin nx.$$

Moreover, differentiating term by term the same series twice with respect to x we find the series

$$\sum_1^\infty (-n^2 f_n) e^{-n^2 t} \sin nx.$$

So, formally, we find that $u_t = u_{xx}$ in $[0, \pi] \times (0, +\infty)$ and that $u(0, t) = u(\pi, t) = 0$. The result is not just formal, for it is possible to show that all these series are uniformly convergent on $[0, \pi] \times [0, +\infty)$, for f is C^1. Finally, since the series converge on $t \geq 0$, u is continuous in $[0, \pi] \times [0, +\infty)$ and $u(x, 0) = f(x)$. □

Remark 15.3.
(i) From (15.16) it follows that $\lim_{t \to +\infty} u(x, t) = 0$. From the physical point of view this means that the temperature decays to zero as $t \to +\infty$, as is in accordance with experience.
(ii) It is possible to show that u is C^k in $[0, L] \times (0, +\infty)$ for all k. In other words, the heat equation has a smoothing effect, in the sense that $u(x, t)$ becomes immediately smooth as long as $t > 0$, even if $u(x, 0) = f(x)$ is merely C^1, or even Hölder continuous. As a consequence, we infer that the "backward heat equation", namely (H) with $t \leq 0$, has no solution if $u(x, 0) = f$ is not C^∞.
(iii) Recalling that $a_n = \frac{2}{\pi} \int_0^\pi f(\vartheta) \sin n\vartheta \, d\vartheta$, see (15.15), we can write (15.16) as

$$u(x, t) = \frac{2}{\pi} \sum_1^\infty \int_0^\pi f(\vartheta) \sin n\vartheta \, d\vartheta \cdot e^{-n^2 t} \sin nx.$$

Since the series is uniformly convergent, we can integrate term by term, yielding, after rearrangements,

$$u(x, t) = \frac{2}{\pi} \int_0^\pi d\vartheta \sum_1^\infty f(\vartheta) \sin nx \cdot \sin n\vartheta \cdot e^{-n^2 t}, \tag{15.17}$$

which provides an integral representation of the solution.

(iv) If we look for solutions of $u_t - cu_{xx} = 0$, in $(0,L) \times (0, +\infty)$, $u(x,0) = f(x)$ for $x \in [0,L]$ and $u(0,t) = u(L,t) = 0$ for $t \geq 0$, we find

$$u(x,t) = \sum_1^\infty f_n e^{-\lambda_n ct} \sin(\sqrt{\lambda_n} x)$$

where $f_n = \frac{2}{L} \int_0^L f(x) \sin(\sqrt{\lambda_n} x)\, dx$ and $\lambda_n = (\frac{n\pi}{L})^2$ are the eigenvalues of $X'' + \lambda X = 0$, $X(0) = X(L) = 0$.

Example 15.3. Let us solve (H) with $f(x) = \sin 3x$.
Here $a_3 = \frac{2}{\pi} \int_0^\pi \sin^2 3x = 1$ whereas $a_n = \frac{2}{\pi} \int_0^\pi \sin 3x \cdot \sin nx = 0$ for all $n \neq 3$. Thus $u(x,t) = e^{-9t} \sin 3x$. ☐

15.2.1 Appendix: further properties of the heat equation

15.2.1.1 Continuous dependence on data
Equation (15.17) can be used to show that the solution of the heat equation depends continuously on the initial data. Precisely, if u_i, $i = 1,2$, denoting the solution of (H) with $f = f_i$, then we find

$$u_1(x,t) - u_2(x,t) = \frac{2}{\pi} \sum_1^\infty \int_0^\pi [f_1(\vartheta) - f_2(\vartheta)] \sin n\vartheta\, d\vartheta \cdot \sin nx e^{-n^2 t}.$$

If $|f_1(\vartheta) - f_2(\vartheta)| < \epsilon$ it follows that

$$|u_1(x,t) - u_2(x,t)| \leq \frac{2}{\pi} \sum_1^\infty \int_0^\pi |f_1(\vartheta) - f_2(\vartheta)|| \sin n\vartheta|\, d\vartheta \cdot \sin nx e^{-n^2 t}$$

$$\leq \epsilon \cdot \frac{2}{\pi} \sum_1^\infty \int_0^\pi |\sin n\vartheta|\, d\vartheta \cdot \sin nx e^{-n^2 t}.$$

Since $\int_0^\pi |\sin n\vartheta|\, d\vartheta \leq \pi$, we find

$$|u_1(x,t) - u_2(x,t)| \leq \epsilon \cdot 2 \sum_1^\infty \sin nx e^{-n^2 t} = K\epsilon,$$

where $K = 2 \sum_1^\infty \sin nx e^{-n^2 t} < \infty$.

15.2.1.2 Neumann boundary conditions

Instead of the Dirichlet boundary condition we can impose other conditions. For example, let us briefly discuss the case in which we seek solutions that satisfy (H) with $u(0,t) = u(\pi, t) = 0$ substituted by $u_x(0,t) = u_x(\pi, t)$. This is called a *Neumann boundary condition*.

The arguments are quite similar to the previous ones. We only indicate the changes: repeating the previous calculations, we find $X''(x) + \lambda X(x) = 0$ and $T'(t) + \lambda T(t) = 0$. The Neumann boundary condition yields $X'(0) = X'(\pi) = 0$. The eigenvalues of $X'' + \lambda X = 0$ are now $\lambda = 0, 1, 2, \ldots$ and the corresponding eigenfunctions are $X_n(x) = A_n \cos nx$. Thus

$$u(x,t) = \sum_0^\infty c_n e^{-n^2 t} \cos nx = c_0 + \sum_1^\infty c_n e^{-n^2 t} \cos nx.$$

Finally, we take into account the initial condition $u(x,0) = f(x)$. Suppose that

$$f(x) = \frac{1}{2}f_0 + \sum_1^\infty b_n \cos nx, \quad b_n = \frac{2}{\pi}\int_0^\pi f(x) \cos nx\,dx, \quad n = 0,1,2,\ldots$$

It follows that

$$u(x,t) = \frac{1}{2}f_0 + \sum_1^\infty b_n e^{-n^2 t} \cos nx.$$

Example 15.4. For example, if $f(x) = 2 + \cos 3x$ then $b_0 = \frac{2}{\pi}\int_0^\pi 2\,dx = 4$ and

$$b_n = \frac{1}{\pi}\int_0^\pi (2 + \cos 3x) \cos nx\,dx = \begin{cases} 1 & \text{if } n = 3, \\ 0 & \text{if } n \neq 0,3, \end{cases}$$

from which we infer that $u(x,t) = 2 + e^{-9t} \cos 3x$. □

15.2.1.3 A uniqueness result

Since the series coefficients c_n in (15.14) are uniquely determined by f, it is clear that (H) has a unique solution *of the form* $u(x,t) = X(x)T(t)$. We want to extend the uniqueness result to *any solution* of (H).

Theorem 15.7. *Given f, problem (H) has a unique solution.*

Proof. Let u, v be two solutions of (H). Setting $w = u - v$ we find that w solves the homogeneous problem

$$\begin{cases} w_t - w_{xx} = 0, & \forall x \in [0,\pi], \ \forall t \geq 0, \\ w(x,0) = 0, & \forall x \in [0,\pi], \\ w(0,t) = w(\pi,t) = 0, & \forall t \geq 0. \end{cases} \tag{15.18}$$

Let us define

$$W(t) = \int_0^{\pi} w^2(x, t)\, dx.$$

Differentiating, and using the equation $w_t - w_{xx} = 0$, we infer

$$W'(t) = 2\int_0^{\pi} w(x, t)w_t(x, t)\, dx = 2\int_0^{\pi} w(x, t)w_{xx}(x, t)\, dx.$$

Integrating by parts, it follows that

$$\int_0^{\pi} w(x, t)w_{xx}(x, t)\, dx = [w(x, t) \cdot w_x(x, t)]_{x=0}^{x=\pi} - \int_0^{\pi} w_x^2(x, t)\, dx$$

$$= w(\pi, t)w_x(\pi, t) - w(0, t)w_x(0, t) - \int_0^{\pi} w_x^2(x, t)\, dx.$$

Using the boundary condition $w(0, t) = w(\pi, t) = 0$ we deduce

$$\int_0^{\pi} w(x, t)w_{xx}(x, t)\, dx = -\int_0^{\pi} w_x^2(x, t)\, dx,$$

and hence

$$W'(t) = -2\int_0^{\pi} w_x^2(x, t)\, dx \le 0.$$

This implies that $W(t)$ is non-increasing, whereby $W(t) \le W(0)$ for $t \ge 0$. Since $w(x, 0) = 0$ we have $W(0) = 0$ and thus

$$\int_0^{\pi} w^2(x, t)\, dx = W(t) \le W(0) = 0.$$

It follows that $w(x, t)$ is identically zero, and this implies that $u(x, t) = v(x, t)$ for all $x \in [0, \pi]$ and all $t \ge 0$, proving the theorem. □

15.3 The vibrating string equation

The *vibrating string equation*, VS equation in short, in its simplest version is given by

$$u_{tt} - c^2 u_{xx} = 0, \tag{VS}$$

where $c \neq 0$ is a given real constant.

The name is related to the fact that (VS) models the vertical vibrations of an elastic string of length L in the absence of external forces. In this application $u(x, t)$ denotes the vertical displacement of the string at time t and point x on the string. Moreover, $c^2 = \frac{\tau}{\rho}$, where ρ is the (constant) mass of the string and τ its (constant) tension.

Sometimes (VS) is called the *D'Alembert equation*, named after the French mathematician J. B. D'Alembert who first studied it.

A solution of (VS) is a C^2 function $u(x, t)$ such that $u_{tt}(x, t) = c^2 u_{xx}(x, t)$ for all $(x, t) \in \mathbb{R}^2$.

Consider the change of variable

$$\begin{cases} \xi = x - ct, \\ \eta = x + ct, \end{cases} \quad \text{that is,} \quad \begin{cases} x = \frac{1}{2}(\xi + \eta), \\ t = \frac{1}{2c}(\eta - \xi). \end{cases}$$

Then one has

$$u_\xi = u_x x_\xi + u_t t_\xi = \frac{1}{2} u_x + \frac{1}{2c} u_t \tag{15.19}$$

and hence

$$u_{\xi\eta} = \frac{\partial u_\xi}{\partial \eta} = \frac{\partial u_\xi}{\partial x} \frac{\partial x}{\partial \xi} + \frac{\partial u_\xi}{\partial y} \frac{\partial y}{\partial \eta} = \frac{1}{2} \frac{\partial u_\xi}{\partial x} - \frac{1}{2c} \frac{\partial u_\xi}{\partial y}. \tag{15.20}$$

From (15.19) we infer

$$\frac{\partial u_\xi}{\partial x} = \frac{1}{2} u_{xx} + \frac{1}{2c} u_{yx},$$

$$\frac{\partial u_\xi}{\partial t} = \frac{1}{2} u_{xy} + \frac{1}{2c} u_{tt}.$$

Substituting into (15.20) we find

$$u_{\xi\eta} = \frac{1}{2}\left(\frac{1}{2} u_{xx} + \frac{1}{2c} u_{yx}\right) - \frac{1}{2c}\left(\frac{1}{2} u_{xy} + \frac{1}{2c} u_{tt}\right).$$

Since u is C^2 the mixed derivatives are equal, that is, $u_{yx} = u_{xy}$. Hence, simplifying,

$$u_{\xi\eta} = \frac{1}{4} u_{xx} - \frac{1}{4c^2} u_{tt}.$$

Finally, recalling that u is a solution of (VS), that is $u_{tt} = c^2 u_{xx}$ we deduce that u satisfies the hyperbolic equation

$$u_{\xi\eta}(\xi, \eta) = 0.$$

From this it follows that $u_\xi(\xi, \eta)$ is independent of η, that is, $u_\xi(\xi, \eta) = U(\xi)$ for some function U. Integrating w. r. t. ξ we find that $u = \int U \, d\xi + q$, with q depending on η. In other words, one has

$$u(\xi, \eta) = p(\xi) + q(\eta),$$

for some functions p, q. Notice that p, q are C^2 functions. Conversely, it is clear that, for any C^2 functions p, q, $u(\xi, \eta) = p(\xi) + q(\eta)$ verifies $u_{\xi\eta} = 0$. Recalling that $\xi = x - ct$, $\eta = x + ct$ we deduce the following theorem.

Theorem 15.8. *All of the solutions of* (VS) *are given by*

$$u(x, t) = p(x - ct) + q(x + ct) \tag{15.21}$$

for any pair of one variable C^2 functions p, q.

We will say that (15.21) is the *general solution* of (VS).

15.3.1 An initial value problem for the vibrating string: D'Alembert's formula

The initial value problem for a second order ordinary differential equation consists of prescribing the solution and its derivative at, say, $t = 0$. Similarly, the initial value problem for the string equation consists of seeking a solution u of (VS) satisfying the initial conditions

$$u(x, 0) = f(x), \tag{VS1}$$
$$u_t(x, 0) = g(x). \tag{VS2}$$

From the physical point of view, this amounts to prescribing the shape of the string and its speed at the initial time $t = 0$. Throughout in the sequel we will assume that f is a C^2 function and g is C^1 defined on $\mathbb{R}$.

Taking $y = 0$ in (15.21) and using (VS1) we find $f(x) = p(x) + q(x)$ and hence $f'(x) = p'(x) + q'(x)$. Moreover, (15.21) also implies $u_t(x, t) = -cp'(x - ct) + cq'(x + ct)$ and hence (VS2) yields $g(x) = -cp'(x) + cq'(x)$. Solving the system

$$\begin{cases} p'(x) + q'(x) = f'(x), \\ -cp'(x) + cq'(x) = g(x), \end{cases}$$

for p', q' we find

$$p'(x) = \frac{1}{2c}(cf'(x) - g(x)) = \frac{1}{2}f'(x) - \frac{1}{2c}g(x),$$
$$q'(x) = \frac{1}{2c}(cf'(x) + g(x)) = \frac{1}{2}f'(x) + \frac{1}{2c}g(x).$$

Integrating and letting G be an antiderivative of g, we find

$$p(x) = \frac{1}{2}f(x) - \frac{1}{2c}G(x),$$

$$q(x) = \frac{1}{2}f(x) + \frac{1}{2c}G(x).$$

Substituting into (15.21) it follows that

$$u(x,t) = \frac{1}{2}[f(x+ct) + f(x-ct)] + \frac{1}{2c}[G(x+ct) - G(x-ct)].$$

Since $G(x) = \int_0^x g(s)\,ds$, we find

$$G(x+ct) - G(x-ct) = \int_0^{x+ct} g(s)\,ds - \int_0^{x-ct} g(s)\,ds = \int_{x-ct}^{x+ct} g(s)\,ds.$$

Notice that u possesses second derivatives u_{tt}, u_{xx} since $f \in C^2$ and $g \in C^1$.
Summarizing, we can state the following existence and uniqueness result.

Theorem 15.9. *If $f(x)$ is C^2 and $g(x)$ is C^1, then the vibrating string equation (VS) has one and only one solution satisfying the initial conditions (VS1) and (VS2), which is given by D'Alembert's formula,*

$$u(x,t) = \frac{1}{2}[f(x+ct) + f(x-ct)] + \frac{1}{2c}\int_{x-ct}^{x+ct} g(s)\,ds.$$

Corollary 15.1. *The solution of (VS) such that $u(x,0) = f(x)$ and $u_t(x,0) = 0$ is given by $u(x,t) = \frac{1}{2}[f(x+ct) + f(x-ct)]$.*

Example 15.5. (a) Solve the ivp $u_{tt} - u_{xx} = 0$, $u(x,0) = x^2$, $u_t(x,0) = \sin x$.
Using D'Alembert's formula, we find

$$u(x,t) = \frac{1}{2}[(x+t)^2 + (x-t)^2] + \frac{1}{2}\int_{x-t}^{x+t} \sin s\,ds$$

$$= x^2 + t^2 + \frac{1}{2}[\cos(x-t) - \cos(x+t)]$$

$$= x^2 + t^2 + \frac{1}{2}[\cos x \cos t + \sin t \sin x - \cos x \cos t + \sin t \sin x]$$

$$= x^2 + t^2 + \sin t \sin x.$$

(b) Solve the ivp $u_{tt} - 4u_{xx} = 0$, $u(x,0) = \sin x$, $u_t(x,0) = 0$.
Using Corollary 15.1 we find

$$u(x,t) = \frac{1}{2}[\sin(x+2t) + \sin(x-2t)]$$

$$= \frac{1}{2}[(\sin x \cdot \cos 2t + \cos x \cdot \sin 2t) + (\sin x \cdot \cos(-2t) - \cos x \cdot \sin(-2t))]$$

$$= \sin x \cdot \cos 2t + \cos x \cdot \sin 2t = \sin(x+2t). \qquad \square$$

15.3.2 Appendix: continuous dependence on the initial conditions

Next, we show that the solution of (VS)–(VS1)–(VS2) depends continuously on the initial data. Let u_i, $i = 1, 2$ be the solutions of (VS)–(VS1)–(VS2) with $f = f_i$, $g = g_i$.

Theorem 15.10. *Suppose that*

$$|f_1(x) - f_2(x)| < \epsilon, \quad |g_1(x) - g_2(x)| < \epsilon, \quad \forall x \in \mathbb{R}. \tag{15.22}$$

Then for all $T > 0$ there exists $K > 0$ such that

$$|u_1(x,t) - u_2(x,t)| < K\epsilon, \quad \forall(x,t) \in \mathbb{R} \times [0,T].$$

Proof. From D'Alembert's formula we infer

$$|u_1(x,t) - u_2(x,t)| \le \frac{1}{2}|f_1(x-ct) - f_2(x-ct)| + \frac{1}{2}|f_1(x+ct) - f_2(x-ct)|$$

$$+ \frac{1}{2c} \int\limits_{x-ct}^{x+ct} |g_1(s) - g_2(s)|\, ds$$

$$\le \epsilon + \frac{1}{2c} \cdot 2ct \cdot \epsilon \le \epsilon + T\epsilon,$$

proving the theorem. $\qquad \square$

Let us consider a function $\varphi(x)$ whose support is $[a,b]$, that is, $\varphi(x) = 0$ if and only if $x \notin (a,b)$.

If u, resp. $\tilde{u}$, denotes the solution of (VS)–(VS1)–(VS2) with $f = \varphi$, resp. $\tilde{f} = f + \varphi$, and $g = 0$, then the D'Alembert formula yields

$$u(x,t) = \frac{1}{2}[f(x+ct) + f(x-ct)],$$

$$\tilde{u}(x,t) = \frac{1}{2}[f(x+ct) + \varphi(x+ct) + f(x-ct) + \varphi(x-ct)].$$

Then

$$\tilde{u}(x,t) - u(x,t) = \frac{1}{2}[\varphi(x+ct) + \varphi(x-ct)].$$

For $t \geq 0$ we set

$$S_1 = \{(x,t) : a < x + ct < b\}, \quad S_2 = \{(x,t) : a < x - ct < b\}, \quad S = S_1 \cap S_2.$$

If $(x,t) \notin S_1$ then $\varphi(x + ct) = 0$; if $(x,t) \notin S_2$ then $\varphi(x - ct) = 0$. Thus if $(x,t) \notin S = S_1 \cap S_2$ then $\tilde{u}(x,t) - u(x,t) = 0$. This means that an initial condition disturbance with support $[a,b]$ does not have any effect outside S; see Fig. 15.3.

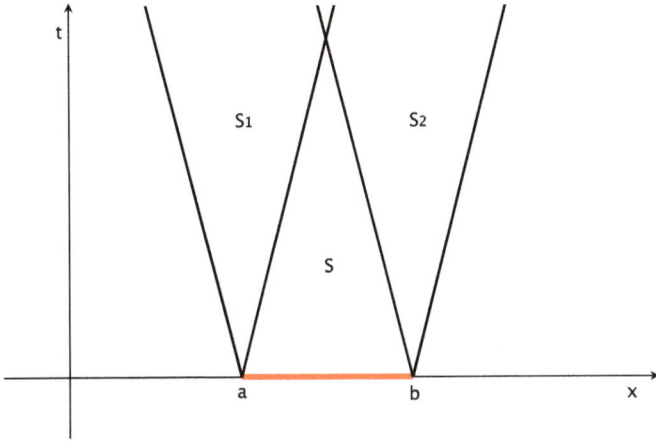

Figure 15.3: Characteristics through $x = a$ and $x = b$.

15.3.3 The method of the characteristics

This is a method which, jointly with D'Alembert's formula, allows us to find further features of u.

Definition 15.1. The straight lines $x \pm ct = k_\pm$, $k_\pm \in \mathbb{R}$, are called the characteristics of $u_{tt} - c^2 u_{xx} = 0$.

From D'Alembert's formula it follows that

$$u(x,t) = \frac{1}{2}[f(k_+) + f(k_-)] + \frac{1}{2c} \int_{x-ct}^{x+ct} g(s)\, ds.$$

If $g = 0$ then $u(x,t) = \frac{1}{2}[f(k_+) + f(k_-)]$, so that u is constant along the characteristics.
    Let $g = 0$. Let there be given $P_1 = (x_1, t_1)$, $P_2 = (x_2, t_2)$, the characteristics passing through P_1, P_2 are $x \pm ct = x_i \pm ct_i$, $i = 1, 2$. Using D'Alembert's formula we find

$$u(x_i, t_i) = \frac{1}{2}[f(x_i - ct_i) + f(x_i + ct_i)] \quad (i = 1, 2). \tag{15.23}$$

This shows that the value of $u(x_i, t_i)$ depends only on the values of f at $x_i \pm ct_i$ which are simply the intersections of the characteristics through P_i with the x axis. They are the endpoints of the interval $I_{P_i} = I_i = [x_i - ct_i, x_i + ct_i]$, $i = 1, 2$, which is called the *domain of dependence* of P_i; see Fig. 15.4.

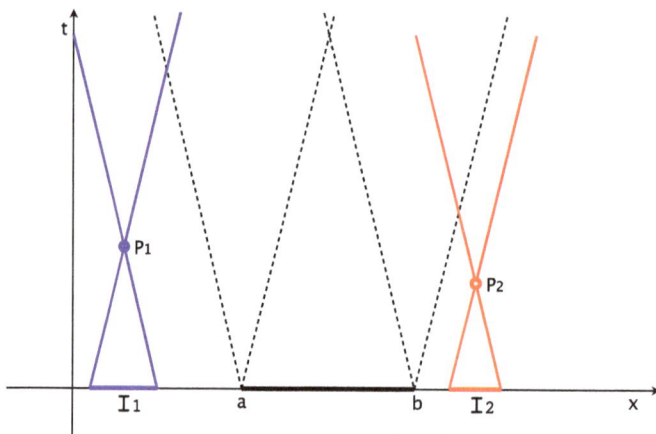

Figure 15.4: The characteristics through P_1 and P_2 and the domains of dependence I_1 of P_1 and I_2 of P_2; the dotted lines are the characteristics through $(a, 0)$ and $(b, 0)$.

Proposition 15.2. *Suppose that f is a function with support $[a, b]$, namely $f(x) = 0$ if and only if $\{x \le a\} \cup \{x \ge b\}$. If $I_{P_i} \cap (a, b) = \emptyset$ then $u(P_i) = 0$. Conversely, if $f(x)$ does not change sign in (a, b) and $u(P_i) = 0$ then $I_{P_i} \cap (a, b) = \emptyset$.*

Proof. If $I_{P_i} \cap (a, b) = \emptyset$ then $f(x_i - ct_i) = f(x_i + ct_i) = 0$, for $x_i - ct_i$ and $x_i + ct_i$ are out of the support of f. Using (15.23) it follows that $u(P_i) = 0$. Conversely, if $u(P_i) = 0$ then (15.23) yields $f(x_i - ct_i) + f(x_i + ct_i) = 0$. Therefore, since f does not change sign in (a, b) and is identically zero outside, we have $f(x_i - ct_i) = f(x_i + ct_i) = 0$. This means that $x_i - ct_i \notin (a, b)$ and $x_i + ct_i \notin (a, b)$ and thus $I_{P_i} \cap (a, b) = \emptyset$. $\square$

Example 15.6. Let $u(x, t)$ be such that $u_{tt} - u_{xx} = 0$, $u(x, 0) = f(x)$, $u_t(x, 0) = 0$, where f is the function given by

$$f(x) = \begin{cases} 0 & \text{if } |x| \ge 1, \\ e^{\left(-\frac{1}{1-|x|^2}\right)} & \text{if } |x| < 1, \end{cases}$$

whose support is the interval $[-1, 1]$. Find $u(0, \frac{1}{2})$, $u(1, 1)$ and $u(2, \frac{1}{2})$.

The characteristics through $A = (0, \frac{1}{2})$ are $x \pm t = \pm\frac{1}{2}$, which give rise to the domain of dependence $I_A = [-\frac{1}{2}, \frac{1}{2}]$ which is contained in the support $[-1, 1]$ of f. Then D'Alembert's formula yields

$$u\left(0, \frac{1}{2}\right) = \frac{1}{2}\left[f\left(\frac{1}{2}\right) + f\left(-\frac{1}{2}\right)\right].$$

Since f is even then $f(\frac{1}{2}) = f(-\frac{1}{2})$ and thus $u(0, \frac{1}{2}) = f(\frac{1}{2}) = e^{-\frac{4}{3}}$.

We refer to Fig. 15.5. The characteristics through $B = (1, 1)$ are $x \pm t = 1 \pm 1$, which give rise to the domain of dependence $I_B = [0, 2]$. Since $x = 0$ is contained in the support $[-1, 1]$ of f, whereas $x = 2$ is not, we have

$$u(1, 1) = \frac{1}{2}[f(0) + \underbrace{f(2)}_{=0}] = \frac{1}{2}f(0) = \frac{1}{2}.$$

The characteristics through $C = (2, \frac{1}{2})$ are $x \pm t = 2 \pm \frac{1}{2}$, which give rise to the domain of dependence $I_C = [-\frac{3}{2}, \frac{5}{2}]$. Since this is disjoint from the support $[-1, 1]$ of f, $u(2, \frac{1}{2}) = 0$.

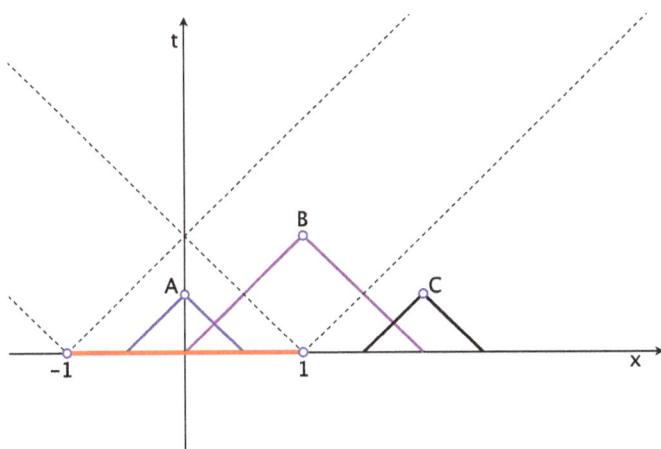

Figure 15.5: The characteristics through A, B, C and the support of f (in red); the dotted lines are the characteristics through $(\pm 1, 0)$.

Notice that in all the region between the lines $x + t = -1$ ($t \geq 0$), the left characteristic through $(-1, 0)$, and $x - t = 1$ ($t \geq 0$), the right characteristic through $(0, 1)$ one has $u(x, t) > 0$, whereas in the complementary region $u(x, t) = 0$. □

Next, let $f = 0$ and suppose that $g(x)$ does not change sign and has support $[a, b]$. Let $I = [\alpha, \beta]$ be the domain of dependence of P, Repeating the previous arguments we can see that D'Alembert's formula yields

$$u(P) = \int_{\alpha}^{\beta} g(s)\, ds.$$

Since $g = 0$ outside (a, b), we infer that

$$u(P) = \int_a^\beta g(s)\, ds = \int_{(a,\beta)\cap(a,b)} g(s)\, ds.$$

Since $g(x)$ does not change sign for $x \in (a,b)$ it follows that $u(P) = 0$ if and only if $I \cap [a,b] = \emptyset$. Precisely, if $(a,\beta) \cap (a,b) \neq \emptyset$ one has

$$u(P) = \int_{(a,\beta)\cap(a,b)} g(s)\, ds \neq 0.$$

In particular, if the support $[a,b]$ of g is completely contained in the domain of dependence of $[a,\beta]$ of P, then $(a,\beta) \cap (a,b) = (a,b)$ and hence

$$u(P) = \int_a^b g(s)\, ds = \text{const.}$$

Notice that if g changes sign then $u(P)$ could be zero, even if $(a,\beta) \cap (a,b) \neq \emptyset$.

Example 15.7. Find $u(1,1)$, where $u_{tt} - u_{xx} = 0$, $u(x,0) = 0$, $u_t(x,0) = g(x)$, and g is an even function with support $[-1,1]$, such that $\int_{-1}^1 g(s)\, ds = k$.

As in Example 15.6, the domain of dependence of $B = (1,1)$ is $[0,2]$. One has $(0,2) \cap (-1,1) = (0,1)$ and thus $u(1,1) = \int_0^1 g(s)\, ds$. Finally, since g is even,

$$u(1,1) = \frac{1}{2}\int_{-1}^1 g(s) = \frac{1}{2}k. \qquad \Box$$

15.3.4 A boundary value problem for the vibrating string: the method of separation of variables

Let us seek solutions of the problem

$$\begin{cases} u_{tt} - u_{xx} = 0, & \forall x \in (0,\pi),\ \forall t > 0, \\ u(x,0) = f(x), & \forall x \in [0,\pi], \\ u_t(x,0) = 0, & \forall x \in [0,\pi], \\ u(0,t) = u(\pi,t) = 0, & \forall t \geq 0. \end{cases} \qquad (15.24)$$

Here, to simplify notation, we have taken $c = 1$.

The boundary condition $u(0,t) = u(\pi,t) = 0$ prescribes the solution at the boundary of the interval $[0,\pi]$. From the physical point of view, a solution of (15.24) describes the displacement of an elastic string which is kept fixed at its endpoints $x = 0$ and $x = \pi$, for all time $t \geq 0$.

To solve (15.24) we will use the method of separation of variables which consists of seeking solutions of (15.24) in the form

$$u(x,t) = X(x)T(t).$$

Since the variable are separated, one finds

$$u_{tt} = X(x)T''(t), \quad u_{xx} = X''(x)T(t).$$

Substituting in the equation $u_{tt} = u_{xx}$ we find $X(x)T''(t) = X''(x)T(t)$. If $X, T \neq 0$ we infer

$$\frac{X''(x)}{X(x)} = \frac{T''(t)}{T(t)}.$$

Since the left hand side is a function of x only, while the right hand side is a function of y only, it follows that they are constant. Calling this constant $-\lambda$, we find

$$\frac{X''(x)}{X(x)} = \frac{T''(t)}{T(t)} = -\lambda, \quad \forall x, y.$$

Thus $u_{tt} = u_{xx}$ is transformed into the following two second order linear ordinary differential equations

$$X''(x) + \lambda X(x) = 0, \tag{15.25}$$
$$T''(t) + \lambda T(t) = 0. \tag{15.26}$$

We now use the boundary condition $u(0, t) = u(L, t)$ to yield

$$X(0) = X(\pi) = 0.$$

The problem

$$X''(x) + \lambda X(x), \quad X(0) = X(\pi) = 0, \tag{15.27}$$

is an eigenvalue problem. We are interested in nontrivial solutions of (15.27), because $X(x) \equiv 0$ yields $u(x, t) = 0$ for all x which does not satisfy the initial value conditions, unless $y = 0$.

Recall that (15.27) has nontrivial solutions if and only if $\lambda = n^2$, $n = 1, 2, \ldots$, given by the *eigenfunctions*

$$X_n(x) = r_n \sin nx, \quad n = 1, 2, \ldots,$$

where $r_n \in \mathbb{R}$; see, e. g., Chapter 13, Section 13.3.

If $\lambda = n^2$, (15.26) becomes

$$T''(t) + n^2 T(t) = 0,$$

whose general solution is

$$T_n(t) = A_n \sin nt + B_n \cos nt, \quad n = 1, 2, \ldots,$$

for constants $A_n, B_n \in \mathbb{R}$.

Thus we find that the function

$$u_n(x, t) = X_n(x)T_n(t) = r_n \sin nx \cdot [A_n \sin nt + B_n \cos nt], \quad n \in \mathbb{N},$$

is a solution of $u_{tt} - u_{xx} = 0$ such that $u(0, t) = u(\pi, t) = 0$ for all $t \geq 0$. The u_n are called the *normal modes* of the string equation.

As a consequence, also any finite sum of normal modes

$$\sum_1^N r_n \sin nx \cdot [A_n \sin nt + B_n \cos nt]$$

gives rise to a solution of the string equation. In general, this superposition of a finite number of normal modes is not sufficient to find a solution satisfying the initial condition $u(x, 0) = f(x)$. For this reason, it might be convenient to consider the infinite series

$$\sum_1^\infty r_n \sin nx \cdot [A_n \sin nt + B_n \cos nt].$$

If this series converges uniformly in $[0, L] \times [0, +\infty)$ and the series obtained by differentiating twice term by term w. r. t. x and y also converge uniformly therein, then it follows that

$$u(x, t) = \sum_1^\infty r_n \sin nx \cdot [A_n \sin nt + B_n \cos nt] \tag{15.28}$$

is a solution of $u_{tt} - u_{xx} = 0$.

Assuming this is the case, we can now find r_n, A_n, B_n by using the initial conditions $u(x, 0) = f(x)$, $u_y(x, 0) = 0$.

First of all, since

$$u_y(x, t) = \sum_1^\infty r_n \sin nx \cdot [nA_n \cos nt + nB_n \sin nt]$$

one has

$$u_t(x,0) = \sum_1^\infty r_n \sin nx \cdot nA_n = \sum_1^\infty nr_n A_n \sin nx.$$

Then $u_t(x,0) = 0$ implies that $r_n A_n = 0$ for all $n \in \mathbb{N}$. Hence, substituting into (15.28), we find

$$u(x,t) = \sum_1^\infty r_n B_n \sin nx \cdot \cos nt.$$

We now use the other initial condition, $u(x,0) = f(x)$.
   Let us suppose that

$$f(x) = \sum_1^\infty a_n \sin nx, \quad a_n := \frac{2}{\pi} \int_0^\pi f(\vartheta) \sin n\vartheta \, d\vartheta.$$

Since

$$u(x,0) = \sum_1^\infty r_n B_n \sin nx,$$

the condition $u(x,0) = f(x)$ implies $r_n B_n = a_n$.
   In conclusion, formally, the solution of (15.24) with $f(x) = \sum_1^\infty a_n \sin nx$, is given by

$$u(x,t) = \sum_1^\infty a_n \sin nx \cdot \cos nt, \quad \text{where } a_n = \frac{2}{\pi} \int_0^\pi f(\vartheta) \sin n\vartheta \, d\vartheta.$$

To make this result not only formal, we should differentiate the preceding series term by term twice w. r. t. x and t and check that these series are uniformly convergent. In general, this requires suitable assumptions on f.
   The situation becomes much simpler if the Fourier coefficients a_n of f are different from zero only for a finite number of terms. Actually in such a case $u(x,t) = \sum_1^\infty a_n \sin nx \cdot \cos nt$ is a finite sum.

Example 15.8. Let us solve (15.24) when f is a trigonometric polynomial as $f(x) = 3 \sin x + \sin 2x$. In this case $a_1 = 3, a_2 = 1$ while $a_n = 0$ for all $n \geq 3$. Then one has

$$u(x,t) = \sum_1^2 a_n \sin nx \cos ny = 3 \sin x \cos t + \sin 2x \cos 2t.$$

Remark 15.4. To solve

$$\begin{cases} u_{tt} - c^2 u_{xx} = 0, & \forall x \in (0,L), \; \forall t > 0, \\ u(x,0) = f(x), & \forall x \in [0,L], \\ u_t(x,0) = 0, & \forall x \in [0,L], \\ u(0,t) = u(L,t) = 0, & \forall t \geq 0, \end{cases}$$

we need to slightly modify the previous calculations as follows.

The components $X(x)$ and $T(t)$ satisfy

$$(i) \quad X'' + \lambda X = 0, \quad X(0) = X(L) = 0, \qquad (ii) \quad T'' + \lambda c^2 T = 0.$$

The eigenvalues of (i) are $\lambda_n = (\frac{n\pi}{L})^2$ with eigenfunctions $X_n = r_n \sin \sqrt{\lambda_n} x$. The solutions of (ii) corresponding to $\lambda = \lambda_n$ are $T_n = A_n \sin(\sqrt{\lambda_n} ct) + B_n \cos(\sqrt{\lambda_n} ct)$.

The rest of the argument is the same. It follows that the solution is given by

$$u(x,t) = \sum_1^\infty a_n \sin(\sqrt{\lambda_n} x) \cdot \cos(\sqrt{\lambda_n} ct), \quad \lambda_n = \left(\frac{n\pi}{L}\right)^2,$$

where $a_n = \frac{2}{L} \int_0^L f(\vartheta) \sin(\sqrt{\lambda_n}\vartheta)\, d\vartheta$.

15.4 Exercises

15.4.1 Exercises on the Laplace equation

1. Find the solution of the Dirichlet problem $\Delta u = 0$ in Ω, $u = h$ on $\partial\Omega$, where h is harmonic.
2. Let u be such that $\Delta u = 0$ in $B = \{(x,y) \in \mathbb{R}^2 : x^2 + y^2 < 1\}$ and $u = x^4 + 2y^2$ on ∂B. Show that $1 \le u(x,y) \le 2$ for all $(x,y) \in \overline{B}$.
3. Let u be such that $\Delta u = 0$ in $\Omega = \{(x,y) \in \mathbb{R}^2 : 2x^2 + y^2 < 1\}$ and $u = 5x^2 + 2y^2$ on $\partial\Omega$. Show that $\frac{5}{2} \le u(x,y) \le 2$ for all $(x,y) \in \overline{\Omega}$.
4. Find the radial harmonic functions in $\mathbb{R}^2 \setminus \{(0,0)\}$.
5. Find the radial functions in $\mathbb{R}^2 \setminus \{(0,0)\}$ such that $\Delta u = x^2 + y^2$.
6. Knowing that the Fourier series of

$$f(\theta) = \begin{cases} -\pi - \theta & \text{for } \theta \in [-\pi, -\frac{1}{2}\pi], \\ \theta & \text{for } \theta \in [-\frac{1}{2}\pi, \frac{1}{2}\pi], \\ \pi - \theta & \text{for } \theta \in [\frac{1}{2}\pi, \pi], \end{cases} \tag{$f1$}$$

is $f(\theta) = \frac{8}{\pi^2}(\sin\theta - \frac{\sin 3\theta}{3^2} + \frac{\sin 5\theta}{5^2} - \cdots)$, solve $\Delta u = 0$ for $r < 1$, $u(1,\theta) = f(\theta)$.
7. Knowing that the Fourier series of $f(\theta) = 1 - |\frac{\theta}{\pi}| (-\pi \le \theta \le \pi)$ is $f(\theta) = \frac{1}{2} + \frac{4}{\pi^2}(\cos\theta + \frac{\cos 3\theta}{3^2} + \frac{\cos 5\theta}{5^2} + \cdots)$, solve $\Delta u = 0$ for $r < 1$, $u(1,\theta) = f(\theta)$.
8. Let u be harmonic in $r < 1$ and such that $u(1,\theta) = f(\theta)$. If f is odd, show that $u(r,0) = u(r,\pi) = 0$.
9. Show that $\Delta u = 0$ for $r < 1$, $u(1,\theta) = f(\theta)$, $u_r(1,\theta) = 0$, has no solution unless $f(\theta) = k$, k constant, in which case the solution is $u \equiv k$.
10. Find $u(r)$ solving $\Delta u = 0$ for $1 < r < R$, $u = a$ for $r = 1$, $u = b$ for $r = R$.

11. Using separation of variables, find the solutions of $\Delta u = 0$ in the rectangle $\Omega = \{(x,y) \in \mathbb{R}^2 : 0 < x < \pi,\ 0 < y < 1\}$ such that $u(x,1) = 0$ for $x \in [0,\pi]$, $u(0,y) = u(1,y) = 0$, for $y \in [0,1]$ and $u(x,0) = \sin x$ for $0 \le x \le \pi$.

12. Using separation of variables, find the solutions of $\Delta u = 0$ in the square $\Omega = \{(x,y) \in \mathbb{R}^2 : 0 < x < \pi,\ 0 < y < \pi\}$, such that $u_x(0,y) = u_x(\pi,y) = 0$ for all y and $u(x,0) = u(x,\pi) = 1 + \cos 2x$.

13. Using separation of variables, solve $\Delta u = 0$ for $x \in \mathbb{R}$, $y \ge 0$, such that $u(x,0) = 0$, $u_y(x,0) = f_n(x) := \sin nx$, $n \in \mathbb{N}$.

14. Let u_a be harmonic in $r < 1$ and such that $u(1,\theta) = a(5 - 8\sin\theta)$. Using the Poisson integral formula, find a such that $u_a(\frac{1}{2}, \frac{\pi}{2}) = 1$.

15. Let u be harmonic in the ball B_R, such that $u(x,y) = (x+y)^2$ on ∂B_R. Find $u(0,0)$.

16. Let u be a harmonic function on $\mathbb{R}^2$. Show that for any $P_0 = (x_0,y_0) \in \mathbb{R}^2$ and $\epsilon > 0$ one has that $u(P_0) = \fint_{\partial B_\epsilon} f$ where B_ϵ is the ball centered in P_0 with radius ϵ and f is the restriction of u on the circle ∂B_ϵ.

17. Show that if there exists a nontrivial solution of $\Delta u = \lambda u$ in Ω and $u = 0$ on $\partial\Omega$, then $\lambda < 0$.

15.4.2 Exercises on the heat equation

1. Solve the heat equation (H) with $f(x) = \sin x$.
2. Solve the heat equation (H) with

$$f(x) = \begin{cases} x & \text{for } x \in [0, \frac{1}{2}\pi], \\ \pi - x & \text{for } x \in [\frac{1}{2}\pi, \pi]. \end{cases}$$

3. Solve the heat equation (H) with $f(x) = \max\{0, \sin 2x\}$.
4. Solve $u_t - 2u_{xx} = 0$, in $(0,1) \times (0,+\infty)$, $u(x,0) = 4\sin(3\pi x)$ for $x \in [0,1]$ and $u(0,t) = u(1,t) = 0$ for $t \ge 0$.
5. Let u be a solution of the heat equation (H). If $f_n = 0$ for all n odd, show that $\frac{\partial u(0,t)}{\partial x} = \frac{\partial u(\pi,t)}{\partial x}$.
6. Letting $c(t) > 0$, solve the following by separation of variables

$$\begin{cases} u_t - c(t)u_{xx} = 0, & \forall x \in [0,\pi],\ \forall t \ge 0, \\ u(x,0) = b\sin x, & \forall x \in [0,\pi], \\ u(0,t) = u(\pi,t) = 0, & \forall t \ge 0. \end{cases}$$

7. Solve by separation of variables

$$\begin{cases} u_t - u_{xx} + 3u = 0, & \forall x \in [0,\pi],\ \forall t \ge 0, \\ u(x,0) = b\sin 2x, & \forall x \in [0,\pi], \\ u(0,t) = u(\pi,t) = 0, & \forall t \ge 0. \end{cases}$$

8. Solve the following Neumann problem for the heat equation

$$\begin{cases} u_t - u_{xx} = 0, & \forall x \in [0, \pi], \ \forall t \geq 0, \\ u(x, 0) = a_1 \cos x + a_2 \cos 2x, & \forall x \in [0, \pi], \\ u_x(0, t) = u_x(\pi, t), & \forall t \geq 0. \end{cases}$$

9. Knowing that

$$(*) \qquad \frac{\sin 6\pi x}{\sin \pi x} = 2 \sum_1^3 \cos 3(2k - 1)x,$$

solve

$$\begin{cases} u_t - u_{xx} = 0, & \forall x \in [0, \pi], \ \forall t \geq 0, \\ u(x, 0) = f(x) := \frac{\sin 6\pi x}{\sin \pi x}, & \forall x \in [0, \pi], \\ u_x(0, t) = u_x(\pi, t), & \forall t \geq 0. \end{cases}$$

10. Solve

$$\begin{cases} u_t - u_{xx} = 0, & \forall x \in [0, \pi], \ \forall t \geq 0, \\ u(x, 0) = |\cos x|, & \forall x \in [0, \pi], \\ u_x(0, t) = u_x(\pi, t), & \forall t \geq 0. \end{cases}$$

11. Solve by separation of variables:

$$\begin{cases} u_t - u_{xx} = 0, & \forall x \in [0, \pi], \ \forall t \geq 0, \\ u(x, 0) = b \sin \frac{3x}{2}, & \forall x \in [0, \pi], \\ u(0, t) = u_x(\pi, t) = 0, & \forall t \geq 0. \end{cases}$$

12. Let $u(x, t)$ be a solution of $u_t - u_{xx} = h(x, t)$ in $A = [0, \pi] \times [0, T]$ for some $T > 0$. Show that if $h(x, t) < 0$ in the interior of A, then u achieves its maximum on the boundary of A.

15.4.3 Exercises on D'Alambert's equation

1. Solve $u_{tt} - u_{xx} = 0$, $u(x, 0) = \sin x$, $u_t(x, 0) = 1$.
2. Solve $u_{tt} - u_{xx} = 0$, $u(x, 0) = x$, $u_t(x, 0) = 2x$.
3. Solve $u_{tt} - 4u_{xx} = 0$, $u(x, 0) = 0$, $u_t(x, 0) = x(\pi - x)$.
4. Show that the solution of $u_{tt} - c^2 u_{xx} = 0$, $u(x, 0) = f(x)$, $u_t(x, 0) = 0$ is even in t.
5. Show that the solution of $u_{tt} - u_{xx} = 0$, $u(x, 0) = 0$, $u_t(x, 0) = g(x)$ is odd in t.
6. Let u be the solution of $u_{tt} - u_{xx} = 0$ in $[0, L] \times [0, +\infty)$, $u(x, 0) = f(x)$, $u_t(x, 0) = 0$. Knowing that $f(\frac{1}{8}L) = f(\frac{3}{8}L) = 2$, find $u(\frac{1}{4}L, \frac{1}{8}L)$.

7. Solve by separation of variables:

$$\begin{cases} u_{tt} - u_{xx} = 0, & \forall x \in (0, \pi), \ \forall t > 0, \\ u(x, 0) = 0, & \forall x \in [0, \pi], \\ u_t(x, 0) = b \sin 3x, & \forall x \in [0, \pi], \\ u(0, t) = u(\pi, t) = 0, & \forall t \geq 0. \end{cases}$$

8. Solve by separation of variables:

$$\begin{cases} u_{tt} - u_{xx} = 0, & \forall x \in (0, \pi), \ \forall t > 0, \\ u(x, 0) = \sin x, & \forall x \in [0, \pi], \\ u_t(x, 0) = \sin x, & \forall x \in [0, \pi], \\ u(0, t) = u(\pi, t) = 0, & \forall t \geq 0. \end{cases}$$

9. Solve by separation of variables:

$$\begin{cases} u_{tt} - u_{xx} = 0, & \forall x \in (0, \pi), \ \forall t > 0, \\ u(x, 0) = \sin 2x, & \forall x \in [0, \pi], \\ u_t(x, 0) = \sin 3x, & \forall x \in [0, \pi], \\ u(0, t) = u(\pi, t) = 0, & \forall t \geq 0, \end{cases}$$

10. Solve by separation of variables:

$$\begin{cases} u_{tt} - u_{xx} = 0, & \forall x \in (0, \pi), \ \forall t > 0, \\ u(x, 0) = 3 + 4 \cos 2x, & \forall x \in [0, \pi], \\ u_t(x, 0) = 0, & \forall x \in [0, \pi], \\ u_x(0, t) = u_x(\pi, t), & \forall t \geq 0. \end{cases}$$

16 The Euler–Lagrange equations in the Calculus of Variations: an introduction

This chapter is intended to serve as an elementary introduction to the Calculus of Variations, one of the most classical topics in mathematical analysis.

For a more complete discussion we refer, e. g., to R. Courant, *Calculus of Variations: With Supplementary Notes and Exercises, Courant Inst. of Math. Sci., N. Y. U., 1962*, or B. Dacorogna, *Introduction to the Calculus of Variations, 2nd Edition, World Scientific, 2008*.

Notation: in this chapter we let x denote the independent variable and $y = y(x)$ the dependent variable.

16.1 Functionals

Given two points in the plane $A = (a, a)$, $B = (b, \beta)$, the length of a smooth curve $y = y(x)$ such that $y(a) = a$, $y(b) = \beta$ is given by

$$\ell[y] = \int_a^b \sqrt{1 + y'^2(x)}\, dx.$$

The map $y \mapsto \ell[y]$ is an example of a *functional*.

In general, given a class of functions Y, a functional is a map defined on Y with values in the set of real numbers $\mathbb{R}$.

We will be mainly concerned with functionals of the form

$$I[y] = \int_a^b L(x, y(x), y'(x))\, dx, \quad y \in Y, \tag{I}$$

where the class Y is given by

$$Y = \{y \in C^2([a, b]) : y(a) = a, y(b) = \beta\} \tag{Y}$$

and the *Lagrangian* $L = L(x, y, p)$ is a function of three variables $(x, y, p) \in [a, b] \times \mathbb{R} \times \mathbb{R}$, such that $L(x, y(x), y'(x))$ is integrable on $[a, b]$, $\forall y \in Y$. In the preceding arclength example, L is given by $L(p) = \sqrt{1 + p^2}$.

To keep the presentation as simple as possible, here and in the sequel we will not deal with the least possible regularity. For example, though $I[y]$ would make sense for $y \in C^1([a, b])$, we take C^2 functions to avoid technicalities in what follows.

The analysis of functionals as (I) is carried out in the *Calculus of Variations*. This is a branch of mathematical analysis dealing with geometrical or physical problems whose

https://doi.org/10.1515/9783111185675-016

solutions are functions that minimize, or maximize, quantities like (I). We will mainly deal with absolute (or global) minima of $I[y]$ on Y, namely $\bar{y} \in Y$ such that

$$I[\bar{y}] \leq I[y], \quad \forall y \in Y.$$

If $\leq$ is replaced by $<$ we say that $\bar{y}$ is a strong minimum. In a similar way we can define local minima as well as local and global maxima of $I[y]$.

Roughly, in the calculus of variations we look for *functions* $y \in Y$ at which (I) achieves a minimum, or maximum, whereas in calculus we search minima or maxima of functions $f : \mathbb{R}^n \mapsto \mathbb{R}$.

16.2 The Euler–Lagrange equation

Theorem 16.1. *Suppose that $L \in C^2([a,b] \times \mathbb{R} \times \mathbb{R})$ and let $\bar{y} \in Y$ be such that*

$$I[\bar{y}] = \min\{I[y] : y \in Y\}. \tag{16.1}$$

Then the function $\bar{y}(x)$ solves the differential equation

$$\frac{d}{dx}\frac{\partial}{\partial p}L(x,y,y') = \frac{\partial}{\partial y}L(x,y,y'), \tag{EL}$$

and the boundary conditions $\bar{y}(a) = \alpha$, $\bar{y}(b) = \beta$.

The equation (EL) is named the *Euler–Lagrange equation* of the functional $I[y]$ on the class Y. Sometime it is written, for short, as $\delta I = 0$, where $\delta I = \frac{d}{dx}L_p - L_y$.

Remark 16.1. Since $\bar{y}$ is the minimum of $I[y]$ on Y, the fact that $I[y]$ might be $+\infty$ on some $y \in Y$ does not play any role (of course, provided $I[y]$ is finite for some $y \in Y$).

Postponing the proof of Theorem 16.1, we now discuss a couple of simple examples.

Example 16.1.
(a) If $L = \sqrt{1+p^2}$ we have the arclength functional $I[y] = \ell[y] = \int_a^b \sqrt{1+y'^2(x)}\, dx$. In this case $L_p = \frac{p}{\sqrt{1+p^2}}$ and hence (EL) becomes

$$\frac{d}{dx}\frac{y'}{\sqrt{1+y'^2}} = 0.$$

With straight calculation we find

$$\frac{y'' \cdot \sqrt{1+y'^2} - y' \cdot \frac{y'y''}{\sqrt{1+y'^2}}}{1+y'^2} = 0 \Longrightarrow \frac{y'' \cdot (1+y'^2) - y'^2 y''}{(1+y'^2)\sqrt{1+y'^2}} = 0 \Longrightarrow y'' = 0.$$

Thus $y(x) = c_1 x + c_2$. Taking into account the boundary conditions $y(a) = \alpha, y(b) = \beta$ we find $y(x) = \alpha + \frac{\beta - \alpha}{b - a} \cdot (x - a)$ which is simply the segment connecting the points (a, α) and (b, β). Notice that this segment is just the (absolute) minimum of the arclength functional.

(b) If $L = \frac{m}{2} p^2 - V(y)$, then $I[y] = \frac{m}{2} \int_a^b y'^2(x)\, dx - \int_a^b V(y(x))\, dx$ and (EL) becomes the nonlinear second order equation

$$\frac{d}{dx} m y'(x) + V'(y) = 0,$$

namely the Newton equation of motion, $my'' + V'(y) = 0$.
Thus a possible minimum or maximum of $I[y]$ on the class Y will solve the boundary value problem

$$my'' + V'(y) = 0, \quad y(a) = \alpha, y(b) = \beta.$$

This remark suggests a way to prove the existence of solutions of a bvp as the previous one: it suffices to show that the corresponding functional $I[y]$ has a minimum (or a maximum) in Y. □

To prove Theorem 16.1 the following lemma is in order.

Lemma 16.1 (Du Bois–Reymond lemma). *Let $\psi \in C([a, b])$ be such that*

$$\int_a^b \psi(x) \cdot z(x)\, dx = 0, \quad \forall z \in C([a, b]) : z(a) = z(b) = 0. \tag{16.2}$$

Then $\psi(x) \equiv 0$.

Proof. By contradiction, let $x_0 \in [a, b]$ be such that $\psi(x_0) \neq 0$: to be specific, let $a < x_0 < b$ and $\psi(x_0) > 0$ (the other cases require small changes). By continuity, there is $\delta > 0$ such that the interval $T = (x_0 - \delta, x_0 + \delta)$ is contained in (a, b) and $\psi(x) > 0$ on T.

Consider a function $\bar{z} \in C([a, b])$ such that $\bar{z}(x) > 0$ for $x \in T$ and zero otherwise; see Fig. 16.1.

Clearly $\bar{z}(a) = \bar{z}(b) = 0$ and hence we can use (16.2) with $z = \bar{z}$, yielding

$$\int_a^b \psi(x) \bar{z}(x)\, dx = 0.$$

Recalling that $\bar{z} \equiv 0$ outside the interval $(x_0 - \delta, x_0 + \delta)$, we deduce

$$\int_a^b \psi(x) \bar{z}(x)\, dx = \int_{x_0 - \delta}^{x_0 + \delta} \psi(x) \bar{z}(x)\, dx = 0.$$

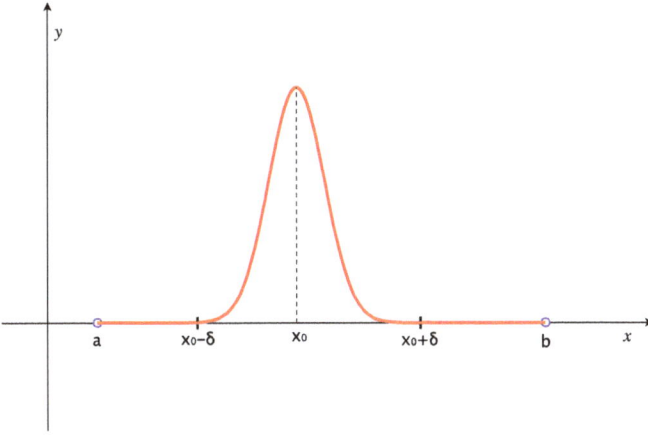

Figure 16.1: Plot of a $\tilde{z}(x)$.

Since both ψ and $\tilde{z}$ are positive on $(x_0 - \delta, x_0 + \delta)$, we find a contradiction, proving the lemma. $\qquad\square$

Proof of Theorem 16.1. We set $Y_0 = \{z \in C^2([a,b]) : z(a) = z(b) = 0\}$. Noticing that $\bar{y} + \epsilon z \in Y, \forall z \in Y_0$, we consider

$$\phi(\epsilon) \stackrel{\text{def}}{=} I[\bar{y} + \epsilon z] = \int_a^b L(x, \bar{y}(x) + \epsilon z(x), \bar{y}'(x) + \epsilon z'(x))\, dx.$$

The function $\phi : \mathbb{R} \mapsto \mathbb{R}$ is continuously differentiable and one has $\phi(0) = I[\bar{y}]$. Since $\bar{y}$ is a minimum of $I[y]$ on the class Y then

$$\phi(\epsilon) = I[\bar{y} + \epsilon z] \geq I[\bar{y}] = \phi(0), \quad \forall \epsilon \in \mathbb{R}.$$

Thus $\epsilon = 0$ is a minimum of ϕ and hence $\phi'(0) = 0$. Let us evaluate $\phi'(\epsilon)$:

$$\phi'(\epsilon) = \frac{d}{d\epsilon} \int_a^b L(x, \bar{y} + \epsilon z, \bar{y}' + \epsilon z')\, dx = \int_a^b \frac{d}{d\epsilon} L(x, \bar{y} + \epsilon z, \bar{y}' + \epsilon z')\, dx$$

$$= \int_a^b [L_y(x, \bar{y} + \epsilon z, \bar{y}' + \epsilon z')z + L_p(x, \bar{y} + \epsilon z, \bar{y}' + \epsilon z')z']\, dx.$$

Taking into account that $z(a) = z(b) = 0$, an integration by parts yields

$$\int_a^b L_p(x, \bar{y} + \epsilon z, \bar{y}' + \epsilon z')z'\, dx = -\int_a^b \frac{d}{dx} L_p(x, \bar{y} + \epsilon z, \bar{y}' + \epsilon z')z\, dx.$$

Then we find

$$
\phi'(\epsilon) = \int_a^b L_y(x, \bar{y} + \epsilon z, \bar{y}' + \epsilon z') z \, dx - \int_a^b \frac{d}{dx} L_p(x, \bar{y} + \epsilon z, \bar{y}' + \epsilon z') z \, dx
$$

$$
= \int_a^b \left[L_y(x, \bar{y} + \epsilon z, \bar{y}' + \epsilon z') - \frac{d}{dx} L_p(x, \bar{y} + \epsilon z, \bar{y}' + \epsilon z') \right] \cdot z \, dx
$$

whereby

$$
\phi'(0) = \int_a^b \left[L_y(x, \bar{y}, \bar{y}') - \frac{d}{dx} L_p(x, \bar{y}, \bar{y}') \right] \cdot z \, dx = 0, \quad \forall z \in Y_0. \tag{16.3}
$$

Using Lemma 16.1 with $\psi(x) = L_y(x, \bar{y}(x), \bar{y}'(x)) - \frac{d}{dx} L_p(x, \bar{y}(x), \bar{y}'(x))$, we infer that $\psi(x) \equiv 0$, namely $L_y(x, \bar{y}(x), \bar{y}'(x)) - \frac{d}{dx} L_p(x, \bar{y}(x), \bar{y}'(x)) = 0, \forall x \in [a.b]$. □

Corollary 16.1. *If L is independent of x, namely $L = L(y, p)$, then (EL) becomes*

$$
y' L_p(y, y') - L(y, y') = k, \quad k \in \mathbb{R}. \tag{EL'}
$$

Proof. Let us evaluate

$$
\frac{d}{dx} [y'(x) L_p(y(x), y'(x)) - L(y(x), y'(x))]
$$

$$
= y'' L_p(y, y') + y' \frac{d}{dx} L_p(y(x), y'(x)) - L_y(y, y') y' - L_p(y, y') y''
$$

$$
= y' \frac{d}{dx} L_p(y(x), y'(x)) - L_y(y, y') y' = y' \cdot \left[\frac{d}{dx} L_p(y(x), y'(x)) - L_y(y, y') \right].
$$

From (EL) we infer that $\frac{d}{dx} [L(y, y') - y' L_p(y, y')] = 0$ and hence (EL') follows. □

Example 16.2. If $L(y, p) = \frac{m}{2} p^2 - V(y)$, we know that (EL) is the Newton equation $m y'' + V'(y) = 0$. On the other hand $L_p = mp$ and hence (EL') becomes

$$
m y'^2 - \frac{m}{2} y'^2 + V(y) = k \quad \Longrightarrow \quad \frac{m}{2} y'^2 + V(y) = k.
$$

Since $\frac{m}{2} y'^2$ is the kinetic energy and $V(y)$ the potential energy, the relationship $\frac{m}{2} y'^2 + V(y) = k$ is simply the principle of conservation of the total energy. □

Remark 16.2. Theorem 16.1 deals with minima of $I[y]$, but the proof makes it clear that (EL) is satisfied by any $\bar{y}$ such that $\frac{d}{d\epsilon} I[\bar{y} + \epsilon z] = 0$ where $z \in Y_0$. These functions are called *stationary points* of the functional I and include local maxima and minima as well as saddles.

Remark 16.3. Given $I[y] = \int_{-1}^{1}(1 - y'^2)^2\, dx$ and $Y = \{y \in C^2([-1,1]) : y(-1) = y(1) = 0\}$, let us show that $\inf_Y I[y] = 0$ but $I[y]$ does not attain the minimum on Y.

To prove this claim, we consider the sequence of functions $y_n \in Y$ such that $y_n(x) \geq 0$ and

$$y_n(x) = \begin{cases} x - 1 & x \in [-1, -\frac{1}{n}], \\ -x + 1 & x \in [\frac{1}{n}, 1]. \end{cases}$$

Since $y'^2 = 1$ on $[-1, -\frac{1}{n}] \cup [\frac{1}{n}, 1]$ we find

$$\int_{-1}^{-\frac{1}{n}}\left(1 - y_n'^2\right) dx = \int_{\frac{1}{n}}^{1}\left(1 - y_n'^2\right) dx = 0.$$

Thus $I[y_n] \to 0$ and $\inf_Y I[y] = 0$. By contradiction, let $I[y]$ have a minimum $\bar{y} \in Y$. Then $\bar{y}$ satisfies the Euler–Lagrange equation, which yields $\bar{y}' = 1$ whereby $\bar{y} = x + c$. Clearly, none of these functions can satisfy both the boundary conditions $y(\pm 1) = 0$.

On the other hand, if we consider the class Y_1 of continuous, piecewise differentiable functions on $[-1, 1]$ with $y(-1) = y(1) = 0$, then all the functions $y \in Y_1$ such that $|y'| = 1$ on the intervals where y is differentiable satisfy $I[y] = 0$ and hence are minima of $I[y]$ on Y_1. Two such functions are plotted in Fig. 16.2.

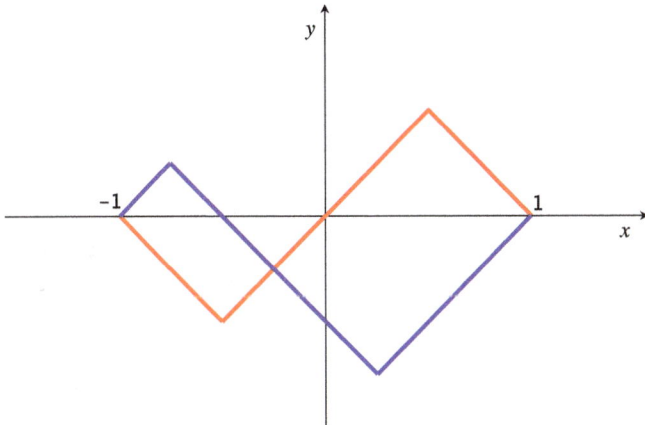

Figure 16.2: Two functions in the class Y_1.

This example highlights that the solution of minimum depends on the class Y. Even more: the functional $I[y]$ could have no minimum on a class Y, whereas it could attain the minimum on a larger class Y_1.

An important problem is to show that $\min_Y I[y]$ is achieved.

We have to be aware that this existence problem is not merely a mathematical curiosity but is relevant, since there are functionals which are bounded from below on Y but $\min_Y I[y]$ is not attained (an example has been discussed in Remark 16.3). The question is far from trivial and cannot be addressed here. It is a branch of the Calculus of Variations named *Direct Methods* and, among other things, has stimulated a lot of research that led to the birth of topics such as functional analysis.

In general, after having transformed the problem under consideration into the search of the minimum of a suitable functional $I[y]$ on a class Y, the arguments are completed by means of two further steps: assuming that $\min_Y I[y]$ exists,

1. one solves the related Euler–Lagrange equation finding a solution $\bar{y}(t)$;
2. one checks that $\bar{y}(t)$ is really the minimum we are looking for.

In what follows we will limit ourselves to discussing the first step only.

In the next two sections we discuss some classical problems that marked the birth of the calculus of variations. Some further problems are proposed as exercises.

16.3 Least time principles

In this section we deal with two celebrated *least time principles*: the fastest descent problem (brachistochrone) and Fermat's principle in optics.

16.3.1 The brachistochrone

The brachistochrone is one of the first problems studied in the Calculus of Variations. After some observations by Galilei, the problem was proposed by the Swiss mathematician Johann Bernoulli in 1696 as a challenge that was solved some time later by Bernoulli himself (see the historical remark at the end on the next section).

Given two points A, B on a vertical plane, we look for the curve of fastest descent on which a point subjected to the gravity only, namely with no friction, falls down from A to B in the *shortest time*. Roughly, we are searching the shape of a runner on which a body falls from A to B is the least time. A simplistic (foolish?) intuition would say that the segment AB, which is the line of shortest distance, is also the curve of the least time. As we will see, this is wrong: the least time curve is an arc of cycloid, called brachistochrone (from Greek: *brákhistos khrónos* = shortest time).

Letting $A = (0,0)$ and $B = (b, \beta)$, $\beta > 0$, we consider a $C^2([0,b])$ function $y = y(x)$ such that $y(0) = 0$, $y(b) = \beta$ (notice that $y \in C^1([a,b])$ would suffice; see the remark in section 16.1).

We start evaluating the time taken by the body to descend from A to B along y. For this it is convenient to introduce curvilinear abscissa $s \in [0, \ell]$, where s is the length of the arc AP and ℓ is the length of the curve $y = y(x), x \in [0, b]$, see Chapter 1, Section 1.3. On a point P on the curve with curvilinear abscissa s the only force acting is the projection of the gravitational force mg on the tangent to the curve at P, namely $mg \sin \theta$, where θ is the angle between this tangent and the vertical axis, as reported in Fig. 16.3 (notice that the y axis is directed downwards).

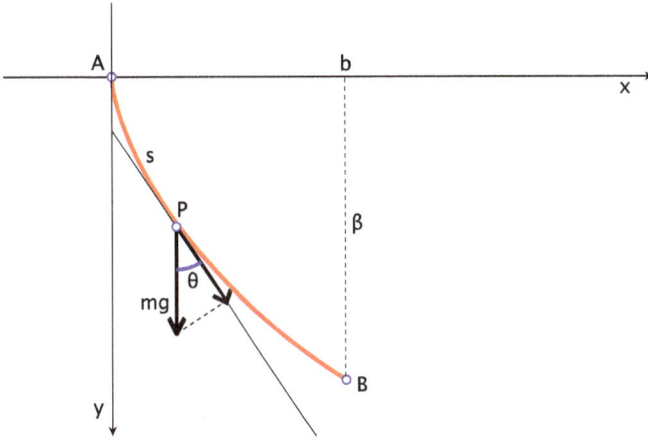

Figure 16.3: The arc AB of equation $y = y(x), x \in [0, b]$.

Then Newton's law yields

$$m\frac{d^2s}{dt^2} = mg \sin \theta.$$

Since from calculus it is known that $\sin \theta = \frac{dy}{ds}$, it follows that

$$\frac{d^2s}{dt^2} = g\frac{dy}{ds} = g\frac{dy}{dt}\frac{dt}{ds}. \tag{16.4}$$

Now, we notice that

$$\frac{d}{dt}\left(\frac{ds}{dt}\right)^2 = 2\frac{ds}{dt} \cdot \frac{d^2s}{dt^2},$$

which, together with (16.4) implies

$$\frac{d}{dt}\left(\frac{ds}{dt}\right)^2 = 2g\frac{dy}{dt}.$$

Integrating this equation with respect to t we find

$$\left(\frac{ds}{dt}\right)^2 = 2gy + c, \quad c \in \mathbb{R}.$$

Assuming that the body is at rest when in A, namely $\frac{ds}{dt} = 0$ at $s = 0$, we find $c = 0$, whereby

$$\dot{s} \stackrel{\text{def}}{=} \frac{ds}{dt} = \sqrt{2gy}.$$

Since the time to reach B is given by $T = \int_0^T dt$, the change of variable $t \mapsto s$ yields

$$T = \int_0^\ell \frac{ds}{\dot{s}} = \int_0^\ell \frac{ds}{\sqrt{2gy}}.$$

Turning back to the cartesian coordinates, we find $ds = \sqrt{1 + y'^2}\, dx$ and thus

$$T = \int_0^b \frac{\sqrt{1 + y'^2}}{\sqrt{2gy}}\, dx. \tag{16.5}$$

In conclusion the brachistochrone is the curve that minimizes the functional

$$I[y] = \int_0^b L(y, y')\, dx, \quad L(y, p) = \frac{\sqrt{1 + p^2}}{\sqrt{2gy}}$$

on the class Y of functions $y \in C^2([0, b])$ such that $y(0) = 0, y(b) = \beta$.

The corresponding Euler–Lagrange equation is $L - y'L_p = k$. Since $L_p = \frac{p}{\sqrt{1+p^2}} \cdot \frac{1}{\sqrt{2gy}}$, we find the equation

$$\sqrt{\frac{1 + y'^2}{2gy}} - \frac{y'^2}{\sqrt{2gy(1 + y'^2)}} = k \implies \frac{1 + y'^2 - y'^2}{\sqrt{2gy(1 + y'^2)}} = k$$

whereby

$$\frac{1}{2gy(1 + y'^2)} = k^2 \implies y(1 + y'^2) = \frac{1}{2gk^2}.$$

Setting $K = \frac{1}{2gk^2}$ we deduce

$$y(1 + y'^2) = K. \tag{16.6}$$

It is convenient to find solutions in a parametric form. With some calculations, which we omit, one finds that (16.6) yields the following equations, as a function of the parameter τ:

$$\begin{cases} x(\tau) = \rho(\tau - \sin\tau), \\ y(\tau) = \rho(1 - \cos\tau), \end{cases} \tag{16.7}$$

which is a cycloid with initial point $(0,0)$.

We take a short detour to recall that a cycloid is a curve traced by a point on a circle of radius ρ that moves in a clockwise direction, without slithering, along $y = 0$; see Fig. 16.4.

Zooming in Fig. 16.4, we plot in Fig. 16.5 a point $P = (x,y)$ on the cycloid and consider the generating circle passing through P. If C is its center, let τ denote the angle $\widehat{PCH}$, as shown in Fig. 16.5. With this notation C has coordinates $(\rho\tau, \rho)$ and a straight trigono-

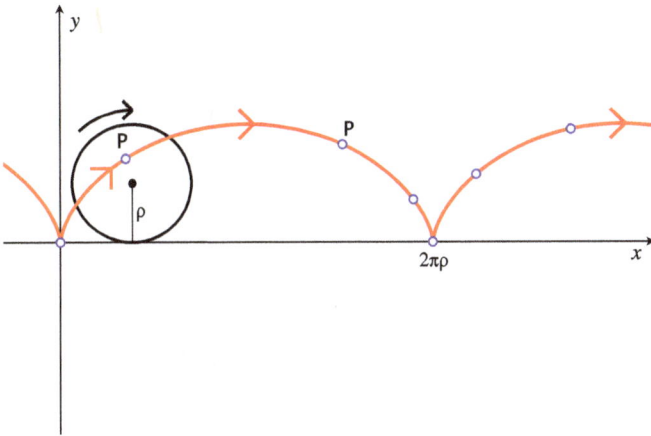

Figure 16.4: The cycloid.

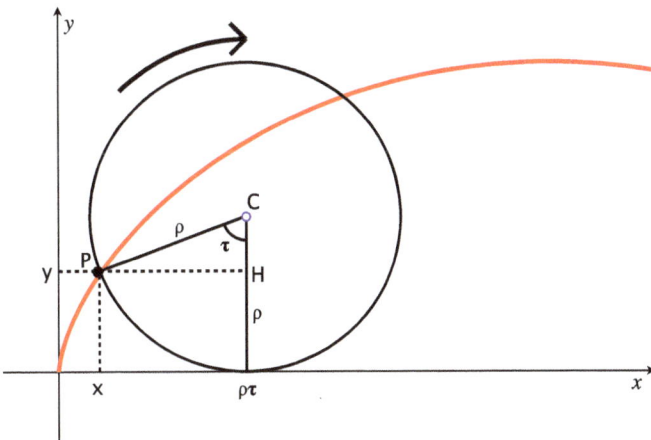

Figure 16.5: Parametric coordinates of the cycloid.

metric calculation yields $x = x(\tau) = \rho\tau - \rho\sin\tau$ and $y = y(\tau) = \rho - \rho\cos\tau$, which are the parametric equations of the cycloid.

Let us come back to the brachistochrone problem. We have shown that the brachistochrone with initial point $A = (0,0)$ is an arc of the cycloid of equation (16.7). Now we have to select the cycloid passing through B. For this we solve $x(\tau^*) = b, y(\tau^*) = \beta$, which yields

$$\begin{cases} \rho(\tau - \sin\tau) = b, \\ \rho(1 - \cos\tau) = \beta. \end{cases}$$

Notice that the function $\tau \mapsto \tau - \sin\tau$ is increasing. Drawing the graphs of $\tau \mapsto \beta(\tau - \sin\tau)$ and $\tau \mapsto b(1 - \cos\tau)$ it is easy to check that there is a unique $\tau^* > 0$ such that

$$\beta(\tau^* - \sin\tau^*) = b(1 - \cos\tau^*).$$

Once we have found $\tau^* > 0$ we infer that $\rho^* = b/(\tau^* - \sin\tau^*)$ (notice that $\tau^* - \sin\tau^* > 0$ since $\tau \mapsto \tau - \sin\tau$ is increasing). Then we infer that the arc of cycloid

$$\begin{cases} x(\tau) = \rho^*(\tau - \sin\tau), \\ y(\tau) = \rho^*(1 - \cos\tau), \end{cases} \qquad \tau \in [0, \tau^*],$$

has endpoints A, B and hence gives rise to the (unique) brachistochrone from A to B.

Remark 16.4. It is possible to show that the brachistochrone is truly the absolute minimum of $I[y]$ on Y.

Example 16.3. Let us compare the time T taken by the body to fall from $A = O = (0,0)$ to $B(\pi, 2)$ along the brachistochrone with the time T_0 taken traveling on the segment OB. We refer to Fig.16.6. By inspection, one shows that solving $2(\tau^* - \sin\tau^*) = \pi(1 - \cos\tau^*)$

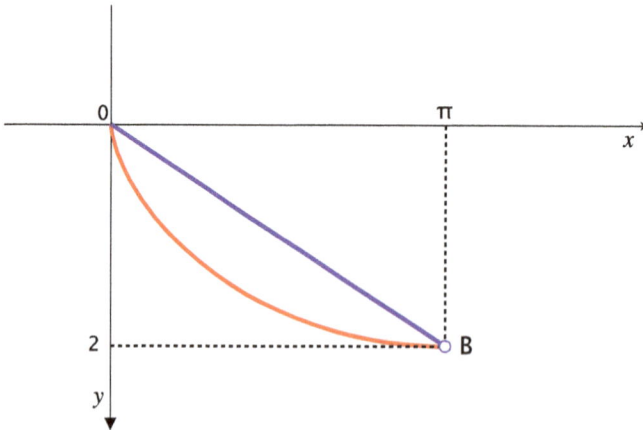

Figure 16.6: The brachistochrone vs. the segment OB.

yields $\tau^* = \pi$ and hence $\rho^* = \pi/(\tau^* - \sin \tau^*) = 1$. Thus the brachistochrone from O to B is given by

$$\begin{cases} x(\tau) = \tau - \sin \tau, \\ y(\tau) = 1 - \cos \tau, \end{cases} \quad \tau \in [0, \pi].$$

Using (16.5) one finds $T = \sqrt{\frac{\rho^*}{g}} \cdot \tau^* = \frac{\pi}{\sqrt{g}}$.

The segment OB is as an inclined plane with height $h = 2$, length $\sqrt{4 + \pi^2}$ and inclination angle α such that $\sin \alpha = \frac{2}{\sqrt{4+\pi^2}}$. From elementary mechanics we know that the velocity of the body sliding down on this inclined plane is given by $v(t) = g \sin \alpha t = \frac{2g}{\sqrt{4+\pi^2}} \cdot t$. We also know by the conservation of the energy that the final velocity is $v = \sqrt{2gh} = 2\sqrt{g}$. Then $\frac{2g}{\sqrt{4+\pi^2}} \cdot T_0 = 2\sqrt{g}$, which yields $T_0 = \frac{\sqrt{4+\pi^2}}{\sqrt{g}}$. Thus $T < T_0$, according to the fact noted before: T is the least descent time. $\qquad \square$

16.3.2 Fermat's principle in geometrical optics

Fermat's principle states that going from one point to another in a medium, the light propagates along the path that takes the shortest time. As for the brachistochrone, one has to be aware that the path minimizes the time, not the length.

To start with let us consider the following elementary cases:
(a) both A and B lie in a medium with constant refraction index n;
(b) the points A, B stay in two different media separated by a plane, each medium having constant refraction index.

Recall that the refractive index n is defined as $n = \frac{c}{v}$, where c and v denote the velocity of light in the vacuum and in the medium, respectively.

Case (a). It is trivial: the velocity in the medium, $v = \frac{c}{n}$, is constant as n is and hence minimizing the

$$\text{time} - \frac{\text{space}}{\text{velocity}}$$

is equivalent to minimizing the length. Therefore the ray travels on the segment AB.

Case (b). The light ray starting from A in the first medium, enters into the second medium via a point C (to be determined) and reaches the point B; see Fig. 16.8.

From (a) we know that in each medium the light ray is a segment. Then the path would be the polygonal path ACB on which the light takes the shortest time. If v_1, v_2 denote the velocity of light in the two media, we find

$$\text{time} = \frac{\text{space}}{\text{velocity}} = \frac{\overline{AC}}{v_1} + \frac{\overline{CB}}{v_2}.$$

Letting $C = (x,0)$ one has $\overline{AC} = \sqrt{(x-a)^2 + a^2}$ and $\overline{CB} = \sqrt{(b-x)^2 + \beta^2}$ and hence it follows that we have to minimize in $[a,b]$ the function

$$T(x) = \frac{\sqrt{(x-a)^2 + a^2}}{v_1} + \frac{\sqrt{(b-x)^2 + \beta^2}}{v_2}.$$

Setting $n_i = \frac{c}{v_i}$, we find $T(x) = \frac{1}{c}[n_1\sqrt{(x-a)^2 + a^2} + n_2\sqrt{(b-x)^2 + \beta^2}]$. The graph of the function $T(x)$ is plotted in Fig. 16.7.

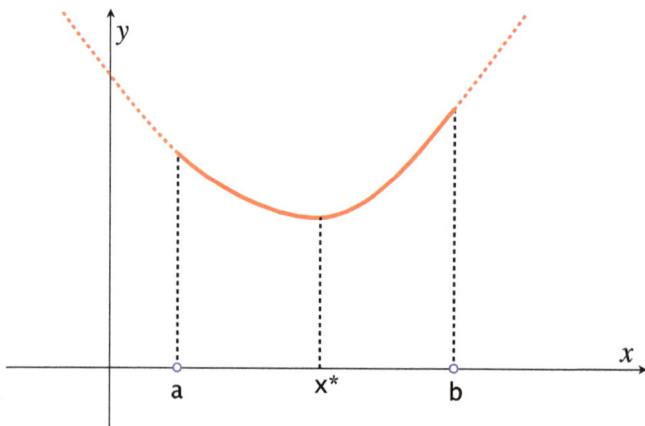

Figure 16.7: Graph of $y = T(x)$.

By a direct calculation we see that $T(x)$ attains the unique absolute minimum at some $x^* \in (a,b)$ such that $T'(x^*) = 0$, which yields

$$n_1\frac{x^* - a}{\sqrt{(x^* - a)^2 + a^2}} - n_2\frac{b - x^*}{\sqrt{(b - x^*)^2 + \beta^2}} = 0.$$

If θ_1, θ_2 denote the angles that the incident light ray and the refracted light ray make with the normal at C to the interface between the two media, see Fig. 16.8, we find

$$\sin\theta_1 = \frac{x^* - a}{\sqrt{(x^* - a)^2 + a^2}}, \quad \sin\theta_2 = \frac{b - x^*}{\sqrt{(b - x^*)^2 + \beta^2}},$$

and thus

$$n_1\sin\theta_1 = n_2\sin\theta_2,$$

which is known as *Snell's law*.

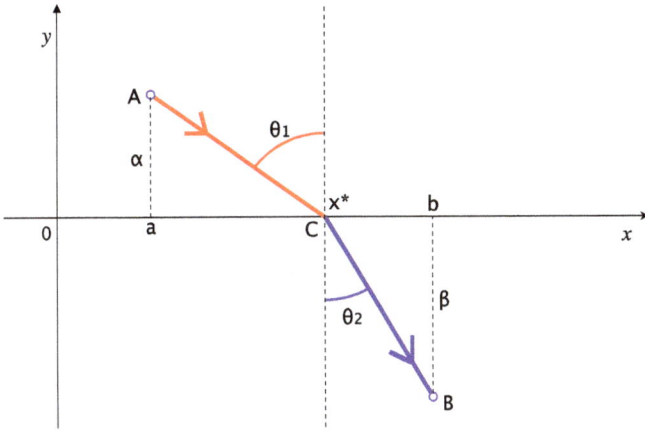

Figure 16.8: Snell's law of refraction.

We now deal with the general case in which light travels in a medium with refractive index $n(y)$. For simplicity, we assume that n depends smoothly on y only.

Given two points $A = (a, \alpha)$ and $B = (b, \beta)$ in the plane (x, y), an argument similar to that presented for the brachistochrone shows that the time taken by the light to travel along a path $y \in Y$ is given by

$$I[y] = \int_a^b n(y)\sqrt{1 + y'^2}\, dx. \tag{16.8}$$

Therefore we have to minimize $I[y]$ on Y. The Lagrangian of the functional (16.8) is given by $L(y, p) = n(y)\sqrt{1 + p^2}$ and then the corresponding Euler–Lagrange equation (EL') becomes

$$L(y, y') - y' L_p(y, y') = n(y)\sqrt{1 + y'^2} - y' \cdot \frac{n(y)y'}{\sqrt{1 + y'^2}}$$

$$= n(y) \cdot \frac{1 + y'^2 - y'2}{\sqrt{1 + y'^2}} = \frac{n(y)}{\sqrt{1 + y'^2}} = k.$$

Let us define $\vartheta(x)$ by setting $\vartheta(x) = \arctan y'(x)$, namely $y'(x) = \tan \vartheta(x)$. With this notation we find

$$\frac{1}{\sqrt{1 + y'^2}} = \frac{1}{\sqrt{1 + \tan^2 \vartheta(x)}} = \cos \vartheta, \quad \cos \vartheta > 0,$$

which yields

$$n(y) \cdot \cos \vartheta = k, \tag{16.9}$$

or, setting $\theta = \frac{1}{2}\pi - \vartheta$,

$$n(y) \cdot \sin \theta = k.$$

Remark 16.5. As for the brachistochrone, it is possible to show that the solution we found before is the absolute minimum of $I[y]$ on Y.

Historical remark. Johann Bernoulli solved the brachistochrone problem using Snell's law, by means of an elegant optical analogy. The body falling from $A = O = (0,0)$ to $B = (b, \beta)$ subjected to the gravitational force only, is compared to a light ray that travels from A to B on a medium with refractive index $n(y) = 1/\sqrt{y}$. Dividing the plane in infinitesimal strips with thickness ϵ, one approximates $1/\sqrt{y}$ with a piecewise constant function, in such a way that we can assume that in the ith strip the refractive index is constant $= n_i$.

Passing through two consecutive strips, Snell's law yields

$$\frac{n_i \sin \theta_i}{n_{i+1} \sin \theta_{i+1}} = 1.$$

Notice that $n_{i+1} < n_i$ implies $\theta_{i+1} > \theta_i$. In this way we find a polygonal path from A to B as shown in Fig. 16.9. Taking the limit as $\epsilon \to 0$, Bernoulli proved that the polygonal path converges to the brachistochrone. See also Exercise 8.

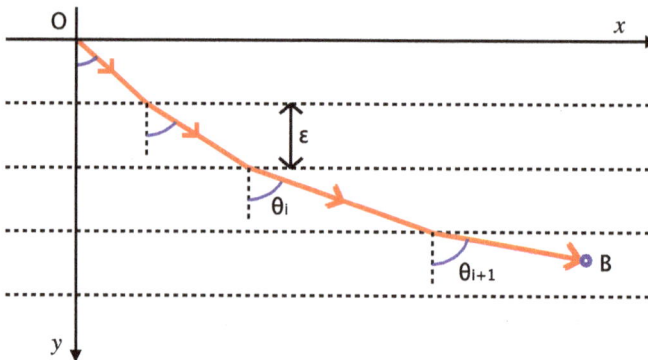

Figure 16.9: Bernoulli's argument to find the brachistochrone using Snell's law.

16.4 Stationary solutions vs. minima

Another question we have to take into account is that (EL) might have more than one solution. In such a case, how can we select the minimal solution if there is any?

The question has been broadly addressed in the past giving rise to several necessary and/or sufficient conditions for the existence of minima of $I[y]$ on a class Y.

In this section we simply state the following necessary condition, due to Legendre.

Theorem 16.2 (Legendre's necessary condition for a minimum). *Let $L \in C^2([a,b] \times \mathbb{R} \times \mathbb{R})$. If $\bar{y}$ is a minimum of the functional $I[y]$ on Y, then*

$$L_{pp}(x, \bar{y}(x), \bar{y}'(x)) \geq 0, \quad \forall x \in [a,b].$$

Example 16.4.

(a) In the Fermat functional (16.8) the lagrangian is $L = n(y)\sqrt{1 + p^2}$. In this case $L_{pp} = n(y) \cdot \frac{1}{(1+p^2)\sqrt{1+p^2}}$ and hence $L_{pp}(\bar{y}, \bar{y}') > 0$ provided $n(\bar{y}) > 0$.

(b) In the brachistochrone problem the lagrangian is $L = \frac{\sqrt{1+p^2}}{\sqrt{2gy}}, y > 0$. Here

$$L_{pp} = \frac{1}{(1+p^2)\sqrt{1+p^2}} \cdot \frac{1}{\sqrt{2gy}} > 0.$$

Therefore in both cases the Legendre necessary condition for minimum is satisfied.

□

Sufficient conditions are more involved and would allow one to show, for example, that the solutions of the brachistochrone and Fermat's problems are really minima, indeed absolute minima, as we already pointed out.

One sufficient condition, simple to state, but maybe difficult to use, is the following: *if $\bar{y} \in Y$ is a solution of $\delta I[y] = 0$ and if $L(x, y, p)$ is a strictly convex function, then $\bar{y}$ is the minimum of $I[y]$ on Y, and the only one*

Example 16.5. Letting $Y = \{y \in C^2([a,b]) : y(a) = y(b) = 0\}$ and $I[y] = \int_a^b [\frac{1}{2}y'^2 + \frac{1}{4}y^4]\, dx$, a solution of $\delta I[y] = 0$ is $\bar{y} = 0$. Since $L(y, p) = \frac{1}{2}p^2 + \frac{1}{4}y^4$ is convex, $\bar{y} = 0$ is the minimum of $I[y]$ on Y and the only one. Hence the boundary value problem $y'' - y^3 = 0, y(a) = y(b) = 0$ has only the trivial solution $\bar{y} = 0$. Notice that the fact that $\bar{y} = 0$ is a minimum of $I[y]$ on Y can be checked directly, since $I[0] = 0 < I[y], \forall y \in Y, y \neq 0$.

□

16.5 On the isoperimetric problem

Another celebrated problem in the calculus of variations is the *isoperimetric problem* which, in the planar case, amounts to finding among the closed curves of given length y, the one that encloses the greatest area.

The problem is equivalent to seeking an arc with endpoints on the x axis and prescribed length such that the area spanned by the arc is maximal. From the variational viewpoint, we look for $y \in Y = \{C^2([a,b]) : y(a) = y(b) = 0\}$ such that $J[y] = \int_a^b \sqrt{1+y'^2}\, dx = \frac{1}{2}y$, which maximizes the area functional $I[y] = \int_a^b y(x)\, dx$.

Some geometrical properties of the solution can be found by elementary arguments. For example Fig. 16.10 shows how we can replace an arc with two bumps by an arc with

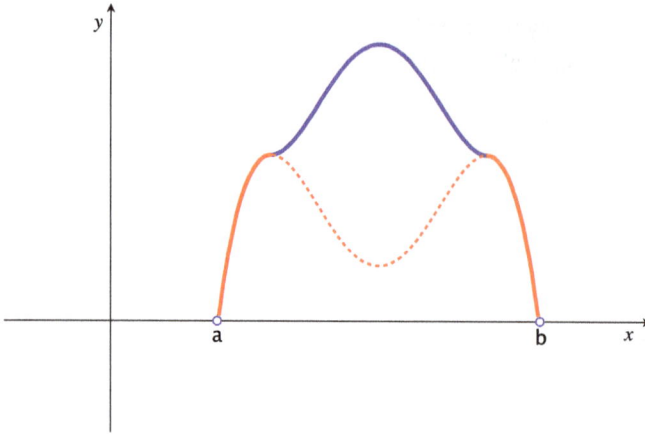

Figure 16.10: Showing that a solution of the isoperimetric problem cannot have more than one bump.

a single bump having the same length and greater area. Thus a solution cannot have more than one bump.

The isoperimetric problem is slightly different from the previous ones since here the test functions $y \in Y$ have to satisfy the constraint $J[y] = \int_a^b \sqrt{1+y'^2}\,dx = \frac{1}{2}\gamma$. This is similar to the problem of finding the maxima of a function $f(x_1,\dots,x_n)$ under the constraint $g(x_1,\dots,x_n) = c$. From calculus we know that if $x = (x_1,\dots,x_n)$ is one of these maxima points then there exists $\lambda \in \mathbb{R}$ (the Lagrange multiplier) such that

$$\nabla f(x) = \lambda \nabla g(x), \quad g(x) = c.$$

Dealing with functionals, the counterpart of the Lagrange multiplier rule is given by the following.

Theorem 16.3. *Given $I[y] = \int L\,dx$ and $J[y] = \int L_1\,dx$ with $L, L_1 \in C^2$, let $\bar{y}$ be such that $I[\bar{y}] = \max\{I[y] : y \in Y \text{ and } J[\bar{y}] = \gamma\}$. Then there exists $\lambda \in \mathbb{R}$ such that $\bar{y}$ satisfies the equation $\delta I[y] = \lambda \cdot \delta J[y]$, namely*

$$\frac{d}{dx}\frac{\partial L}{\partial p} - \frac{\partial L}{\partial y} = \lambda\left[\frac{d}{dx}\frac{\partial L_1}{\partial p} - \frac{\partial L_1}{\partial y}\right]. \tag{16.10}$$

Let us apply the previous theorem to the isoperimetric problem. Since we have $I[y] = \int_a^b y\,dx$ and $J[y] = \int_a^b \sqrt{1+y'^2}\,dx$, equation (16.10) becomes

$$1 + \lambda \cdot \frac{d}{dx}\frac{y'}{\sqrt{1+y'^2}} = 0.$$

Integrating we find

$$x + \lambda \cdot \frac{y'}{\sqrt{1+y'^2}} = c \Rightarrow \lambda \cdot \frac{y'}{\sqrt{1+y'^2}} = c - x.$$

Adding $-x$ to both sides and then squaring, we obtain

$$\lambda^2 \frac{y'^2}{1+y'^2} = (c-x)^2 \Rightarrow y'^2 = \frac{(c-x)^2}{\lambda^2 - (c-x)^2} \Rightarrow y' = \frac{c-x}{\sqrt{\lambda^2 - (c-x)^2}}.$$

Integrating, we obtain

$$y = -\sqrt{\lambda^2 - (c-x)^2} + c' \Rightarrow (y-c')^2 + (x-c)^2 = \lambda^2,$$

which is the equation of a circle centered at (c, c') and radius λ. The values of c, c' and λ are found by using the boundary conditions $y(a) = y(b) = 0$ and the constraint $J[y] = \frac{1}{2}\gamma$.

As for the brachistochrone and for the Fermat principles, one checks that Legendre's necessary condition for a minimum is satisfied. Moreover, it is possible to see that the half circle really encloses the greatest area among all the smooth curves from $(a, 0)$ to $(b, 0)$, with given length.[1]

Example 16.6. If $a = -1$, $b = 1$ and $\gamma = \pi$, the boundary conditions $y(a) = y(b) = b$ become $y(\pm 1) = 0$ and yield the system

$$\begin{cases} c'^2 + (-1 - c)^2 = \lambda^2, \\ c'^2 + (1 - c)^2 = \lambda^2, \end{cases}$$

whereby $c = c' = 0$. Hence $y^2 + x^2 = \lambda^2$. As for λ, we use the constraint $J[y] = \int_{-1}^{1} \sqrt{1+y'^2}\, dx = \pi$, with $y = \sqrt{\lambda^2 - x^2}$ to infer

$$\pi = \int_{-1}^{1} \left(1 + \frac{x^2}{\lambda^2 - x^2}\right) dx = \lambda^2 \int_{-1}^{1} \frac{dx}{\lambda^2 - x^2} = \lambda \cdot \left[\operatorname{arctanh}\left(\frac{x}{k}\right)\right]_{-1}^{1} = \pi\lambda,$$

whereby $\lambda = 1$. Thus we find the half circle $x^2 + y^2 = 1$. ☐

The general isoperimetric problem was completely solved only in the middle of the last century by the Italian mathematician Ennio De Giorgi.

---

1 A myth relates the isoperimetric problem to queen Dido: after her arrival on the Tunisian coast, the queen received a piece of land, from the local king, which could have been enclosed by a bull's skin. Dido cut the skin in very thin strips and enclosed the greatest area with a half circle. In that location Carthage was founded.

16.6 The Sturm–Liouville eigenvalue problem revisited

In section 13.3 of Chapter 13 we studied the Sturm–Liouville eigenvalue problem which, using the notation introduced in this chapter, can be written as

$$(r(x)y')' + \lambda p(x)y = 0, \quad y(a) = y(b) = 0. \tag{EP}$$

Here we want to revisit this problem from the point of view of the Calculus of Variations.

In the sequel we assume that $r, p \in C^2([a, b])$, and $r, p > 0$ on (a, b). Moreover, keeping the notations of Chapter 13, we let λ_1 denote the least eigenvalue of (EP).

Given the functionals

$$I[y] = \int_a^b r(x)y'^2(x)\, dx, \quad J[y] = \int_a^b p(x)y^2(x)\, dx,$$

defined on $y \in Y = \{y \in C^2([a, b]) : y(a) = y(b) = 0\}$, consider the variational problem

$$\min_{u \in S} I[u], \quad \text{where } S = \{u \in Y : J[u] = 1\},$$

which is simply the problem of finding $\min_Y I[y]$ constrained by the condition $J[y] = 1$.

It is possible to show that the previous minimum is attained at some $\bar{u} \in S$. Then, according to Theorem 16.3, $\bar{u}$ satisfies (16.10), namely

$$\frac{d}{dx}(r \cdot \bar{u}') = \bar{\lambda}(-p \cdot \bar{u}) \Rightarrow (r \cdot \bar{u}')' + \bar{\lambda} p \cdot \bar{u} = 0,$$

for some Lagrange multiplier $\bar{\lambda}$. Multiplying the preceding equation by $\bar{u}$ and integrating by parts it is easy to check that $\bar{\lambda} = I[\bar{u}]$. This shows the following.

Lemma 16.2. $\bar{\lambda} = I[\bar{u}]$ is an eigenvalue of (EP), a corresponding eigenfunction being $\bar{u}$.

Furthermore, the following Lemma holds.

Lemma 16.3. If Λ is any eigenvalue of (EP) and $\phi(x)$ is a corresponding eigenfunction such that $\int_a^b p\phi^2\, dx = 1$, then $\Lambda = I[\phi]$. In particular, taking $\Lambda = \bar{\lambda}$, one has $\bar{\lambda} = I[\bar{u}]$.

Proof. The eigenfunction ϕ satisfies the boundary value problem

$$(r(x)\phi')' + \Lambda \cdot p(x)\phi = 0, \quad \phi(a) = \phi(b) = 0.$$

Multiplying the equation by ϕ and integrating on $[a, b]$ we get

$$\int_a^b (r(x)\phi')'\phi\, dx + \Lambda \int_a^b p(x)\phi^2\, dx = 0.$$

Taking into account that $\phi(a) = \phi(b) = 0$, an integration by parts yields

$$\int_a^b (r(x)\phi')'\phi\, dx = -\int_a^b r(x)\phi'^2\, dx,$$

whereby

$$-\int_a^b r(x)\phi'^2\, dx + \Lambda \int_a^b p(x)\phi^2\, dx = 0, \quad \text{i.e.,} \quad \Lambda \int_a^b p(x)\phi^2\, dx = \int_a^b r(x)\phi'^2\, dx.$$

Since $\int_a^b p(x)\phi^2\, dx = 1$ we finally infer

$$\Lambda = \int_a^b r(x)\phi'^2\, dx = I[\phi]. \qquad \square$$

Remark 16.6. The condition $\int_a^b p\phi^2\, dx = 1$ is not restrictive. Actually, if $\int_a^b p\phi^2\, dx = k \neq 1$, it suffices to take $\tilde{\phi} = \phi/\sqrt{k}$ (notice that $k > 0$ for $p > 0$ and $\phi \neq 0$) which is also an eigenfunction.

Theorem 16.4. $\bar{\lambda} = \lambda_1$, *the least eigenvalue of* (EP).

Proof. We know that $\bar{\lambda}$ is an eigenvalue of (EP); see Lemma 16.2. Let Λ be any eigenvalue of (EP) and denote by ϕ a corresponding eigenfunction such that $\int_a^b p\phi^2\, dx = 1$. We will prove the lemma by showing that $\bar{\lambda} \leq \Lambda$. Actually, by Lemma 16.3 one has $\Lambda = I[\phi]$ as well as $\bar{\lambda} = I[\bar{u}]$. Then, recalling that $I[\bar{u}] = \min_{u \in S} I[u]$, we find $\bar{\lambda} = I[\bar{u}] \leq I[\phi] = \Lambda$. $\square$

Next we introduce the *the Rayleigh quotient* defined for $y \in Y, y \neq 0$, by

$$\mathcal{R}(y) = \frac{\int_a^b r(x)y'^2(x)\, dx}{\int_a^b p(x)y^2(x)\, dx} = \frac{I[y]}{J[y]}.$$

It is immediate to check that $\mathcal{R}(y)$ is homogeneous of degree 0, in the sense that

$$\mathcal{R}(ay) = \mathcal{R}(y), \quad \forall a \in \mathbb{R}, \quad a \neq 0, \quad \forall y \in Y, y \neq 0. \tag{16.11}$$

Theorem 16.5. $\min_{u \in S} I[u] = \min_{y \in Y} \mathcal{R}(y)$ *and hence* $\lambda_1 = \min_{y \in Y} \mathcal{R}(y)$.

Proof. If $\bar{u} \in S$ is such that $I[\bar{u}] = \min_{u \in S} I[u]$ then

$$I[\bar{u}] \leq I[u], \quad \forall u \in Y \text{ such that } J[u] = 1.$$

For any $y \in Y, y \neq 0$, the positive number $a = 1/\sqrt{J[u]}$ is such that

$$J[\alpha y] = \int_a^b p(x) \cdot (\alpha y)^2 \, dx = \alpha^2 \int_a^b p(x)y^2 \, dx = \alpha^2 J[y] = 1.$$

Thus $\alpha y \in S$ and hence

$$I[\bar{u}] \le I[\alpha y]. \tag{16.12}$$

Recalling that $J[\alpha y] = 1$ we get

$$I[\alpha y] = \frac{I[\alpha y]}{J[\alpha y]} = \mathcal{R}(\alpha y) = \mathcal{R}(y), \tag{16.13}$$

where, in the last equality, we have used (16.11).

From (16.12) and (16.13) we infer $I[\bar{u}] \le \mathcal{R}(y)$. Since this holds for any $y \in Y$, it follows that

$$I[\bar{u}] = \min_{u \in S} I[u] \le \min_{y \in Y} \mathcal{R}(y).$$

Conversely, if $\bar{y} \in Y$ is such that $\mathcal{R}(\bar{y}) = \min_Y \mathcal{R}(y)$ then $\mathcal{R}(\bar{y}) \le \min_{u \in S} I[u]$. The argument is quite similar to the previous one and is omitted.

Finally, $\lambda_1 = \min_{y \in Y} \mathcal{R}(y)$ follows immediately from Theorem 16.4. □

As an application, we prove an elegant inequality involving the first eigenvalue λ_1.

Theorem 16.6 (Poincaré inequality). *If λ_1 is the first eigenvalue of* (EP) *then one has*

$$\lambda_1 \cdot \int_a^b p(x)y^2 \, dx \le \int_a^b r(x)y'^2 \, dx, \quad \forall y \in Y.$$

Proof. The previous theorem yields $\lambda_1 = \min_{y \in Y} \mathcal{R}(y)$, namely

$$\lambda_1 \le \frac{\int_a^b r(x)y'^2 \, dx}{\int_a^b p(x)y^2 \, dx}, \quad \forall y \in Y.$$

From this the Poincaré inequality immediately follows. □

The next example shows a possible application of the Poincaré inequality.

Example 16.7. Let λ_1 denote the first eigenvalue of $y'' + \lambda p(x)y = 0, y(a) = y(b) = 0$. Prove that the nonlinear boundary value problem

$$y'' + \lambda p(x)y - y^3 = 0, \quad y(a) = y(b) = 0, \tag{16.14}$$

has only the trivial solution, provided that $\lambda \le \lambda_1$.

Solution. Let y be a nontrivial solution of (16.14). Multiplying the equation by y and integrating we find

$$\int_a^b yy''\,dx + \lambda \int_a^b p(x)y^2\,dx = \int_a^b y^4\,dx > 0,$$

for $y(x) \neq 0$.

Now an integration by parts yields

$$-\int_a^b y'^2\,dx + \lambda \int_a^b p(x)y^2\,dx > 0 \quad \Longrightarrow \quad \int_a^b y'^2\,dx < \lambda \int_a^b p(x)y^2\,dx.$$

Since $\lambda \leq \lambda_1$ we infer $\int_a^b y'^2\,dx < \lambda_1 \int_a^b p(x)y^2\,dx$, which is in contradiction with the Poincaré inequality. □

16.7 Exercises

1. Find the Euler–Lagrange equation of $I[y] = \int_a^b L\,dx$, where $L(y,p) = F(y)G(p)$ with $F, G \in C^2(\mathbb{R})$.
2. Let $I[y] = \int_0^\pi [\frac{1}{2}y'^2 - \frac{1}{4}y^4]\,dx$ and $Y = \{y \in C^2([0,\pi]) : y(0) = y(\pi) = 0\}$. Show that $\inf_Y I[y] = -\infty$.
3. Find the Euler–Lagrange equation of $I[y] = \int_0^\pi [\frac{1}{2}y'^2 - \frac{\lambda}{2}y^2]\,dx$ on $Y = \{y \in C^2([0,\pi]) : y(0) = y(\pi) = 0\}$ and solve it.
4. Given $I[y] = \int_0^\pi [\frac{1}{2}y'^2 - \frac{\lambda}{2}y^2 + \frac{1}{4}y^4]\,dx$, $y \in Y = \{y \in C^2([0,\pi]) : y(0) = y(\pi) = 0\}$, show:
   (a) $I[y]$ is bounded below on Y, namely $\exists C \in \mathbb{R}$ such that $I[y] \geq C, \forall y \in Y$,
   (b) $\inf_Y I[y] < 0$ provided $\lambda > 1$,
   (c) knowing that this infimum is achieved at $\bar{y}$, what is the boundary value problem solved by $\bar{y}$?
5. Let $Y = \{y \in C^2([0,2]) : y(0) = 0, y(2) = 2\}$ and $I[y] = \int_0^2 \sqrt{\frac{1+y'^2}{2gy}}\,dx$. Solve $\delta I = 0$ on Y.
6. Show that the time T taken by a body to descend from $A = (0,0)$ to $B = (1,1)$ along the Brachistochrone is less than or equal to $2/\sqrt{g}$, where g is the gravitational acceleration.
7. Compare the time taken by a point to travel from $A = (0,0)$ to $B = (2\pi,0)$ on the Brachistochrone and on the line $y = 0$.
8. Consider a medium with refractive index $n(y) = 1/\sqrt{2gy}$ and check if the light rays are Brachistochrone curves.
9. Finding the shape of a heavy chain with length γ, suspended at the points $(-1,1)$, $(1,1)$, leads to finding a curve $y = y(x)$ that minimizes $I[y] = \int_{-1}^1 y\sqrt{1+y'^2}\,dx$ on

$Y = \{y \in C^2([a,b]) : y(-1) = y(1) = 1, \ \int_{-1}^{1} \sqrt{1+y'^2}\, dx = y\}$. Find the solution when $y > 2$.

10. As first noted by Newton, the problem of finding the shape of the solid of revolution with least resistance when it moves in a fluid with constant velocity parallel to the axis of revolution leads to looking for the minimum of $I[y] = \int_a^b y \cdot \frac{y'^3}{1+y'^2}\, dx$ on $Y = \{y \in C^2([a,b]) : y(a) = y(b) = 0\}$. Find the Euler–Lagrange equation of $I[y]$.

Solutions

Solutions to Chapter 2

2. Solve $x' + (\ln 2)x = 2$.

   Solution. An integrating factor is $\mu(t) = e^{(\ln 2)t}$. Thus we find $(xe^{(\ln 2)t})' = 2e^{(\ln 2)t}$. Integrating both sides we get

   $$xe^{(\ln 2)t} = \int 2e^{(\ln 2)t}\, dt + c = \frac{2}{\ln 2} \cdot e^{(\ln 2)t} + c$$

   and hence

   $$x(t) = ce^{-(\ln 2)t} + \frac{2}{\ln 2}.$$

4. Solve $x' - 4t = 2x$, $x(0) = 2$.

   Solution. Rearranging $x' - 4t = 2x$, we have $x' - 2x = 4t$. Then $(e^{-2t}x)' = 4te^{-2t}$, $(e^{-2t}x) = 2te^{-2t} - e^{-2t} + c$, $x(t) = 2t - 1 + ce^{2t}$, $x(0) = 2$ implies $2 = -1 + c$, $c = 3$; so $x(t) = -2t - 1 + 3e^{2t}$.

6. Solve $x' - t = x + 1$, $x(0) = 1$

   Solution. Rewriting the equation as $x' - x = t + 1$, we have $(e^{-t}x)' = te^{-t} + e^{-t}$; hence $(e^{-t}x) = -te^{-t} - e^{-t} - e^{-t} + k$ and $x = -t - 2 + ke^{t}$ from which we get the desired solution $x = -t - 2 + 3e^{t}$.

8. Solve $x' + x = \sin t + \cos t$.

   Solution. $(e^{t}x)' = e^{t}\sin t + e^{t}\cos t$, $x(t) = \sin t + ce^{-t}$.

10. Solve $tx'' + x' = t + 2$.

    Solution. Letting $z = x'$ and solving for z, we have $z = \frac{1}{2}t + 2 + \frac{c}{t}$. Integrating both sides we obtain $x = \frac{1}{4}t^2 + 2t + c\ln t + k$.

12. Show that if $k < 0$ then there is no solution of $x' + kx = h$ such that $\lim_{t\to+\infty} x(t)$ is finite, but the constant one.

    Solution. The general solution is $x(t) = ce^{-kt} + \frac{h}{k}$. Thus, if $k < 0$ and $c \neq 0$ one has $\lim_{t\to+\infty} x(t) = \pm\infty$, depending on the sign of c.

14. Show that for any $x_0 \in \mathbb{R}$ the solution of $x' - x = h$, $x(0) = x_0$ is such that $\lim_{t\to-\infty} x(t) = -h$.

    Solution. The general solution is $x(t) = ce^{t} - h$. For $t = x = 0$ we obtain $c = h + x_0$ and hence the solution of the ivp is $x(t) = (h + x_0)e^{t} - h$. Thus $\lim_{t\to-\infty} x(t) = -h$.

16. * Explain why the boundary value problem $x' + kx = 0$, $x(1) = 2$, $x(3) = -1$ has no solution. This shows that, unlike the initial value problems, boundary value problems for first order homogeneous equations don't always have solutions.

    Solution. The general solution is $x(t) = ce^{-kt}$. Thus, $x(1) = 2 \Rightarrow 2 = ce^{-k}$ and hence $c = 2e^{k} > 0$. On the other hand, $x(3) = -1 \Rightarrow -1 = ce^{-3k}$ and hence $c = -e^{3k} < 0$, contradiction.

https://doi.org/10.1515/9783111185675-017

18. Solve $x' + (\cot t)x = \cos t, t \neq n\pi$.

Solution. $(x \sin t)' = \sin t \cos t, x(t) = \frac{1}{2} \sin t + c \csc t$.

20. Show that the boundary value problem $x' + p(t)x = q(t), x(t_1) = x_1, x(t_2) = x_2, p(t)$ and $q(t)$ continuous, cannot have more than one solution.

Solution. This is obvious because if $x(t)$ and $y(t)$ are two solutions such that $x(t_1) = y(t_1)$, then by the Uniqueness Theorem, $x(t) \equiv y(t)$.

22. Find the largest interval in which the solution to the following Cauchy Problems exist.

1. $x' + \frac{1}{t^2-4}x(t) = 0, x(-1) = 10$.

2. $x' + \frac{1}{t^2-4}x(t) = 0, x(3) = 1$.

3. $x' + \frac{1}{t^2-4}x(t) = 0, x(-10) = 1$.

Solution. 1. $(-2, 2)$, 2. $(2, \infty)$, 3. $(-\infty, -2)$.

24. (a) Show that if $x_1(t)$ is a solution of $x' + p(t)x = f(t)$ and $x_2(t)$ is a solution of $x' + p(t)x = g(t)$, then $x_1 + x_2$ is a solution of $x' + p(t)x = f(t) + g(t)$.

(b) Use the method in part (a) to solve $x' + x = 2e^t - 3e^{-t}$.

Solution.

(a). $(x_1' + px_1) + (x_2' + px_2) = f + g \Rightarrow (x_1' + x_2') + p(x_1 + x_2) = f + g$.

(b). $x' + x = 2e^t \Rightarrow x_1 = e^t + ce^{-t}$ and $x' + x = -3e^{-t} \Rightarrow x_2 = -3te^{-t} + ke^{-t}$. Then $x(t) = x_1(t) + x_2(t)$.

26. * Find a number k such that $x' + kx = 0, x(1) = 2, x(2) = 1$ has a solution.

Solution. $x(t) = ce^{-kt} \Rightarrow 2 = ce^{-k}$ and $1 = ce^{-2k} \Rightarrow \frac{ce^{-k}}{ce^{-2k}} = 2$; and $k = \ln 2$. Hence $x(t) = ce^{(-\ln 2)t}$ and $x(1) = 2 \Rightarrow 2 = ce^{-\ln 2} \Rightarrow c = 4$. Hence $x' + (\ln 2)x = 0$ has the solution $x(t) = 4e^{(-\ln 2)t}$ that satisfies the required boundary value conditions.

28. * Let $x_1(t)$ be any solution of (2.1) $x' + p(t)x = q(t)$. Recall that $x(t) = ce^{-\int_{t_0}^t p(t)\,dt}$ is the general solution of (2.2) $x' + p(t)x = 0$. Show that $x(t) = x_1(t) + ce^{-\int_{t_0}^t p(t)\,dt}$ is the general solution of the nonhomogeneous equation (2.1).

Solution. First show, by substitution, that $x(t) = x_1 + ce^{-\int_{t_0}^t p(t)\,dt}$ is a solution of (2.1). Then let $z(t)$ be any solution of (2.1) and let t_0 be any number in its domain of definition. Let $z(t_0) = k$. Show that there exists $\bar{c}$ ($\bar{c} = z(t_0) - x_1(t_0)$) such that $x(t) = x_1(t) + \bar{c}e^{-\int_{t_0}^t p(t)\,dt}$ is a solution of (2.1) and $x(t_0) = k$, which would imply, by the uniqueness theorem, that $x(t) \equiv z(t)$.

30. Explain why the function $x(t) = e^t - 1$ cannot be a solution of any homogeneous equation $x' + p(t)x = 0$, where $p(t)$ is some continuous function.

Solution. We note that $x(0) = 0$. Also the trivial solution $z(t) \equiv 0$ satisfies the initial condition $z(0) = 0$. Therefore, by the uniqueness of solutions, we must have $x(t) \equiv 0$ for all t, which is not true, since, for example, $x(1) = e - 1 \neq 0$.

32. Explain why $x(t) = t^2 - 1$ cannot be a solution of a homogeneous equation $x' + p(t)x = 0, p(t)$ continuous.

Solution. $x(1) = 0$ implies $x(t) \equiv 0$.

Solutions to Chapter 3

2. Verify that for any $(t_0, x_0) \in \mathbb{R}^2$ the local existence and uniqueness theorem applies to the ivp $x' = \ln(1 + x^4)$, $x(t_0) = x_0$.

Solution. The function $f(t, x) = \ln(1+x^4)$ is continuous for all $x \in \mathbb{R}$ with continuous $f_x(t, x) = \frac{4x^3}{1+x^4}$.

4. Find the domain of definition of the solution to the ivp $x' = x^2$, $x(0) = a$.

Solution. If $a = 0$, the solution is $x \equiv 0$. Let $a \neq 0$. As in Example 3.1 we see that $x = \frac{1}{c-t}$ solve $x' = x^2$, for all constant c. The initial condition $x(0) = a$ yields $a = \frac{1}{c}$, namely $c = \frac{1}{a}$. So $x = \frac{1}{\frac{1}{a}-t} = \frac{a}{1-at}$. In particular, these solutions are defined for $t \neq \frac{1}{a}$.

Moreover, taking into account that the solutions are continuous, we have to take only the upper branches of the hyperbolas if $a > 0$ and the lower branches if $a < 0$.

Hence:
- if $a = 0$ the ddf is $\mathbb{R}$,
- if $a > 0$ the ddf is $(-\infty, \frac{1}{a})$,
- if $a < 0$ the ddf is $(\frac{1}{a}, +\infty)$.

6. Show that the Cauchy problem $x' = x^{1/3}$, $x(0) = 0$, does not have a unique solution.

Solution. One solution is $x = 0$. Another solution for $t \geq 0$ is given by $x(t) = \left(\frac{2}{3}t\right)^{3/2}$.

8. Show that the solutions of $x' = \sqrt{1+x^2}$ are defined on all $t \in \mathbb{R}$.

Solution. The continuous function $f(x) = \sqrt{1+x^2}$ has bounded derivative $f'(x) = \frac{x}{\sqrt{1+x^2}}$.

10. Show that the solution of the Cauchy problem $x' = \sin x$, $x(0) = 1$, is such that $0 < x(t) < \pi$ and is increasing.

Solution. $x(t)$ is defined for all $t \in \mathbb{R}$, for $f(x) = \sin x$ is continuously differentiable and $f'(x) = \cos x$ is bounded. Moreover $x = k\pi$, $k = 0, \pm1, \pm2, \ldots$ are the equilibria solutions. Since $0 < x(0) = 1 < \pi$ then $0 < x(t) < \pi$. Since $\sin x > 0$ for $0 < x < \pi$, then $x'(t) > 0$ and thus $x(t)$ is increasing.

12. Let $f(x)$ be an even $C^1(\mathbb{R})$ function and let $x_0(t)$ be a solution of $x' = f(x)$ such that $x(0) = 0$. Prove that $x_0(t)$ is an odd function.

Solution. Setting $z(t) = -x_0(-t)$, we find $z'(t) = x_0'(-t) = f(x_0(-t)) = f(-z(t))$. Since f is even, then $f(-z) = f(z)$ and thus $z' = f(z)$. Moreover, $z(0) = -x_0(0) = 0$. Then both z and x_0 are solutions to the ivp $x' = f(x)$, $x(0) = 0$, and hence, by uniqueness, $z(t) = x_0(t)$. Recalling that $z(t) = -x_0(-t)$ we find $-x_0(-t) = x_0(t)$, namely $x_0(-t) = -x_0(t)$ which means that x_0 is an odd function.

14. Solve $x' = |x| - 1$, $x(0) = 0$.

Solution. For $x \geq 0$, resp. $x \leq 0$, the equation becomes $x' = x - 1$, resp. $x' = -x - 1$. Considering separately the two equations, we apply Theorem 3.2 to infer that the ivp has a unique solution defined for all $t \in \mathbb{R}$. The equation has two equilibria $x = \pm1$ and hence the solution $x(t)$ of the ivp satisfies $-1 < x(t) < 1$ for all t. Since $x(0) = 0$ and $x'(0) = |x(0)| - 1 = -1$, the solution $x(t)$ is negative for $t > 0$ and positive for $t < 0$. Thus $x(t)$ satisfies $x' = -x - 1$ for $t > 0$ and $x' = x - 1$ for $t < 0$. Solving the two linear equation as learned in Chapter 2, we obtain $x(t) = -1 + ce^{-t}$ for $t \geq 0$ and $x(t) = 1 + ce^t$ for $t \leq 0$. For $t = x = 0$ we find $c = 1$ in the former case and $c = -1$ in the latter case. In conclusion, the solution is given by

$$x(t) = \begin{cases} -1 + e^{-t} & (t \geq 0), \\ 1 - e^t & (t \leq 0). \end{cases}$$

Since $x'(0+) = x'(0-) = -1$ the above function is of class C^1.

16. Find the $\lim_{t \to +\infty} x(t)$, where $x(t)$ is the solution of the Cauchy problem $x' = 1 - e^x$, $x(0) = 1$.

    Solution. Since $1 - e^x < 0$ provided $x > 0$, the solution is decreasing. Moreover $x = 0$ is the (unique) equilibrium solution and thus $0 < x(t) < 1$ for all $t > 0$. It follows that $\lim_{t \to +\infty} x(t) = 0$.

18. Let $x(t)$ be the solution of the Cauchy problem $x' = \ln(1 + x^2)$, $x(0) = 0$. Show that $\lim_{t \to +\infty} x(t) = +\infty$ and $\lim_{t \to -\infty} x(t) = -\infty$.

    Solution. The solution $x(t)$ is defined for all $t \in \mathbb{R}$, for $f(x) = \ln(1 + x^2)$ has bounded derivative $f'(x) = \frac{2x}{1+x^2}$. Since $\ln(1 + x^2) > 0$, then $x(t)$ is increasing. Moreover $x'' = f'(x)x' = \frac{2x}{1+x^2} \cdot \ln(1 + x^2)$. Thus, $x''(t) > 0 \iff x(t) > 0$. It follows that $x(t)$ is convex for $t > 0$ and concave for $t < 0$ which in turn implies that $\lim_{t \to +\infty} x(t) = +\infty$ and $\lim_{t \to -\infty} x(t) = -\infty$.

20. Study the convexity of the solutions to $x' = x^3 - 1$.

    Solution. Differentiating we get $x'' = 3x^2x' = 3x^2(x^3 - 1)$. Hence $x''(t) > 0$ iff $x(t) > 1$. Notice that the equation has a unique equilibrium solution $x = 1$ and all the other solutions are either greater than 1 or smaller than 1. Thus the solutions such that $x(t) > 1$ are convex, whereas the solutions such that $x(t) < 1$ are concave.

22. Study the convexity of the solutions to $x' = x(1 + x)$.

    Solution. $x' = x + x^2 \Rightarrow x'' = x' + 2xx' \Rightarrow x'' = x + x^2 + 2x(x + x^2) = x + 3x^2 + 2x^3 = x(1 + 3x + 2x^2)$. Solving $x(1 + 3x + 2x^2) > 0$ we find that $x''(t) > 0$ whenever $-1 < x(t) < -\frac{1}{2}$ or $x(t) > 0$. The equation has two equilibria $x = 0$ and $x = -1$; and $x' > 0$ for $\{x < -1\} \cup \{x > 0\}$, whereas $x' < 0$ for $-1 < x < 0$. It follows that the solutions such that $x(t) > 0$ or $x(t) < -1$ are (increasing and) convex, whereas the solutions such that $-1 < x(t) < 0$ (are decreasing and) change their concavity at the unique t^* such that $x(t^*) = -\frac{1}{2}$.

24. Find the locus of maxima of $x' = x^3 - t$.

    Solution. At maxima one has $x' = 0 \Rightarrow x^3 = t$. Moreover $x''(t) = 3x^2(t)x'(t) - 1 = -1$. Thus the locus of maxima is the curve $x^3 = t$.

26. Show that the solution of $x' = x^4$, $x(0) = x_0 \neq 0$, is strictly convex if $x_0 > 0$ and strictly concave if $x_0 < 0$. Extend the result to $x' = x^p$, $x(0) = x_0 \neq 0$, $p \geq 1$.

    Solution. All the solutions of the ivp are either positive or negative, for $x_0 \neq 0$. Moreover $x'' = 4x^3x' = 4x^7 > 0 \iff x > 0$. If $x' = x^p$, then $x'' = px^{p-1}x^p = px^{2p-1}$. So $x'' > 0 \iff x^{2p-1} > 0 \iff x > 0$, since $2p - 1$ is an odd integer.

28. Let $x(t)$ be the solution to $x' = f(t,x), x(0) = x_0 > 0$, where f is smooth for $(t,x) \in \mathbb{R}^2$. If $x(t)$ is defined on $[0,+\infty)$ and $f(t,x) > x$, show that $\lim_{t\to+\infty} x(t) = +\infty$.
Solution. $f > x \Rightarrow x' > x \Rightarrow x' - x > 0$. Multiplying by e^{-t} we get $(e^{-t}x)' > 0$. Integrating from 0 to t it follows that $e^{-t}x(t) - x(0) > 0 \Rightarrow x(t) > x(0)e^t = x_0 e^t$. Since $x_0 > 0$, it follows that $\lim_{t\to+\infty} x(t) = +\infty$.

30. Find a function $f(x)$ such that

$$x(t) = \begin{cases} e^t & \text{if } t \geq 0, \\ t+1 & \text{if } t \leq 0, \end{cases}$$

is the solution of the Cauchy problem $x' = f(x), x(0) = 1$.
Solution. Notice that x is continuously differentiable, for

$$x'(t) = \begin{cases} e^t & \text{if } t \geq 0, \\ 1 & \text{if } t \leq 0, \end{cases}$$

and the two functions are equal at $t = 0$. This also shows that $x' = f(x)$, where

$$f(x) = \begin{cases} x & \text{if } x \geq 1, \\ 1 & \text{if } x \leq 1. \end{cases}$$

Solutions to Chapter 4

Solutions to Section 4.3

2. Solve $x' = e^{-2x}e^t$.
   Solution. $e^{2x} dx = e^t dt$ implies $\frac{1}{2}e^{2x} = e^t + c$ or $e^{2x} = 2e^t + k, k = 2c$ arbitrary constant.

4. Solve $x' = \frac{t(x+1)}{t+1}$.
   Solution. $\frac{dx}{x+1} = \frac{t\,dt}{t+1} = (1 - \frac{1}{t+1})\,dt$ implies $\ln|x+1| = t - \ln|t+1| + c$.

6. Solve $x' = x(x+1)$.
   Solution. As in the previous exercise, $x(t) \equiv 0$ is one solution; otherwise,

$$\int \frac{dx}{x(x+1)} = \int dt + c \Rightarrow \int \frac{dx}{x(x+1)} = t + c.$$

Since $\frac{1}{x(x+1)} = \frac{1}{x} - \frac{1}{x+1}$ it follows that

$$\int \frac{dx}{x(x+1)} = \int \frac{1}{x} - \int \frac{1}{x+1} = \ln|x| - \ln|x+1| = \ln\left|\frac{x}{x+1}\right|.$$

Thus

$$\ln\left|\frac{x}{x+1}\right| = t + c \Rightarrow \left|\frac{x}{x+1}\right| = e^{t+c} = e^c \cdot e^t,$$

which is the general solution in implicit form.

8. Solve $x' = \frac{2t^p+1}{3x^q+1}$, p, q positive numbers.

    Solution. Separating the variables and integrating we find

    $$\int (3x^q + 1)\,dx = \int (2t^p + 1)\,dt + c \Rightarrow \frac{3}{q+1}x^{q+1} + x = \frac{2}{p+1}t^{p+1} + t + c.$$

10. Solve $e^{t+2x}x' = e^{2t-x}$.

    Solution. $e^t e^{2x} x' = e^{2t} e^{-x} \Rightarrow e^{3x}\,dx = e^t\,dt \Rightarrow e^{3x} = 3e^t + k.$

12. Solve $\sqrt{1-t^2}\,x' = x$.

    Solution. Separating the variables and integrating we find

    $$\int \frac{dx}{x} = \int \frac{dt}{\sqrt{1-t^2}} + c \Rightarrow \ln|x| = \arcsin t + c.$$

14. Solve $x' = \frac{x+1}{1+t^2}$.

    Solution.

    $$\int \frac{dx}{x+1} = \int \frac{dt}{1+t^2} \Rightarrow \ln|1+x| = \arctan t + c.$$

16. Solve $x' = 4t\sqrt{x}$, $x > 0$, $x(0) = 1$.

    Solution. Separating the variables and integrating we find

    $$\int \frac{dx}{\sqrt{x}} = 4\int t\,dt + c \Rightarrow 2\sqrt{x} = 2t^2 + c.$$

    Moreover, $x(0) = 1 \Rightarrow c = 2$. Hence, $\sqrt{x} = t^2 + 1$ whereby $x(t) = (t^2 + 1)^2$.

18. Solve $x' = 4t^3\sqrt{x}$, $x \geq 0$, $x(0) = 1$.

    Solution.

    $$\int \frac{dx}{\sqrt{x}} = 4\int t^3\,dt + c \Rightarrow 2\sqrt{x} = t^4 + c.$$

    Then $c = 2$ and $x(t) = \left(\frac{1}{2}t^4 + 1\right)^2$.

20. Solve $x' = x(2 - x)$, first with the initial value $x(0) = -1$ and then with the initial value $x(0) = 1$.

    Solution. The general solution is

    $$x(t) = \frac{2ce^{2t}}{1 + ce^{2t}}$$

    and $x(0) = 1 \Rightarrow c = 1$, $x(0) = -1 \Rightarrow c = -1$.

22. Solve $e^{2x-t}\,dx = 2e^{x+t}\,dt$.

    Solution. $e^{2x}e^{-t}\,dx = 2e^x e^t\,dt \Rightarrow e^x\,dx = 2e^{2t}\,dt \Rightarrow e^x = e^{2t} + c.$

Solutions to Section 4.5

2. Solve $x' = \frac{x}{t} - (\frac{x}{t})^2$.

   Solution. Setting $x = tz$, the problem becomes $tz' + z = z - z^2 \Rightarrow tz' = -z^2$. Solving this separable equation we find

   $$\int \frac{dz}{z^2} = -\int \frac{dt}{t} \Rightarrow -\frac{1}{z} = -\ln|t| + c \Rightarrow z = \frac{1}{\ln|t| - c} \Rightarrow \frac{x}{t} = \frac{1}{\ln|t| - c} \Rightarrow x = \frac{t}{\ln|t| - c}.$$

4. Solve $x' = \frac{x}{t} + \tan(\frac{x}{t})$.

   Solution. $x = tz \Rightarrow tz' + z = z + \tan z \Rightarrow tz' = \tan z$. Separating variables and integrating, we have

   $$\int \frac{\cos z \, dz}{\sin z} = \int \frac{1}{t} \, dt,$$

   or $\ln|\sin z| = \ln|t| + c$ whereby $|\sin z| = e^c|t| \Rightarrow |\sin \frac{x}{t}| = e^c|t| = k|t|$, k a positive constant.

6. Solve $x' = \frac{x^2 + tx + t^2}{t^2}$.

   Solution. $x = tz \Rightarrow tz' + z = z^2 + z + 1 \Rightarrow tz' = z^2 + 1 \Rightarrow \frac{dz}{z^2 + 1} = \frac{dt}{t}$. Integrating

   $$\int \frac{dz}{z^2 + 1} = \int \frac{dt}{t} \Rightarrow \arctan z = \ln|t| + c.$$

   Solving for z we get $z = \tan(\ln|t| + c)$ and thus

   $$x = tz = t \cdot \tan(\ln|t| + c), \quad t \neq 0.$$

8. Solve $x' = \frac{x^2 + 2t^2}{tx}$.

   Solution. In differential form, we have $(tx) \, dx = (x^2 + 2t^2) \, dt$. Substituting $x = tz$ and $dx = t \, dz + z \, dt$, and simplifying yields $z \, dz = \frac{2}{t} \, dt$. Integrating and making the substitution $z = \frac{x}{t}$, we obtain

   $$x^2 = t^2 \ln t^4 + ct^2.$$

10. Solve $x' = e^{\frac{x}{t}} + \frac{x}{t}$.

    Solution. Letting $x = tz$ and simplifying, we get $tz' = e^z$, or $e^{-z} \, dz = \frac{dt}{t}$. Integrating and substituting $\frac{x}{t}$ for z, we have

    $$e^{\frac{-x}{t}} + \ln t = c.$$

Solutions to Section 4.7

2. Solve the Bernoulli equation $x' = \frac{x}{t} + 3x^3$.

   Solution. $y = x^{-2} \Rightarrow y' = -2 \cdot \frac{y}{t} - 6 \Rightarrow y' + 2 \cdot \frac{y}{t} = -6 \Rightarrow y = \frac{c}{t^2} - 2t \Rightarrow x = y^{\frac{-1}{2}} = (\frac{t^2}{c - 2t^3})^{\frac{1}{2}}.$

4. Solve $x' + x = x^2$.

    Solution. Let $y = x^{-1}$. Then $x' = -x^2 y'$ and $x = x^2 y$ and hence $y' - y = -1$. Solving for y, we have $y = 1 + ce^t$ and $x = \frac{1}{1+ce^t}$.

6. Solve the Bernoulli equation $x' = -\frac{x}{t} + x^{-2}$, $x(1) = 1$.

    Solution. Substituting $y = x^3$, $x' = \frac{1}{3}x^{-2}y'$, $x = x^{-2}y$ and simplifying, we obtain $y' + \frac{3}{t}y = 3$. Solving this equation, we get $y = ct^{-3} + \frac{3}{4}t \Rightarrow x = (\frac{3t^4+4c}{4t^3})^{1/3}$.

    $x(1) = 1 \Rightarrow 1 = \frac{3+4c}{4} \Rightarrow 4c = 1 \Rightarrow x = (\frac{3t^4+1}{4t^3})^{1/3}$.

8. Solve $x' - tx^2 = x$.

    Solution. Writing it as $x' - x = tx^2$, we see that this is a Bernoulli equation. Letting $z = x^{-1}$ leads to $z' + z = -t$. Solving this linear equation, we have $z(t) = 1 - t + ce^{-t}$ and consequently $x = \frac{1}{1-t+ce^{-t}}$.

10. Solve $x' - x = x^{\frac{1}{3}}$.

    Solution. Let $y = x^{\frac{2}{3}}$. Then $x = x^{\frac{1}{3}}y$ and $x' = \frac{3}{2}x^{\frac{1}{3}}y'$. Substituting and simplifying, we have $y' - \frac{2}{3}y = \frac{2}{3}$. Solving for y, we have $y = -1 + ce^{\frac{2}{3}t}$, yielding $x = (-1 + ce^{\frac{2}{3}t})^{\frac{3}{2}}$.

Solutions to Section 4.9

2. Solve the Clairaut equation $x = tx' - e^{x'}$ and find the singular solution.

    Solution. The general solution is $x = ct - e^c$. The system (4.15) becomes

    $$\begin{cases} x = ct - e^c, \\ t = e^c. \end{cases}$$

    Solving $e^c = t$ we get $c = \ln t$, $t > 0$. Hence the singular solution can be given in cartesian form as $x = t \ln t - t$, $t > 0$. See Fig. 4.4.

4. Solve the Clairaut equation $x = tx' + \frac{1}{3(x')^3}$ and find the singular solution.

    Solution. $x = tc + \frac{1}{3c^3}$, $c \neq 0$. Singular solution:

    $$\begin{cases} x = ct + \frac{1}{3c^3} \\ t = \frac{1}{c^4} \end{cases} \Rightarrow \begin{cases} x = c \cdot \frac{1}{c^4} + \frac{1}{3c^3} = \frac{4}{3c^3} \\ t = \frac{1}{c^4} > 0 \end{cases}$$

    $$\Rightarrow t = \left|\frac{3}{4}x\right|^{4/3} > 0 \text{ or } x = \pm\frac{4}{3}t^{3/4}, \quad t > 0.$$

6. Solve the Cauchy problem $x = tx' + \frac{1}{x'}$, $x(1) = b$.

    Solution. (The equation is similar to Exercise 55 of the old book.) The general integral is $x = ct + \frac{1}{c}$, $c \neq 0$. Singular solution:

    $$\begin{cases} x = ct + \frac{1}{c} \\ t = \frac{1}{c^2} > 0 \end{cases} \Rightarrow x = c \cdot \frac{1}{c^2} + \frac{1}{c} = \frac{2}{c} \Rightarrow \frac{1}{c} = \frac{x}{2} \Rightarrow t = \frac{1}{4}x^2, \quad (t > 0)$$

The initial condition yields $b = c + \frac{1}{c} \Rightarrow c^2 - bc + 1 = 0 \Rightarrow c_{1,2} = \frac{b \pm \sqrt{b^2-4}}{2}$. Thus for $|b| < 2$ the ivp has no solution, for $|b| > 2$ it has 2 solutions: $x = c_{1,2}\,t + \frac{1}{c_{1,2}}$, whereas for $b = 2$, resp $b = -2$, the ivp has one solution which is the line $x = t + 1$, resp. $x = -t - 1$, and a second solution which is the singular solution $t = \frac{1}{4}x^2$.

Solutions to Chapter 5

2. Solve $x^2 + ye^x + (y + e^x)y' = 0$.
    Solution. Writing the equation in the form $(x^2 + ye^x)\,dx + (y + e^x)\,dy = 0$, we see that it is exact because $M_y = e^x = N_x$. Using Method 2, we have $\int_0^x (x^2 + ye^x)\,dx + \int_0^y (y+1)\,dy = \frac{1}{3}x^3 + ye^x - y + \frac{1}{2}y^2 + y = \frac{1}{3}x^3 + ye^x + \frac{1}{2}y^2 = c$.

4. Solve the initial value problem $(3x^2y + 2xy^2)\,dx + (x^3 + 2x^2y - 6)\,dy = 0, y(1) = -1$.
    Solution. We let $y = 0$ in the first term and integrate, obtaining $F(x,y) = \int_0^x (x^3 + 2x^2y - 6)\,dy = x^3y + x^2y^2 - 6y$. Thus the general solution is given by $x^3y + x^2y^2 - 6y = k$. Substituting the initial values $x = 1, y = -1$ yields $k = 6$ and the desired solution is $x^3y + x^2y^2 - 6y = 6$.

6. Solve $(12x^5 - 2y)\,dx + (6y^5 - 2x)\,dy = 0$.
    Solution. Using Method 1, we have $\int_0^x 12x^5\,dx + \int_0^y (6y^5 - 2x)\,dy = 2x^6 + y^6 - 2xy = c$.

8. Find the number a such that $(x^3 + 3axy^2)\,dx + (x^2y + y^4)\,dy = 0$ is exact and solve it.
    Solution. Setting $M_y = N_x$, we have $6axy = 2xy \Rightarrow a = \frac{1}{3}$. Hence $(x^3 + xy^2)\,dx + (x^2y + y^4)\,dy = 0$, which is exact, with the general solution given by $\frac{1}{4}x^4 + \frac{1}{2}x^2y^2 + \frac{1}{5}y^5 = c$.

10. Solve $2xy^3 + 1 + (3x^2y^2)y' = 0, y(1) = 1$.
    Solution. The equation written as $(2xy^3 + 1)\,dx + (3x^2y^2)\,dy = 0$ is exact and the general solution is given by $\int_0^x dx + \int_0^y (3x^2y^2)\,dy = x + x^2y^3 = k$. The initial condition $y(1) = 1$ implies $k = 2$ and the desired solution is $x + x^2y^3 = 2$.

12. Find the solution of $(x^2 - 1)\,dx + y\,dy = 0$ passing through $(-1, b)$ with $b > 0$ and show that it can be given in the form $y = y(x)$.
    Solution. The general solution is $\frac{1}{3}x^3 - x + \frac{1}{2}y^2 = c$. Since c is arbitrary constant, we can write the equation as $3y^2 + 2x^3 - 6x = c$. The initial values $x = -1, y = b$ imply that $c = 3b^2 + 4$ and hence $3y^2 + 2x^3 - 6x = 3b^2 + 4$. Solving for y we find $y = \pm \frac{1}{\sqrt{3}}\sqrt{3b^2 + 4 + 6x - 2x^3}$. For $x = -1$ one has $y(-1) = \pm|b|$. Since $b > 0$ we take the positive square root yielding $y(x) = \frac{1}{\sqrt{3}}\sqrt{3b^2 + 4 + 6x - 2x^3}$.

14. Solve $y\,dx - 3x\,dy = 0$.
    Solution. The equation is not exact. Checking $\frac{N_x - M_y}{M} = \frac{-4}{y}$, we see that there is an integrating function $\mu(y)$ given by $\mu(y) = \frac{1}{y^4}$. Multiplying the equation by $\mu(y) = \frac{1}{y^4}$, we obtain the exact equation $\frac{1}{y^3}\,dx - \frac{3x}{y^4}\,dy = 0$. Let $x_0 = 0, y_0 = 1$. Using Method 1, we have $\int_0^x dx - \int_1^y 3xy^{-4}\,dy = \frac{x}{y^3} = c$. The given equation is also separable and can be solved by separating the variables.

16. Solve $\frac{2y^2 + 1}{x}\,dx + 4y \ln x\,dy = 0$.

Solution. First we notice that the term $\frac{2y^2+1}{x}$ prohibits us from using $x_0 = 0$ as the lower limit. So, let us use $x_0 = 1$, $y_0 = 1$. Then, using Method 1, we let $y = 1$ in the first integral and integrate, obtaining

$$F(x,y) = \int_1^x \frac{3}{x}\,dx + \int_1^y 4y\ln x\,dy = 3\ln x + 2y^2\ln x - 2\ln x = \ln x + 2y^2\ln x = c.$$

18. (A) Solve $(y + x)\,dx + dy = 0$ by finding an integrating factor $\mu(x)$.

(B) Explain whether or not it also has an integrating factor that is a function of y only.

Solution.

(A) Since $\frac{M_y - N_x}{N} = 1$, $\mu(x) = e^{\int 1\,dx} = e^x$. Multiplying $(y+x)\,dx+dy = 0$ by e^x, we get the exact equation $(e^x y + x e^x)\,dx + e^x\,dy = 0$. Solving it, we have $F(x,y) = xe^x - e^x + e^x y$.

(B) Since $\frac{N_x - M_y}{M} = \frac{0-1}{y+x}$ is not a function of y only, then there is no integrating factor $\mu(y)$.

20. Solve $y(\cos x + \sin^2 x)\,dx + \sin x\,dy = 0$ in two ways, first as a separable equation and then as an exact equation.

Solution. The equation can be written as

$$\frac{dy}{y} + \left[\frac{\cos x + \sin^2 x}{\sin x}\right]dx = \left[\frac{\cos x}{\sin x} + \sin x\right]dx,$$

and integrating it yields

$$\ln|y| + \ln|\sin x| - \cos x = c,$$

or

$$y\sin x e^{-\cos x} = c.$$

To solve it as an exact equation, let us try to find an integrating factor that is a function if x only. We have $\frac{M_y - N_x}{N} = \sin x$. Hence $e^{-\cos x}$ is an integrating factor. Multiplying the given equation by $e^{-\cos x}$, we obtain the exact equation

$$(y\cos x\, e^{-\cos x} + y\sin^2 x\, e^{-\cos x})\,dx + e^{-\cos x}\sin x\,dy = 0.$$

Now if we use Method 1 and let $y = 0$ in the first part, we have only one integral to deal with. So, we have

$$F(x,y) = \int_0^x e^{-\cos x}\sin x\,dy = ye^{-\cos x}\sin x = c,$$

the same as in part A.

22. Solve $(y + \frac{1}{2}xy^2 + x^2y) dx + (x + y) dy = 0$.

Solution. It is not exact. Since $\frac{M_y - N_x}{N} = \frac{1+xy+x^2-1}{x+y} = \frac{xy+x^2}{x+y} = x$, then an integrating factor is $\mu = e^{\frac{1}{2}x^2}$. Solving the exact equation

$$e^{\frac{1}{2}x^2}\left(y + \frac{1}{2}xy^2 + x^2y\right) dx + e^{\frac{1}{2}x^2}(x + y) dy = 0$$

we find $e^{\frac{1}{2}x^2}(xy + \frac{1}{2}y^2) = c$.

24. Solve $(3y + x) dx + x dy = 0$.

Solution. It is not exact. Since $\frac{M_y - N_x}{N} = \frac{3-1}{x} = \frac{2}{x}$, then an integrating factor is $\mu(x) = e^{\int \frac{2}{x}} = e^{\ln x^2} = x^2$. Solving the exact equation $(3yx^2 + x^3) dx + x^3 dy = 0$ we find $\int_0^x x^3 dx + \int_0^y x^3 dy = \frac{1}{4}x^4 + x^3y = c$.

26. Does the equation $(x - 2y) dx + (xy + 1) dy = 0$ have an integrating factor that is a function of y only? Explain.

Solution. No, because $\frac{N_x - M_y}{M} = \frac{y+2}{x-2y}$ is not a function of y alone.

28. Consider the exact equation $(xy^2 - 1) dx + (x^2y + 1) dy = 0$. Solve it by taking the lower limits as $(2, 1)$, i. e. $x_0 = 2$ and $y_0 = 1$. Then find the particular solution satisfying the initial condition $y(1) = -1$.

Solution. $F(x, y) = \int_2^x (x - 1) dx + \int_1^y (x^2y + 1) dy = \frac{1}{2}x^2y^2 + y - x - 1$. Thus the general solution is described by $\frac{1}{2}x^2y^2 + y - x = c$. To find the particular solution, we substitute the initial values $x_0 = 1, y_0 = -1$, yielding $c = -\frac{3}{2}$ and the particular solution $\frac{1}{2}x^2y^2 - x + y = -\frac{3}{2}$.

Solutions to Chapter 6

Solutions to Section 6.2.2

2. Show that $x_1 = 2 \sin t$ and $x_2 = \sin 2t$ are linearly independent.

Solution. Suppose that $c_1(2 \sin t) + c_2(\sin 2t) = 0$. Then $= 2c_1 \sin t + 2c_2 \sin t \cos t = 0$. If we let $t = \frac{\pi}{2}$, we obtain $2c_1 = 0$ and hence $c_1 = 0$. If we let $t = \frac{\pi}{4}$, we obtain $2c_2 \frac{\sqrt{2}}{2} \frac{\sqrt{2}}{2} = 0$ and hence $c_2 = 0$.

4. Show that $x_1 = \sin^2 t$ and $x_2 = \cos^2 t$ are linearly independent.

Solution. Suppose that $c_1 \sin^2 t + c_2 \cos^2 t = 0$ for all t. Then we have

$$c_1 \sin^2 t + c_2(1 - \sin^2 t) = 0 \Rightarrow (c_1 - c_2) \sin^2 t + c_2 = 0.$$

If $c_1 - c_2 = 0$, then $c_2 = 0$ and we have $c_1 \sin^2 t = 0$ which implies that $c_1 = 0$. If $c_1 - c_2 \neq 0$, then $\sin^2 t = \frac{-c_2}{c_1-c_2}$, which is absurd since the left side is a variable and the right side is a constant.

6. Let $y(t)$ be any function, $y(t) \neq 0$ for $t \in \mathbb{R}$. Show that if $x_1(t)$ and $x_2(t)$ are linearly independent, then so are $y(t)x_1(t)$ and $y(t)x_2(t)$.

Solution. $c_1 y(t)x_1(t) + c_2 y(t)x_2(t) = 0 \Rightarrow y(t)(c_1 x_1(t) + c_2 x_2(t)) = 0$. Since $y(t) \neq 0$, we must have $c_1 x_1(t) + c_2 x_2 = 0$. Since x_1 and x_2 are linearly independent, we must have $c_1 = c_2 = 0$.

8. Show that if $p \neq q$ then $u_1 = t^p$ and $u_2 = t^q$ cannot be solutions of the same second order equation $L[x] = 0$.

    Solution. Evaluate the Wronskian $W(u_1, u_2; t) = \begin{vmatrix} u_1 & u_2 \\ u_1' & u_2' \end{vmatrix} = \begin{vmatrix} t^p & t^q \\ pt^{p-1} & qt^{q-1} \end{vmatrix} = qt^{p+q-1} - pt^{p+q-1} = (q-p)t^{p+q-1}$. We find $W(u_1, u_2)(0) = 0$ whereas $W(u_1, u_2)(1) = q - p \neq 0$. Thus they cannot be solutions of any second order linear equation since Abel's Theorem implies that the Wronskian of two solutions is either always 0 or never 0.

10. Solve $W(t+1, x(t)) = t + 1, x(0) = 1$.

    Solution. The problem asks to solve the initial value problem

    $$(t+1)x' - x = t + 1, \quad x(0) = 1.$$

    The solution is $x(t) = (t+1)[\ln|t+1| + 1]$.

12. Let u_1, u_2 be solutions of $L[x] = 0$ on I such that $u_1(t_0) = u_2(t_0) = 0$ for some $t_0 \in I$. Show that they are linearly dependent and hence all their zeros are in common.

    Solution. One has $W(u_1, u_2)(t_0) = u_1(t_0)u_2'(t_0) - u_1'(t_0)u_2(t_0) = 0$. Therefore, $W(u_1, u_2) \equiv 0$ and hence u_1 and u_2 are linearly dependent. Thus $u_1 = \lambda u_2$ and all their zeros are in common.

14. Solve $W(t, x) = 1, x(1) = 2$.

    Solution. Solving $tx' - x = 1, x(1) = 2$, yields $x = 3t - 1$.

16. Let x_1, x_2 be solutions of $x'' - x' + q(t)x = 0$, p, q continuous. If $W(x_1, x_2)(2) = 5$, find $W(x_1, x_2)(3)$.

    Solution. $W(x_1, x_2)(t) = ce^t$ and since $W(x_1, x_2)(2) = 5$, then $ce^2 = 5$ and hence $c = 5e^{-2}$. So, $W(x_1, x_2)(t) = 5e^{t-2}$ yields $W(x_1, x_2)(3) = 5e$.

Solutions to Section 6.3.4

2. Find the general solution of $x'' + 2x' + 10x = 0$.

    Solution. $C(\lambda) = \lambda^2 + 2\lambda + 10 = 0 \Rightarrow \lambda = -1 \pm 3i$. Thus $x = e^{-t}(c_1 \sin 3t + c_2 \cos 3t)$.

4. Solve $x'' + 3x' = 0, x(0) = 1, x'(0) = -1$.

    Solution. $C(\lambda) = \lambda^2 + 3\lambda = 0 \Rightarrow \lambda = 0, \lambda = -3$. Thus the general solution is $x = c_1 + c_2 e^{-3t}$. To find the solution to the initial value problem, we set $x(0) = c_1 + c_2 = 1$ and $x'(0) = -3c_2 = -1$. Thus $c_2 = \frac{1}{3}$ and $c_1 = 1 - \frac{1}{3} = \frac{2}{3}$ obtaining $x(t) = \frac{2}{3} + \frac{1}{3}e^{-3t}$.

6. Find the general solution of $9x'' + 6x' + x = 0$.

    Solution. $C(\lambda) = 9\lambda^2 + 6\lambda + 1 = 0 \Rightarrow \lambda = -\frac{1}{3}$ (double). Thus $x = e^{-\frac{1}{3}t}(c_1 + c_2 t)$.

8. Show that the nonconstant solutions of $x'' - kx' = 0, x(0) = 0, x'(0) = k$ are increasing or decreasing.

    Solution. The general solution is $x = c_1 + c_2 e^{kt}$. The initial conditions yield $c_1 + c_2 = 0$, $kc_2 = k$ whereby $x = -1 + e^{kt}$. Thus $x(t)$ is increasing if $k > 0$ and decreasing if $k < 0$.

10. Find $k \in \mathbb{R}$ such that all solutions of $x'' + 2kx' + 2k^2 x = 0$ tend to 0 as $t \to +\infty$.

Solution. $C(\lambda) = \lambda^2 + 2k\lambda + 2k^2 = 0 \Rightarrow \lambda = -k \pm ik$. Then the general solution is $x = e^{-kt}(c_1 \sin kt + c_2 \cos kt)$. If $k > 0$ we find $\lim_{t \to +\infty} x(t) = 0$ for any c_1, c_2. If $k \le 0$ the nontrivial solutions do not tend to 0 as $\to +\infty$.

12. Find $x(t)$ such that $x'' - 2x' = 0$, $\lim_{t \to -\infty} x(t) = 1$ and $x'(2) = 1$.

Solution. The general solution of the equation is $x = c_1 + c_2 e^{2t}$. Moreover, $\lim_{t \to -\infty} x(t) = 1$ implies $c_1 = 1$ whereby $x(t) = 1 + c_2 e^{2t}$. The condition $x'(2) = 1$ yields $2c_2 e^4 = 1 \Rightarrow c_2 = \frac{1}{2} e^{-4}$. Thus $x(t) = 1 + \frac{1}{2} e^{-4} e^{2t} = 1 + \frac{1}{2} e^{2t-4}$.

14. Show that all the solutions x of $x'' + 2x' + x = 0$ are such that $\lim_{t \to +\infty} x(t) = 0$.

Solution. Since $C(\lambda) = \lambda^2 + 2\lambda + 1 = 0$ has the double root $\lambda = -1$, the general solution of the equation is $x = e^{-t}(c_1 + c_2 t)$. It is clear that $\lim_{t \to +\infty} x(t) = 0 \ \forall c_1, c_2 \in \mathbb{R}$.

16. Solve the bvp $x'' - 6x' + 9x = 0$, $x(-1) = 1$, $x(1) = 0$.

Solution. Since $C(\lambda) = \lambda^2 - 6\lambda + 9 = 0$ has the double root $\lambda = 3$, the general solution of the equation is $x = e^{3t}(c_1 + c_2 t)$. The boundary conditions $x(-1) = 1$, $x(1) = 0$ yield the system

$$\begin{cases} e^{-3}(c_1 - c_2) = 1, \\ e^{3}(c_1 + c_2) = 0. \end{cases}$$

Solving for c_1, c_2 we find $c_2 = -c_1$ and $2c_1 = e^3$ whereby $c_1 = \frac{1}{2} e^3$, $c_2 = -\frac{1}{2} e^3$. Therefore the solution is $x = \frac{1}{2} e^3 e^{3t}(1 - t) = \frac{1}{2} e^{3t+3}(1 - t)$.

18. Let $a \ne 0$. Show that the bvp $x'' + 3x' - 4x = 0$, $x(-a) = x(a) = 0$ has only the trivial solution.

Solution. Since the roots of $C(\lambda) = \lambda^2 + 3\lambda - 4 = 0$ are $\lambda = 1, -4$, the general solution of the equation is $x = (c_1 e^t + c_2 e^{-4t})$. The boundary conditions $x(-a) = 0$, $x(a) = 0$ yield the system

$$\begin{cases} c_1 e^{-a} + c_2 e^{4a} = 0, \\ c_1 e^{a} + c_2 e^{-4a} = 0. \end{cases}$$

The determinant of this system is given by

$$\begin{vmatrix} e^{-a} & e^{4a} \\ e^{a} & e^{-4a} \end{vmatrix} = e^{-5a} - e^{5a} \ne 0, \quad \forall a \ne 0.$$

Thus, by the Cramer rule, we find $c_1 = c_2 = 0$, which gives rise to the trivial solution.

20. Solve $x'' - x = t$.

Solution. The general solution of the homogeneous equation is $x = c_1 e^t + c_2 e^{-t}$. We can see by inspection that a particular solution of the nonhomogeneous equation is given by $y = -t$; otherwise we may use Variation of Parameters or the method of Undetermined Coefficients. Thus, the general solution is $x = c_1 e^t + c_2 e^{-t} - t$.

22. Solve $x'' - 4x' + 3x = \sin t$.

    Solution. The solution is given by $x = c_1 e^t + c_2 e^{3t} + y$, where $y'' - 4y' + 3y = \sin t$.
    Taking $y = A \sin t + B \cos t$ we find $y' = A \cos t - B \sin t, y'' = -A \sin t - B \cos t$. Then
    $-A \sin t - B \cos t - 4(A \cos t - B \sin t) + 3(A \sin t + B \cos t) = \sin t \Rightarrow (2A + 4B) \sin t +$
    $(2B - 4A) \cos t = \sin t \Rightarrow 2A + 4B = 1, -4A + 2B = 0 \Rightarrow A = \frac{1}{10}, B = \frac{1}{5}$. Thus
    $y = \frac{1}{10} \sin t + \frac{1}{5} \cos t$ and $x = c_1 e^t + c_2 e^{3t} + \frac{1}{10} \sin t + \frac{1}{5} \cos t$.

24. Solve $x'' + 4x = \sin \omega t$.

    Solution. The solution is given by $x = c_1 \sin 2t + c_2 \cos 2t + y$, where $y'' + 4y = \sin \omega t$.
    We distinguish 2 cases:

    (1) $\omega \neq 2$ (nonresonant case). Taking $y = A \sin \omega t + B \cos \omega t$, we find $y'' = -A\omega^2 \sin \omega t - B\omega^2 \cos \omega t$. Thus $y'' + 4y = -A\omega^2 \sin \omega t - B\omega^2 \cos \omega t + 4(A \sin \omega t + B \cos \omega t) = (-A\omega^2 + 4A) \sin \omega t + (-B\omega^2 + 4B) \cos \omega t$. Then $y'' + 4y = \sin \omega t$ implies

    $$4A - A\omega^2 = 1, \quad 4B - B\omega^2 = 0 \Rightarrow A = \frac{1}{4 - \omega^2}, \quad B = 0.$$

    In conclusion, $x = c_1 \sin 2t + c_2 \cos 2t + \frac{1}{4-\omega^2} \sin \omega t$.

    (2) $\omega = 2$ (resonant case). Taking $y = At \sin 2t + Bt \cos 2t$, we find $y' = A \sin 2t + B \cos 2t + 2At \cos 2t - 2Bt \sin 2t$ whereby

    $$y'' = 2A \cos 2t - 2B \sin 2t + 2A \cos 2t - 2B \sin 2t - 4At \sin 2t - 4Bt \cos 2t$$
    $$= (-4B - 4At) \sin 2t + (4A - 4Bt) \cos 2t$$

    Then

    $$y'' + 4y = (-4B - 4At) \sin 2t + (4A - 4Bt) \cos 2t + 4(At \sin 2t + Bt \cos 2t)$$
    $$= -4B \sin 2t + 4A \cos 2t$$

    and $y'' + 4y = \sin 2t$ yields $A = 0, B = -\frac{1}{4}$. Hence $y = -\frac{1}{4}t \cos 2t$ and

    $$x = c_1 \sin 2t + c_2 \cos 2t - \frac{1}{4}t \cos 2t.$$

26. Find the general solution of $x'' - x = t^3 + 2t$.

    Solution. We solve this problem in two parts. First we find the solution x_1 of $x'' - x = 2t$ and then the solution x_2 of $x'' - x = t^3$. We may use the method of Undetermined Coefficients or simply observe that $x_1 = -2t$ solves $x'' - x = 2t$. Then using the method of Undetermined Coefficients or Variation of parameters, we find that $x_2 = -t^3 - 6t$ solves $x'' - x = t^3$. Consequently, $x_1 + x_2 = -t^3 - 8t$ solves $x'' - x = t^3 + 2t$. Since the general solution of $x'' - x = 0$ is given by $y(t) = c_1 e^t + c_2 e^{-t}$, then the general solution of the given nonhomogeneous equation is $-t^3 - 8t + c_1 e^t + c_2 e^{-t}$.

Solutions to Section 6.4.1

2. Solve $t^2 x'' - tx' + x = 0$.

 Solution. Substituting $t = e^s$, we obtain $\ddot{x} - 2\dot{x} + x = 0$. Since $\lambda = 1$ is a repeated root of the characteristic equation, the general solution is given by $x(s) = c_1 e^s + c_2 s e^s$, and $x(t) = c_1 t + c_2 t \ln t$.

4. Solve $t^2 x'' - 5tx' + 5x = 0$.

 Solution. Substitution of $t = e^s$ changes the equation to $\ddot{x} - 6\dot{x} + 5x = 0$. Roots of he characteristic equation are $\lambda_1 = 1$ and $\lambda_2 = 5$. Thus $x(s) = c_1 e^s + c_2 e^{5s}$ and $x(t) = c_1 t + c_2 t^5$.

6. Solve

$$t^2 x'' + tx' + x = 0, \quad x(1) = 1, \quad x'(1) = -1.$$

 Solution. Letting $t = e^s$, we obtain $\ddot{x} + x = 0$. The general solution is $x(s) = c_1 \sin s + c_2 \cos s$ and $x(t) = c_1 \sin(\ln t) + c_2 \cos(\ln t)$. Now, $x(1) = 1$ implies $c_1 \sin 0 + c_2 \cos(0) = 1$ which yields $c_2 = 1$. Calculating x', we have $x'(t) = \frac{c_1}{t} \cos(\ln t) - \frac{1}{t} \sin(\ln t)$. Therefore, $x'(1) = -1$ implies $c_1 = -1$ and the desired solution is

$$x(t) = -\sin(\ln t) + \cos(\ln t).$$

Solutions to Chapter 7

2. Show that if $u_1, u_2, \ldots, u_k$, $2 \le k < n$, are linearly dependent, then any larger set $u_1, u_2, \ldots, u_k, u_{k+1}, \ldots, u_n$ is linearly dependent.

 Solution. By assumption, there exist constants $c_1, c_2, \ldots, c_k$, not all zero, such that $c_1 u_1 + c_2 u_2 + \cdots + c_k u_k \equiv 0$. We can now let $c_{k+1} = \cdots = c_n = 0$ and obtain $c_1 u_1 + \cdots + c_n u_n \equiv 0$, where not all the c_i's are zero.

4. Solve $x''' - 4x' = 0$.

 Solution. The roots of $\lambda^3 - 4\lambda = 0$ are $\lambda = 0, \pm 2$ and hence $x = c_1 + c_2 e^{2t} + c_3 e^{-2t}$.

6. Solve $x''' - 2x'' + x' - 2x = 0$.

 Solution. By observation we see that $\lambda = 2$ is a root of the characteristic equation. Hence $\lambda^3 - 2\lambda^2 + \lambda - 2 = (\lambda - 2)(\lambda^2 + 1)$. Therefore the roots of $\lambda^3 - 2\lambda^2 + \lambda - 2 = 0$ are given by $\lambda = 2$ and $\lambda = \pm i$, yielding $x = c_1 e^{2t} + c_2 \sin t + c_3 \cos t$.

8. Solve $x'''' - 3x'' + 2x = 0$.

 Solution. The roots of the biquadratic equation $\lambda^4 - 3\lambda^2 + 2 = 0$ are $\lambda = \pm 1$ and $\lambda = \pm \sqrt{2}$. Thus $x = c_1 e^t + c_2 e^{-t} + c_3 e^{\sqrt{2}t} + c_4 e^{-\sqrt{2}t}$.

10. Solve $x'''' - 4x''' = 0$.

 Solution. The roots of $\lambda^4 - 4\lambda^3 = 0$ are $\lambda = 0$ (triple) and $\lambda = 4$. Thus $x = c_1 + c_2 t + c_3 t^2 + c_4 e^{4t}$.

12. Solve $x''' - x = e^t$.

Solution. Since $\lambda^3 - 1 = (\lambda - 1)(\lambda^2 + \lambda + 1)$, then the roots of $\lambda^3 - 1 = 0$ are $\lambda = 1$ and $\lambda = \frac{-1 \pm i\sqrt{3}}{2}$. Then the general solution of the homogeneous equation is $x = c_1 e^t + e^{-\frac{1}{2}t}(c_2 \sin \frac{\sqrt{3}}{2}t + c_3 \cos \frac{\sqrt{3}}{2}t) + v$. Since we are in the resonant case, we look for $v = Ate^t$. We find $v''' - v = 3Ae^t + Ate^t - Ate^t = e^t \Rightarrow A = \frac{1}{3}$. Then $x = c_1 e^t + e^{-\frac{1}{2}t}(c_2 \sin \frac{\sqrt{3}}{2}t + c_3 \cos \frac{\sqrt{3}}{2}t) + \frac{1}{3}te^t$.

14. Solve the ivp $x''' + x'' = 0$, $x(0) = 1$, $x'(0) = 1$, $x''(0) = 0$.

Solution. The roots of $\lambda^3 + \lambda^2 = 0$ are $\lambda = 0$ (double) and $\lambda = -1$. Then the general solution is $x = c_1 + c_2 t + c_3 e^{-t}$. The initial conditions yield the system

$$\begin{cases} c_1 + c_3 = 1 \\ c_2 - 3c_3 = 1 \\ c_3 = 0 \end{cases} \Rightarrow c_1 = c_2 = 1, c_3 = 0$$

Thus $x = 1 + t$.

16. Prove that the solutions of a third order equation cannot be all oscillatory.

Solution. The characteristic equation is a third order polynomial and hence it has at least one real root λ. The corresponding solution $x = c_1 e^{\lambda t}$ is not oscillatory.

18. Show that for every $b \ne 0$ there is a one-parameter family of nontrivial solutions of $x'''' - b^2 x'' = 0$ which tend to zero at $t \to +\infty$.

Solution. The characteristic equation is $\lambda^4 - b^2\lambda^2 = 0$. Solving it, we find $\lambda = 0$ (double) and $\lambda = \pm b$. Thus $x = c_1 + c_2 t + c_2 e^{bt} + c_4 e^{-bt}$. If $b > 0$ we take $c_1 = c_2 = c_3 = 0$, yielding $x = c_4 e^{-bt}$ which $\to 0$ as $t \to +\infty$ for any $c_4 \ne 0$. If $b < 0$ we take $c_1 = c_2 = c_4 = 0$ yielding $x = c_3 e^{bt}$ which $\to 0$ as $t \to +\infty$ for any $c_3 \ne 0$.

20. Solve $x''' - x'' = t$ by the method of Variation of Parameters.

Solution. Solving

$$\begin{cases} v_1' + v_2' t + v_3' e^t = 0, \\ v_2' + v_3' e^t = 0, \\ v_3' e^t = t, \end{cases} \tag{1}$$

we obtain $v_1 = \frac{1}{3}t^3 - \frac{1}{2}t^2$, $v_2 = -\frac{1}{2}t^2$, $v_3 = -te^{-t} - e^{-t}$. The general solution is $x(t) = c_1 + c_2 t + c_3 e^t + \frac{-1}{6}t^3 - \frac{1}{2}t^2$.

22. Show that $t^3 x''' + 3t^2 x'' + tx' + x = 0$, $x(1) = x'(1) = x''(1) = 1$ has exactly one one-parametric solution that tends to 0, as $t \to \infty$, and two oscillating ones.

Solution. Transforming the equation via $t = e^s$, we obtain

$$\dddot{x} - 3\ddot{x} + 2\dot{x} + 3(\ddot{x} - \dot{x}) + \dot{x} + x = \dddot{x} + x = 0.$$

The characteristic equation $\lambda^3 + 1 = (\lambda + 1)(\lambda^2 - \lambda + 1) = 0$ has roots $\lambda = -1, \lambda = \frac{1}{2} \pm \frac{\sqrt{3}}{2}i$ and the general solution is $x(s) = c_1 e^{-s} + c_2 e^{\frac{1}{2}s} \sin \frac{\sqrt{3}}{2}s + c_3 e^{\frac{1}{2}s} \cos \frac{\sqrt{3}}{2}s$, or

$$x(t) = c_1 t^{-1} + c_2 t^{\frac{1}{2}} \sin\left(\frac{\sqrt{3}}{2}\ln t\right) + c_3 t^{\frac{1}{2}} \cos\left(\frac{\sqrt{3}}{2}\ln t\right).$$

The assertion of the problem is now easy to verify.

24. Solve $t^3x''' + 2t^2x'' - 4tx' + 4x = 0$, $x(1) = x'(1) = 1$, $x''(1) = 0$.

Solution. Transforming via $t = e^s$, we obtain

$$\dddot{x} - \ddot{x} - 4\dot{x} + 4x = 0.$$

The characteristic equation is $\lambda^3 - \lambda^2 - 4\lambda + 4 = \lambda^2(\lambda - 1) - 4(\lambda - 1) = (\lambda - 1)(\lambda^2 - 4) = (\lambda - 1)(\lambda + 2)(\lambda - 2) = 0$, yielding the roots $\lambda_1 = 1$, $\lambda_2 = 2$, $\lambda_3 = -2$ and the general solution $x(s) = c_1e^s + c_2e^{2s} + c_3e^{-2s}$, or

$$x(t) = c_1t + c_2t^2 + c_3t^{-2}.$$

Now, substituting the initial values, we obtain the required solution

$$x(t) = t$$

26. Solve $x''' - x'' + x' - x = 0$.

Solution. The characteristic equation is $C(\lambda) = \lambda^3 - \lambda^2 + \lambda - 1 = (\lambda_1 - 1)(\lambda^2 + 1) = 0$, whose solutions are $\lambda_1 = 1$, $\lambda_2 = i$, $\lambda_3 = -i$. Thus the general solution is

$$x = c_1e^t + c_2\cos t + c_3\sin t.$$

28. Evaluate the Wronskian $W(2, \sin^2 t, t^7e^t, e^{t^2}, \cos^2 t, t^3, t^5 - 1)$.

Solution. The subset $\{2, \sin^2 t, \cos^2 t\}$ consists of linearly dependent functions. Therefore 2, $\sin^2 t$, t^7e^t, e^{t^2}, $\cos^2 t$, t^3, $t^5 - 1$ are linearly dependent and hence their Wronskian is identically zero.

Solutions to Chapter 8

2. Use the eigenvalue method to find the general solution of

$$\begin{cases} x' = 4x + 6y, \\ y' = x + 3y. \end{cases}$$

Solution. The system is equivalent to

$$\bar{x}' = \begin{pmatrix} 4 & 6 \\ 1 & 3 \end{pmatrix}\bar{x}$$

where

$$\bar{x} = \begin{pmatrix} x \\ y \end{pmatrix}.$$

The eigenvalues are $\lambda = 1$, $\lambda = 6$. Corresponding to $\lambda = 1$, $(A - \lambda I)\bar{v}_1 = 0$ yields two equations, $3\bar{v}_1 + 6\bar{v}_2 = 0$ and $\bar{v}_1 + 2\bar{v}_2 = 0$, which are the same equation. So let $v_2 = 1$. Then $v_1 = -2$ implies

$$\bar{x}_1 = \left(\begin{array}{c} -2e^t \\ e^t \end{array} \right).$$

Similarly, corresponding to $\lambda = 6$, we obtain

$$\bar{x}_2 = \left(\begin{array}{c} 3e^{6t} \\ e^{6t} \end{array} \right).$$

$W(\bar{x}_1, \bar{x}_2)(0) = -5$ shows that they form a fundamental set of solutions, and the general solution is given by $\bar{x}(t) = c_1\bar{x}_1 + c_2\bar{x}_2$.

4. Solve

$$\bar{x}' = \left(\begin{array}{cc} 1 & 1 \\ 2 & 2 \end{array} \right)\bar{x}.$$

Solution. The eigenvalues are $\lambda = 0$, $\lambda = 3$ with corresponding eigenvectors

$$\bar{v}_1 = \left(\begin{array}{c} 1 \\ -1 \end{array} \right), \quad \bar{v}_2 = \left(\begin{array}{c} 1 \\ 2 \end{array} \right).$$

The general solution is

$$\bar{x} = c_1\bar{x}_1 + c_2\bar{x}_2 = c_1\left(\begin{array}{c} 1 \\ -1 \end{array} \right) + c_2\left(\begin{array}{c} e^{3t} \\ 2e^{3t} \end{array} \right) = \left(\begin{array}{c} c_1 + c_2 e^{3t} \\ -c_1 + 2c_2 e^{3t} \end{array} \right).$$

6. Solve problem 5 by converting the system to a linear scalar equation suggested by Theorem 8.5.

Solution. We have to solve the system of two equations $x' = 3x + y$ and $y' = -x + y$. The corresponding linear equation from Theorem 8.2 is $x'' - 4x + 4 = 0$ and $\lambda = 2$ is repeated root of the corresponding characteristic equation. Hence two linearly independent solutions are $x_1 = e^{2t}$, $x_2 = te^{2t}$. Corresponding to $x_1 = e^{2t}$, we solve $x' = 3x + y$ for y and obtain $y_1 = 2e^{2t} - 3e^{2t} = -e^{2t}$, yielding

$$\bar{x}_1 = \left(\begin{array}{c} e^{2t} \\ -e^{2t} \end{array} \right).$$

Similarly, corresponding to $x_2 = te^{2t}$, we obtain $y_2 = e^{2t} + 2te^{2t} - 3te^{2t} = e^{2t} - te^{2t}$ and hence

$$\bar{x}_2 = \left(\begin{array}{c} te^{2t} \\ e^{2t} - te^{2t} \end{array} \right).$$

The general solution is

$$\bar{x} = \begin{pmatrix} c_1 e^{2t} + c_2 t e^{2t} \\ -c_1 e^{2t} + c_2(e^{2t} - t e^{2t}) \end{pmatrix}$$

yielding

$$\bar{x}_p = \begin{pmatrix} e^{2t} - t e^{2t} \\ -2e^{2t} + t e^{2t} \end{pmatrix}.$$

8. Find the inverse of

$$\begin{pmatrix} a & 0 & 0 \\ 0 & a & 0 \\ 0 & 0 & a \end{pmatrix}.$$

Solution.

$$A^{-1} = \begin{pmatrix} \frac{1}{a} & 0 & 0 \\ 0 & \frac{1}{a} & 0 \\ 0 & 0 & \frac{1}{a} \end{pmatrix}.$$

10. Show that any system

$$\begin{cases} x_1' = ax_1 + bx_2, \\ x_2' = cx_1 + dx_2. \end{cases}$$

has a nontrivial constant solution, $x_1 = k_1, x_2 = k_2$, if and only if the determinant

$$\begin{vmatrix} a & b \\ c & d \end{vmatrix} = 0.$$

Solution. By Cramer's rule, the system

$$\begin{cases} ak_1 + bk_2 = 0, \\ ck_1 + dk_2 = 0 \end{cases}$$

has a nontrivial solution if and only if $\begin{vmatrix} a & b \\ c & d \end{vmatrix} = 0$.

12. First explain why the equation below has a nontrivial constant solution and then solve it.

$$\bar{x}' = \begin{pmatrix} 1 & -1 \\ 2 & -2 \end{pmatrix} \bar{x}.$$

Solution. It has a constant solution because

$$\begin{vmatrix} 1 & -1 \\ 2 & -2 \end{vmatrix} = 0.$$

The eigenvalues are $\lambda = 0, -1$. The corresponding solutions are

$$\bar{x}_1(t) = \begin{pmatrix} 1 \\ 1 \end{pmatrix}, \quad \bar{x}_2(t) = \begin{pmatrix} -e^{-t} \\ -2e^{-t} \end{pmatrix}.$$

14. Use the eigenvalue method to find the solution to the initial value problem

$$\bar{x}' = \begin{pmatrix} 1 & 1 \\ -1 & 1 \end{pmatrix} \bar{x}, \quad \bar{x}(0) = \begin{pmatrix} 1 \\ -1 \end{pmatrix}.$$

Solution. The eigenvalues are $\lambda = 1 \pm i$. To find an eigenvector corresponding to $\lambda = 1 + i$, we set up $(A - \lambda I)\bar{v} = 0$, which yields two equations $-iv_1 + v_2 = 0$ and $-v_1 - iv_2 = 0$. These two equations are essentially the same, so we choose $v_1 = 1$ which gives $v_2 = i$. Hence, using Euler's equation,

$$\bar{x} = e^{(1+i)t}\bar{v} = e^{(1+i)t}\begin{pmatrix} 1 \\ i \end{pmatrix} = \begin{pmatrix} e^t \cos t \\ -e^t \sin t \end{pmatrix} + i \begin{pmatrix} e^t \sin t \\ e^t \cos t \end{pmatrix}$$

yielding two real solutions

$$\bar{x}_1 = \begin{pmatrix} e^t \cos t \\ -e^t \sin t \end{pmatrix}, \quad \bar{x}_2 = \begin{pmatrix} e^t \sin t \\ e^t \cos t \end{pmatrix}.$$

The general solution is

$$\bar{x} = \begin{pmatrix} c_1 e^t \cos t + c_2 e^t \sin t \\ -c_1 e^t \sin t + c_2 e^t \cos t \end{pmatrix}.$$

Substituting $t = 0$ and solving for c_1, c_2, we find the required solution $\bar{x}_p$ as

$$\bar{x}_p = \begin{pmatrix} e^t \cos t - e^t \sin t \\ -e^t \sin t - e^t \cos t \end{pmatrix}.$$

16. Show that

$$\bar{x}' = \begin{pmatrix} a & b \\ c & d \end{pmatrix} \bar{x}.$$

cannot have periodic solutions if b and c are either both positive or both negative.

Solution. After converting the system to the scalar equation

$$x'' - (a + d)x' + (ad - bc) = 0,$$

we see that the discriminant of the characteristic equation is

$$(a + d)^2 - 4(ad - bc) = (a - d)^2 + 4bc$$

and hence if $bc > 0$ then there are two distinct real roots.

18. Solve

$$\bar{x}' = \begin{pmatrix} 1 & -1 & 1 \\ 1 & 2 & -2 \\ 2 & 1 & -1 \end{pmatrix} \bar{x}.$$

Solution. Eigenvalues are: $0, 0, 2$. One solution corresponding to 0 is

$$\bar{x}_1 = \begin{pmatrix} 0 \\ 1 \\ 1 \end{pmatrix}.$$

A second solution can be found as $\bar{v}_2 + t\bar{v}_1 =$, where $\bar{v}_2$ is the extending eigenvector, yielding

$$\bar{x}_2 = \begin{pmatrix} \frac{1}{3} \\ t \\ -\frac{1}{3} + t \end{pmatrix}.$$

The solution corresponding to $\lambda = 2$ is

$$\bar{x}_3 = \begin{pmatrix} 2e^{2t} \\ -e^{2t} \\ e^{2t} \end{pmatrix}.$$

The general solution is

$$\bar{x}(t) = c_1 \begin{pmatrix} 0 \\ 1 \\ 1 \end{pmatrix} + c_2 \begin{pmatrix} \frac{1}{3} \\ t \\ -\frac{1}{3} + t \end{pmatrix} + c_3 \begin{pmatrix} 2e^{2t} \\ -e^{2t} \\ e^{2t} \end{pmatrix}.$$

20. Find a fundamental set of solutions for

$$\bar{x}' = \begin{pmatrix} 3 & 4 & 1 \\ 1 & 3 & -2 \\ 0 & 0 & 3 \end{pmatrix} \bar{x}.$$

Solution. There are 3 distinct eigenvalues, $\lambda = 1, 3, 5$, which yield 3 distinct linearly independent solutions

$$\bar{x}_1 = \begin{pmatrix} 2e^t \\ -e^t \\ 0 \end{pmatrix}, \quad \bar{x}_2 = \begin{pmatrix} 8e^{3t} \\ -e^{3t} \\ 4e^{3t} \end{pmatrix}, \quad \bar{x}_3 = \begin{pmatrix} 2e^{5t} \\ e^{5t} \\ 0 \end{pmatrix}.$$

Solutions to Chapter 9

2. Using the phase plane analysis, show that the solution of $x'' - 4x^3 = 0$, $x(0) = 0$, $x'(0) = -1$ is increasing, convex for $t < 0$ and concave for $t > 0$. Sketch a qualitative graph of the solution.
   Solution. The energy is $E(x,y) = \frac{1}{2}y^2 - x^4$, where $y = x'$. The energy is constant: $\frac{1}{2}y^2 - x^4 = c$. To evaluate c we use the initial conditions $x(0) = 0$, $x'(0) = -1$ yielding $c = 1/2$. Since $E(x(t), y(t)) = \frac{1}{2}$ and $(x(0), y(0)) = (0, -1)$ then $x(t), y(t)$ belongs to the lower branch E_- of $\{E = \frac{1}{2}\}$ (marked in blue in Figure 9.4 below). On E_- we have that $y(t) \leq y(0) = -1$. Thus $x'(t) = y(t) < 0$ for all t, which implies that $x(t)$ is decreasing, whereby $t < 0 \Rightarrow x(t) > 0$ and $t > 0 \Rightarrow x(t) < 0$. Moreover, from the equation we infer that $x'' = 4x^3$ and hence

$$t < 0 \Rightarrow x(t) > 0 \Rightarrow x''(t) > 0 \Rightarrow x(t) \text{ is convex;}$$
$$t > 0 \Rightarrow x(t) < 0 \Rightarrow x''(t) < 0 \Rightarrow x(t) \text{ is concave.}$$

   A sketch of the graph of the solution $x(t)$ is reported in the Figure 9.4 below, right side.

4. Show that the solution of $x'' - x + 4x^3 = 0$, $x(0) = 1$, $x'(0) = 0$, is periodic.
   Solution. Here $V(x) = -\frac{1}{2}x^2 + x^4$, $E = \frac{1}{2}y^2 - \frac{1}{2}x^2 + x^4$, $c = E(1,0) = \frac{1}{2}$ and
   (i) $V(x) \leq \frac{1}{2} \Leftrightarrow -\frac{1}{2}x^2 + x^4 \leq \frac{1}{2} \Leftrightarrow 2x^4 - x^2 - 1 \leq 0 \Leftrightarrow -1 \leq x \leq 1$;
   (ii) $V'(1) = 3$ and $V'(-1) = -3$.
   Therefore $E = \frac{1}{2}$ is a smooth closed bounded curve with no singular point, since the only singularity is $(0, 0)$. Thus $x(t)$ is periodic.

6. Let $x_a(t)$ be the solution of the nonlinear harmonic oscillator $x'' + \omega^2 x - x^3 = 0$, $x_a(0) = 0$, $x_a'(0) = a$. Find $a > 0$ such that $x_a(t)$ is periodic.
   Solution. From the energy relationship $\frac{1}{2}y^2 + \frac{1}{2}\omega^2 x^2 - \frac{1}{4}x^4 = c$ it follows that the energy of x_a is given by $c = \frac{1}{2}[x_a'(0)]^2 = \frac{1}{2}a^2$. It is known (see Subsection 9.3.1.1) that the nonlinear harmonic oscillator has a periodic solution whenever $0 < c < \frac{1}{4}\omega^2$. Thus $x_a(t)$ is periodic whenever $0 < \frac{1}{2}a^2 < \frac{1}{4}\omega^2$, namely $0 < a < |\omega|$.

8. Let $x_a(t), y_a(t)$ be the solution of the Hamiltonian system

$$\begin{cases} x' = ax + y, \\ y' = -x - ay, \end{cases}$$

   such that $x(0) = 0$, $y(0) = 1$. Find a such that $x_a(t), y_a(t)$ is periodic.

Solution. We see that $H(x,y) = \frac{1}{2}x^2 + axy + \frac{1}{2}y^2$. Since $H(0,1) = \frac{1}{2}$ then we find $x^2 + 2axy + y^2 = 1$. The equation $x^2 + 2axy + y^2 = 1$ is equivalent to $(x + ay)^2 + (1-a^2)y^2$. Hence, the conic is an ellipse if $a^2 < 1$. Otherwise, the conic is a hyperbola for $a^2 > 1$ or it is a pair of straight lines if $a^2 = 1$. Thus $x_a(t), y_a(t)$ is periodic whenever $a^2 < 1$.

10. Find k such that the equation $x'' = k^2 x - x^3$ has a homoclinic $x(t)$ to 0 such that $\max_{\mathbb{R}} x(t) = 2$.

   Solution. As in the previous exercise one has that $\max_{\mathbb{R}} x(t)$ is the positive solution x_0 of $E(x,0) = 0$. In this case we have $E(x,0) = -\frac{1}{2}k^2x^2 - \frac{1}{4}x^4 = 0$ and hence $x_0^2 = 2k^2$. Thus, if $x_0 = 2$ we find $4 = 2k^2 \Rightarrow k = \pm\sqrt{2}$.

12. Show that for $k < 0$ the equation $x'' = kx - x^3$ has no homoclinic to 0.

   Solution. The energy $E = \frac{1}{2}y^2 - \frac{1}{2}kx^2 + \frac{1}{4}x^4$ has a local proper maximum at $(0,0)$, provided $k < 0$. Thus in the phase plane $(0,0)$ is surrounded by closed curves $E = \epsilon > 0$ small enough, and there cannot be homoclinics to 0.

14. Show that for $p \geq 1$ the equation $x'' + x^p = 0$ has no heteroclinic.

   Solution. The energy $E = \frac{1}{2}y^2 + \frac{1}{p+1}x^{p+1}$ has a unique singular point $x = y = 0$. Using (‡) at page 158 we can conclude that the given equation has no heteroclinic.

16. Find a such that $x'' + 2x - 2x^3 = 0$, $x(0) = 0$, $x'(0) = a$, is a heteroclinic.

   Solution. $E = \frac{1}{2}y^2 + x^2 - \frac{1}{2}x^4$, $E(0,a) = \frac{1}{2}a^2$. To be a heteroclinic, $E_c = \{E = c\}$ has to contain the singular points $(\pm 1, 0)$. Thus $c = E(\pm 1, 0) = \frac{1}{2}$. It follows that $\frac{1}{2} = \frac{1}{2}a^2 \Rightarrow a = \pm 1$.

18. * Let $T = T(a)$ be the period of the periodic solution of the nonlinear harmonic oscillator $x'' + \omega^2 x - x^3 = 0$ such that $x(0) = 0$, $x'(0) = a > 0$. Show that $\lim_{a\to 0} T(a) = \frac{2\pi}{\omega}$.

   Solution. Since $E(0,a) = \frac{1}{2}a^2$ we infer that $y = \pm\sqrt{a^2 - \omega^2 x^2 + \frac{1}{2}x^4}$. Letting $x_0 > 0$ be such that $E(x_0, 0) = \frac{1}{2}a^2$, we find $\frac{1}{2}a^2 = a^2 - \omega^2 x_0^2 + \frac{1}{2}x_0^4$ which gives $a^2 = \omega^2 x_0^2 - \frac{1}{2}x_0^4$.
   Next, we have

$$T(a) = 4\int_0^{x_0} \frac{dx}{y} = 4\int_0^{x_0} \frac{dx}{\sqrt{a^2 - \omega^2 x^2 + \frac{1}{2}x^4}} = 4\int_0^{x_0} \frac{dx}{\sqrt{\omega^2 x_0^2 - \frac{1}{2}x_0^4 - \omega^2 x^2 + \frac{1}{2}x^4}}.$$

The change of variable $x \to x_0 \xi$ yields

$$T(a) = 4\int_0^1 \frac{x_0\, d\xi}{\sqrt{\omega^2 x_0^2 - \frac{1}{2}x_0^4 - \omega^2 x_0^2\xi^2 + \frac{1}{2}x_0^4\xi^4}}.$$

Canceling x_0 we get

$$T(a) = 4\int_0^1 \frac{d\xi}{\sqrt{\omega^2 - \frac{1}{2}x_0^2 - \omega^2\xi^2 + \frac{1}{2}x_0^2\xi^4}}.$$

As $a \to 0$, we have that $x_0 \to 0$ and hence

$$\lim_{a \to 0} T(a) = 4 \int_0^1 \frac{dx}{\sqrt{\omega^2 - \omega^2 \xi^2}} = \frac{4}{\omega} \int_0^1 \frac{dx}{\sqrt{1 - \xi^2}} = \frac{4}{\omega} \cdot \arcsin 1 = \frac{2\pi}{\omega}.$$

20. * Show that there exists a unique $\lambda > 0$ such that the boundary value problem $x'' + \lambda x^3 = 0$, $x(0) = x(\pi) = 0$, has a positive solution.

Solution. Consider the solution such that $x(0) = 0$, $x'(0) = 1$, whose energy is $E = \frac{1}{2}$. In the phase plane any $E = \frac{1}{2}$ is a closed bounded curve with no singularity and hence it carries a periodic solution with period $T = T(\lambda)$. Finding a positive solution of the given bvp amounts to finding λ such that $T(\lambda) = 2\pi$. We evaluate $T(\lambda)$ as in the previous exercise. From $E(x_0, 0) = E(0, 1)$ we infer that $\frac{1}{4} \lambda x_0^4 = \frac{1}{2} \implies \lambda x_0^4 = 2$. We have

$$T(\lambda) = 4 \int_0^{x_0} \frac{dx}{y} = 4 \int_0^{x_0} \frac{dx}{\sqrt{1 - \frac{1}{2} \lambda x^4}} = 4 \int_0^1 \frac{x_0 \, d\xi}{\sqrt{1 - \frac{1}{2} \lambda x_0^4 \xi^4}}.$$

Since $x_0 = (2/\lambda)^{1/4}$, we find $\frac{1}{2} \lambda x_0^4 = \frac{1}{2} \lambda \cdot \frac{2}{\lambda} = 1$. Then

$$T(\lambda) = 4 \int_0^1 \frac{x_0 \, d\xi}{\sqrt{1 - \frac{1}{2} \lambda x_0^4 \xi^4}} = 4 x_0 \int_0^1 \frac{d\xi}{\sqrt{1 - \xi^4}} = 4 \left(\frac{2}{\lambda} \right)^{1/4} \int_0^1 \frac{d\xi}{\sqrt{1 - \xi^4}}.$$

Setting $A = \int_0^1 \frac{d\xi}{\sqrt{1 - \xi^4}}$, the equation $T(\lambda) = 2\pi$ yields

$$4 \left(\frac{2}{\lambda} \right)^{1/4} \cdot A = 2\pi \implies \left(\frac{2}{\lambda} \right)^{1/4} = \frac{\pi}{2A} \implies \lambda = 2 \cdot \left(\frac{2A}{\pi} \right)^4.$$

Solutions to Chapter 10

2. Establish the stability of the equilibrium of

$$\begin{cases} x' = -y + ax, \\ y' = x + ay, \end{cases}$$

in terms of a.

Solution. The coefficient matrix is

$$\begin{pmatrix} a & -1 \\ 1 & a \end{pmatrix}$$

whose eigenvalues are $\lambda = a \pm i$. Thus:
1. if $a < 0$ then $x = 0$ is an asymptotically stable focus;
2. if $a > 0$ then $x = 0$ is an unstable focus;
3. if $a = 0$ then $x = 0$ is a stable center.

4. Given the 3×3 system

$$\begin{cases} x' = -x + y, \\ y' = -x - y, \\ z' = y - z, \end{cases}$$

show that $(0, 0, 0)$ is a stable equilibrium.

Solution. The matrix of the system is

$$A = \begin{pmatrix} -1 & 1 & 0 \\ -1 & -1 & 0 \\ 0 & 1 & -1 \end{pmatrix}.$$

Its eigenvalues are the solution of

$$\begin{vmatrix} -1 - \lambda & 1 & 0 \\ -1 & -1 - \lambda & 0 \\ 0 & 1 & -1 - \lambda \end{vmatrix} = 0$$

namely, $(-1 - \lambda)[(-1 - \lambda)^2 + 1] = 0 \Rightarrow \lambda = -1$ and $\lambda = -1 \pm i$. Thus we have asymptotic stability.

6. Establish the stability of the equilibrium $(0, 0)$ of the system

$$\begin{cases} x'' = -2x + y, \\ y'' = x. \end{cases}$$

Solution. We can either write the system as a first order system in x, x', y, y' or we can transform the system to the fourth order equation $x'''' = -2x'' + y'' = -2x'' + x \Rightarrow$ $x'''' + 2x'' - x = 0$. The characteristic equation of this latter equation is $\lambda^4 + 2\lambda^2 - 1 = 0$ whose roots are $\pm\sqrt{-1 \pm \sqrt{2}}$. Since $\sqrt{-1 + \sqrt{2}} > 0$, we have instability.

8. Show that the origin is a center for the harmonic oscillator $x'' + \omega^2 x = 0$, $\omega \neq 0$.

Solution. Roots of the characteristic equation are $\pm i\omega$.

10. Using stability by linearization, show that the trivial solution of $x'' + 2x' + x + x^2 = 0$ is stable.

Solution. The linearized equation at $x = 0$ is $y'' + 2y' + y = 0$. In this case, $\lambda = -1$ is a double root and $x = 0$ is stable.

12. Show that $x = 0$ is an unstable solution of $x'''' - 4x'' + x = 0$.

Solution. The roots of the characteristic equation $C(\lambda) = \lambda^4 - 4\lambda^2 + 1$ are

$$\lambda = \pm\sqrt{2 \pm \sqrt{3}}.$$

Since the roots $+\sqrt{2 \pm \sqrt{3}}$ are positive, we have instability.

14. Show that the solution $x = 0$ of $x'' + \omega^2 x \pm x^{2k} = 0$, $\omega \neq 0$, $k \in \mathbb{N}$, is stable provided $k \geq 1$.

    Solution. $x = 0$ is a strict local minimum of the potential $U(x) = \frac{1}{2}\omega^2 x^2 \pm \frac{1}{2k+1}x^{2k+1}$.

16. Study the stability of the equilibria $k\pi$ of the pendulum equation $x'' + \sin x = 0$.

    Solution. The potential is $U(x) = -\cos x$ which has a strict minimum at $2n\pi$, $n \in \mathbb{Z}$. Thus they are stable. To establish the nature of $\xi_n = (2n + 1)\pi$ we consider the linearized problem $y'' + (\cos \xi_n)y = 0$, namely $y'' - y = 0$. Since the roots of the characteristic equation $\lambda^2 - 1 = 0$ are $\lambda = \pm 1$, we infer that $\xi_n = (2n+1)\pi$, $n \in \mathbb{Z}$, is a saddle, namely a hyperbolic equilibrium.

18. Using stability by linearization, show that if $g'(0) = 0$ then $x = 0$ is unstable for $x'' + g(x') - x = 0$.

    Solution. The linearized equation at $x = 0$ is $y'' + g'(0)y' - y = y'' - y = 0$. The roots of the corresponding characteristic equation are $\lambda = \pm 1$. Thus $x = 0$ is unstable.

20. Study the stability of the solution $x = 0$ of $x'' - \mu(1 - x^2)x' - x = 0$.

    Solution. The linearized equation at $x = 0$ is $y'' - \mu y' - y = 0$. For any $\mu \in \mathbb{R}$ the characteristic equation $\lambda^2 - \mu\lambda - 1 = 0$ has a positive root. Hence $x = 0$ is unstable.

Solutions to Chapter 11

2. Find the recursive formula and the first five nonzero terms of the series solution of $x'' + tx' = 0$.

   Solution. Substituting

   $$\sum_0^{+\infty} a_k t^k$$

   yields

   $$\sum_2^{+\infty} k(k-1)a_k t^{k-2} + \sum_1^{+\infty} ka_k t^k$$

   $$= \sum_0^{+\infty} (k+1)(k+2)a_{k+2}t^k + \sum_1^{+\infty} ka_k t^k = 0.$$

   For $k = 0$, we obtain $2a_2 = 0$ and for $k \geq 1$, we have the recursive formula

   $$a_{k+2} = -\frac{ka_k}{(k+2)(k+1)}, \quad k = 1, 2, \ldots$$

   It follows: $a_2 = a_4 = a_6 = \cdots = 0$ whereas $a_3 = -\frac{a_1}{3 \cdot 2}$, $a_5 = -\frac{3a_3}{5 \cdot 4} = \frac{3a_1}{5 \cdot 4 \cdot 3 \cdot 2}$, $a_7 = -\frac{5a_5}{7 \cdot 6} = -\frac{5 \cdot 3a_1}{7 \cdot 6 \cdot 5 \cdot 4 \cdot 3 \cdot 2}$, and so on. Hence

   $$x(t) = a_0 + a_1\left(t - \frac{t^3}{3!} + \frac{3t^5}{5!} - \frac{5 \cdot 3t^7}{7!} + \cdots\right).$$

4. Use the infinite series method to find the function that solves $x'' - 4x = 0$, $x(0) = 1$, $x'(0) = 2$.
Solution. Substituting

$$x(t) = \sum_0^\infty a_n t^n$$

we obtain the recursive formula

$$a_{n+2} = \frac{4a_n}{(n+1)(n+2)}, \quad n \geq 0.$$

We also have $a_0 = 1$, $a_1 = 2$, yielding $a_n = \frac{2^n}{n!}$ and

$$x(t) = \sum_0^\infty \frac{(2t)^n}{n!} = e^{2t}.$$

6. Use the series solution method to find the function that solves the initial value problem $x' - 2tx = 0$, $x(0) = 1$.
Solution. Taking derivative and adjusting indices, we obtain

$$\sum_0^\infty (n+1)c_{n+1}t^n - \sum_1^\infty 2c_{n-1}t^n = 0.$$

For $n = 0$, we obtain $c_1 = 0$. Also, by assumption, $c_0 = 1$. For $n \geq 1$, we have

$$c_{n+1} = \frac{2c_{n-1}}{n+1}.$$

Therefore, For $n \geq 0$, we obtain $c_{2n+1} = 0$ and $c_{2n} = \frac{1}{n!}$ and hence

$$x(t) = 1 + t^2 + \frac{t^4}{2} + \frac{t^6}{3!} + \frac{t^8}{4!} \cdots = \sum_0^\infty \frac{(t^2)^n}{n!} = e^{t^2}.$$

8. Find the general series solution of $x'' - tx' = 0$.
Solution. Taking derivatives of $x_k = \sum_0^\infty a_k t^k$ and adjusting indices, we obtain

$$\sum_0^\infty (k+1)(k+2)a_{k+2}t^k - \sum_1^\infty ka_k t^k = 0,$$

yielding $a_2 = 0$ and

$$a_{k+2} = \frac{ka_k}{(k+2)(k+1)}, \quad k = 1, 2, \ldots$$

Hence

$$x(t) = a_0 + a_1\left(t + \frac{t^3}{3!} + \frac{3t^5}{5!} + \frac{5 \cdot 3t^7}{7!} + \cdots\right)$$

10. Find the infinite series solution for $(1-t)x' - x = 0$, $x(0) = 1$ and identify the function represented by the series.

    Solution. Substituting the series

    $$x(t) = \sum_0^\infty c_n t^n$$

    we obtain

    $$\sum_1^\infty nc_n t^{n-1} - \sum_1^\infty nc_n t^n - \sum_0^\infty c_n t^n = \sum_0^\infty (n+1)c_{n+1} t^n - \sum_1^\infty nc_n t^n - \sum_0^\infty c_n t^n = 0.$$

    For $n = 0$, we have $c_1 - c_0 = 0 \Rightarrow c_1 = c_0 = 1$. For $n \geq 2$, we have the recursive formula

    $$c_{n+1} = \frac{(n+1)c_n}{n+1} = c_n.$$

    Hence

    $$x(t) = \sum_0^\infty t^n = \frac{1}{1-t}.$$

12. Show that $t = 0$ is a strict maximum of J_0.

    Solution. Differentiating $J_0(t) = \sum_0^\infty \frac{(-1)^n}{(n!)^2}(\frac{t}{2})^{2n}$ term by term, it follows that $J'(0) = 0$ and $J_0''(0) = -\frac{1}{2}$.

14. Show that if $a \neq 0$ is such that $J_0(a) = 0$ then $J'(a) \neq 0$.

    Solution. Otherwise $J'(a) = 0$ and hence, from the equation $a^2 J_0''(a) + a J_0'(a) + a^2 J_0(a) = 0$ it follows that $J_0''(a) = 0$. Differentiating $t^2 J_0''(t) + t J_0'(t) + t^2 J_0(t) = 0$ we find

    $$2t J_0''(t) + t^2 J_0'''(t) + J_0'(t) + t J_0''(t) + 2t J_0(t) + t^2 J_0'(t) = 0$$

    whereby $J_0'''(a) = 0$. Repeating the calculation we get $\frac{d^k J_0}{dt^k}(a) = 0$ for any $k \in \mathbb{N}$. Thus $J_0(t) \equiv 0$, a contradiction.

16. Show that $(t J_1(t))' = t J_0(t)$.

    Solution.

    $$t J_1 = \frac{t^2}{2}\sum_0^\infty \frac{(-1)^n}{n! \cdot (n+1)!}\left(\frac{t}{2}\right)^{2n} = 2\left(\frac{t}{2}\right)^2 \sum_0^\infty \frac{(-1)^n}{n! \cdot (n+1)!}\left(\frac{t}{2}\right)^{2n}$$

    $$= 2\sum_0^\infty \frac{(-1)^n}{n! \cdot (n+1)!}\left(\frac{t}{2}\right)^{2n+2}$$

Thus

$$(tJ_1)' = 2\sum_0^\infty \frac{(-1)^n}{n!\cdot(n+1)!}\cdot\frac{2n+2}{2}\left(\frac{t}{2}\right)^{2n+1} = 2\sum_0^\infty \frac{(-1)^n(n+1)}{n!\cdot(n+1)!}\left(\frac{t}{2}\right)^{2n+1}$$

$$= 2\cdot\frac{t}{2}\sum_0^\infty \frac{(-1)^n}{n!\cdot n!}\left(\frac{t}{2}\right)^{2n} = tJ_0.$$

18. Show that $x(t) = tJ_1(t)$ solves $tx'' - x' + tx = 0$.
Solution. $x' = J_1 + tJ_1'$ and $x'' = 2J_1' + tJ_1''$. Thus

$$tx'' - x' + tx = 2tJ_1' + t^2J_1'' - J_1 - tJ_1' + t^2J_1 = t^2J_1'' + tJ_1' + (t^2-1)J_1 = 0,$$

since the last equation is the Bessel equation of order 1.

20. Find the equation satisfied by $J_0(\frac{1}{2}t^2)$.
Solution. Let $x(t) = J_0(\frac{1}{2}t^2)$. Then $x'(t) = tJ_0'(\frac{1}{2}t^2)$ and $x'' = J_0'(\frac{1}{2}t^2) + t^2J_0''(\frac{1}{2}t^2)$. Since $J_0''(s) = -J_0'(s)/s - J_0(s)$, namely $(s = \frac{1}{2}t^2)$ $J_0''(\frac{1}{2}t^2) = -2J_0'(\frac{1}{2}t^2)/t^2 - J_0(\frac{1}{2}t^2)$, we find

$$x''(t) = J_0'(t^2/2) + t^2\cdot[-2J_0'(t^2/2)/t^2 - J_0(t^2/2)]$$
$$= J_0'(t^2/2) - 2J_0'(t^2/2) - t^2J_0(t^2/2) = -J_0'(t^2/2) - t^2J_0(t^2/2).$$

Since $J_0'(\frac{1}{2}t^2) = x'(t)/t$, it follows that $x(t)$ satisfies $x'' + \frac{x'}{t} + t^2x = 0$.

22. Solve the problem $t^2x'' + tx' + t^2x = 0$ such that $x(0) = b$ and $x(a) = 0$, where $a > 0$ is a zero of J_0.
Solution. The general solution of the equation is given by $x(t) = c_1J_0(t) + c_2Y_0(t)$. Since $J_0(0) = 1$ whereas $Y_0(t) \to -\infty$ as $t \to 0+$, then $x(0) = b$ implies $c_2 = 0$ and $c_1 = b$. Therefore $x(t) = bJ_0(t)$, which automatically yields $x(a) = bJ_0(a) = 0$.

24. Solve the Euler equation $t^2x'' - 2x = 0$ by means of the Frobenius method.
Solution. Here $p = q_1 = 0$ and $q = -2$. The indicial equation is $\mathcal{I}(r) = r(r-1) - 2 = r^2 - r - 2 = 0$ with roots $\rho_1 = -1, \rho_2 = 2$. Moreover, for $k \in \mathbb{N}$ one has $\mathcal{I}(\rho_1 + k) = \mathcal{I}(k-1) = (k-1)(k-2) - (k-1) - 2 = k^2 - 4k + 1 \neq 0$, since the roots of $s^2 - 4s + 1 = 0$ are $s = 2\pm\sqrt{3} \notin \mathbb{Z}$. Similarly, $\mathcal{I}(\rho_2 + k) = \mathcal{I}(k+2) = k^2 + 2k + 2 \neq 0$ for $k \in \mathbb{N}$ (actually $\mathcal{I}(k+2) > 0$). Then $A_k = -\frac{q_1A_{k-1}}{\mathcal{I}(\rho_1+k)} = 0$ and $B_k = -\frac{q_1B_{k-1}}{\mathcal{I}(\rho_2+k)} = 0$ for $k = 1, 2,\ldots$ and thus

$$x = A_0t^{\rho_1} + B_02t^{\rho_2} = \frac{A_0}{t} + B_0t^2.$$

26. Find a solution of $t^2x'' - tx' + (1-t)x = 0$ such that $x(0) = x_0$.
Solution. Here $p = -1, q = 1, q_1 = -1$. The indicial equation is $\mathcal{I}(r) = r(r-1) - r + 1 = r^2 - 2r + 1 = 0$ with roots $\rho_1 = \rho_2 = 1$. Since $\mathcal{I}(1+k) = (k+1)^2 - 2(k+1) + 1 = k^2$ and $q_1 = -1$, then (11.5) yields

$$A_k = (-1)^k\frac{(-1)^kA_0}{\mathcal{I}(1+1)\mathcal{I}(1+2)\cdots\mathcal{I}(1+k)} = \frac{A_0}{2^2\cdot3^2\cdots k^2} = \frac{A_0}{(k^2)!}.$$

Or, we can directly use equation (11.4) to infer that for $k = 1, 2, \ldots$

$$\begin{cases} -A_0 + A_1 & = 0, \\ -A_1 + 2^2 A_2 & = 0, \\ -A_2 + 3^2 A_3 & = 0, \\ \vdots & \vdots \\ -A_{k-1} + k^2 A_k & = 0. \end{cases}$$

Thus we find $A_1 = A_0$, $A_2 = \frac{A_1}{2^2} = \frac{A_0}{2^2}$, $A_3 = \frac{A_2}{3^2} = \frac{A_0}{3^2 2^2} = \frac{A_0}{(3!)^2}$, $A_4 = \frac{A_3}{4^2} = \frac{A_0}{4^2 3^2 2^2} = \frac{A_0}{(4!)^2}$, $A_5 = \frac{A_4}{5^2} = \frac{A_0}{5^2 4^2 3^2 2^2} = \frac{A_0}{(5!)^2}$, and in general $A_k = \frac{A_0}{(k!)^2}$.

Thus a solution of the equation is given by

$$x(t) = A_0 \cdot \left(1 + t + \frac{t^2}{(2!)^2} + \frac{t^3}{(3!)^2} + \cdots \right) = A_0 \cdot \sum_0^\infty \frac{t^k}{(k!)^2}.$$

The condition $x(0) = x_0$ yields $A_0 = x_0$ and hence we find

$$x(t) = x_0 \cdot \sum_0^\infty \frac{t^k}{(k!)^2}.$$

Solutions to Chapter 12

2. Find $\mathcal{L}[e^{-\frac{1}{2}t}]$.

   Solution.

   $$\mathcal{L}[e^{-\frac{1}{2}t}] = \frac{1}{s + \frac{1}{2}} = \frac{2}{2s + 1}, \quad \text{by } (L2).$$

4. Find $\mathcal{L}[e^{2t} \sinh t]$.

   Solution. $(L7) \rightarrow \mathcal{L}[\sinh t] = \frac{1}{s^2 - 1}$. Property 9 yields $\mathcal{L}[e^{2t} \sinh t] = \frac{1}{(s-2)^2 - 1}$.

6. Find $\mathcal{L}[e^{3t} \cos 2t]$.

   Solution. By $(L9)$ we have $\mathcal{L}[e^{3t} \cos 2t] = \mathcal{L}[\cos 2t](s-3)$. Since $\mathcal{L}[\cos 2t](s) = \frac{s}{s^2 + 4}$, we get

   $$\mathcal{L}[\cos 2t](s - 3) = \frac{s - 3}{(s - 3)^2 + 4}.$$

8. If $\mathcal{L}[f(t)] = F(s) = \frac{4}{s^2 - 16}$, find $f(t) = \mathcal{L}^{-1}[F(s)]$.

Solution. We note that $F(s)$ can be written as

$$F(s) = \frac{4}{s^2 - 16} = \frac{\frac{1}{2}}{s - 4} - \frac{\frac{1}{2}}{s + 4}.$$

Therefore, $f(t) = \frac{1}{2}e^{4t} - \frac{1}{2}e^{-4t}$.

10. Find $\mathscr{L}[e^{-4t}\sin 3t]$.

Solution. Using $(L9)$, we have $\mathscr{L}[e^{-4t}\sin 3t] = F(s+4)$, where $F(s) = \mathscr{L}[\sin 3t]$. Since $\mathscr{L}[\sin 3t] = \frac{3}{s^2+9}$, then $F(s+4) = \frac{3}{(s+4)^2+9}$.

12. Find $f(t)$ that such that $\mathscr{L}[f(t)] = \frac{s+1}{s^2+s-2}$.

Solution.

$$\mathscr{L}[f(t)] = \frac{s+1}{s^2+s-2} = \frac{\frac{2}{3}}{s-1} + \frac{\frac{1}{3}}{s+2}.$$

implies that $f(t) = \frac{2}{3}e^{t} + \frac{1}{3}e^{-2t}$.

14. * Show that $\mathscr{L}[J_0] = \frac{1}{\sqrt{1+s^2}}$, where J_0 is the Bessel function of order 0, such that $J_0(0) = 1, J_0'(0) = 0$.

Solution. Notice that $J_0 \in \mathscr{C}$, since J_0 is continuous and bounded. Moreover, J_0 satisfies $tJ_0'' + J_0' + tJ_0 = 0$. Taking the Laplace transform, one has $\mathscr{L}[tJ_0''] + \mathscr{L}[J_0'] + \mathscr{L}[tJ_0] = 0$, whereby

$$-\frac{d}{ds}\mathscr{L}[J_0''] + \mathscr{L}[J_0'] - \frac{d}{ds}\mathscr{L}[J_0] = 0. \tag{†}$$

Setting $X = \mathscr{L}[J_0]$, we find

$$\mathscr{L}[J_0''] = s^2\mathscr{L}[J_0] - J_0(0)s - J_0'(0) = s^2X - J_0(0)s - J_0'(0)$$
$$\mathscr{L}[J_0'] = s\mathscr{L}[J_0] - J_0(0) = sX - J_0(0).$$

Thus (†) yields

$$-\frac{d}{ds}(s^2X - J_0(0)s - J_0'(0)) + sX - J_0(0) + \frac{d}{ds}X = 0,$$

namely $-2sX - s^2X' + J_0(0) + sX - J_0(0) - X' = 0$. Rearranging, we find

$$(1+s^2)X' + sX = 0.$$

Solving this separable equation for X we obtain

$$\frac{X'}{X} = -\frac{s}{1+s^2} \Rightarrow \ln|X| = -\frac{1}{2}\ln(1+s^2) = \ln(1+s^2)^{-1/2},$$

where we have used the known fact that $X(0) = \int_0^{+\infty} J_0 = 1$. Finally, $\ln|X| = \ln\frac{1}{\sqrt{1+s^2}}$ yields

$$X = \frac{1}{\sqrt{1+s^2}}.$$

16. Find $\mathscr{L}^{-1}[\frac{2s+3}{s^2}]$.

Solution. $\mathscr{L}^{-1}[\frac{2s+3}{s^2}] = \mathscr{L}^{-1}[\frac{2}{s} + \frac{3}{s^2}] = 2\mathscr{L}^{-1}[\frac{1}{s}] + 3\mathscr{L}^{-1}[\frac{1}{s^2}] = 2 + 3t$.

18. Find $\mathscr{L}^{-1}[\frac{1}{s^4+s^2}]$.

Solution. Using partial fractions, we have: $\frac{1}{s^4+s^2} = \frac{1}{s^2(s^2+1)} = \frac{1}{s^2} - \frac{1}{s^2+1} \Rightarrow \mathscr{L}^{-1}[\frac{1}{s^4+s^2}] = \mathscr{L}^{-1}[\frac{1}{s^2} - \frac{1}{s^2+1}] = \mathscr{L}^{-1}[\frac{1}{s^2}] - \mathscr{L}^{-1}[\frac{1}{s^2+1}] = t - \sin t$.

20. Using the Laplace transform, solve the ivp $x' - x = 1$, $x(0) = 2$.

Solution. Letting $X = \mathscr{L}[x]$, we have

$$sX - 2 - X = \mathscr{L}[1] = \frac{1}{s} \Rightarrow (s-1)X = \frac{2s+1}{s},$$

$$\Rightarrow X = \frac{2s+1}{s(s-1)} = \frac{2}{s-1} + \frac{1}{s(s-1)} = \frac{2}{s-1} + \frac{1}{s-1} - \frac{1}{s} = 3\frac{1}{s-1} - \frac{1}{s},$$

yielding the solution $x(t) = \mathscr{L}^{-1}[X] = 3e^t - 1$.

22. Use the Laplace transform to solve $x'' - 2x' + x = 0$, $x(0) = 0$, $x'(0) = 1$.

Solution. Applying the Laplace transform, we find

$$(s^2X - 1) - 2(sX) + X = 0$$

which yields

$$X = \frac{1}{(s-1)^2}.$$

Therefore, $x(t) = te^t$ (see L4).

24. Using the Laplace transform solve the ivp $x'' + x = 0$, $x(0) = x'(0) = 1$.

Solution. Applying the Laplace transform, we find $\underbrace{s^2X - s - 1}_{\mathscr{L}[x'']}+X = 0$ whereby $X = \frac{s+1}{1+s^2} = \frac{s}{1+s^2} + \frac{1}{1+s^2}$. Then

$$x = \mathscr{L}^{-1}\left[\frac{s}{1+s^2}\right] + \mathscr{L}^{-1}\left[\frac{1}{1+s^2}\right] = \cos t + \sin t.$$

26. Using the Laplace transform, solve the ivp $4x'' - 12x' + 5x = 0$, $x(0) = 1$, $x'(0) = k$, depending upon $k \in R$.

Solution. Applying the Laplace transform, we find

$$4 \cdot \underbrace{(s^2X - s - k)}_{\mathscr{L}[x'']} - 12 \cdot \underbrace{(sX - 1)}_{\mathscr{L}[x']} + 5X = 0$$

and rearranging, $(4s^2 - 12s + 5)X - 4s - 4k + 12 = 0$. It follows $X = \frac{4s+4k-12}{4s^2-12s+5}$. Seeking A, B such that

$$\frac{4s + 4k - 12}{4s^2 - 12s + 5} = \frac{A}{2s-1} + \frac{B}{2s-5},$$

we get

$$A(2s - 5) + B(2s - 1) = 4s + 4k - 12 \Rightarrow 2(A + B)s - 5A - B = 4s + 4k - 12$$

and hence the system

$$\begin{cases} A + B = 2, \\ -5A - B = 4k - 12. \end{cases}$$

Solving, we find $A = 2 - B$ and $5A - (2 - A) = 4k - 12$, whereby $-4A = 4k - 10 \Rightarrow A = -k + \frac{5}{2}$, $B = 2 - A = k - \frac{1}{2}$.

Therefore, $X = \frac{-k+\frac{5}{2}}{2s-1} + \frac{k-\frac{1}{2}}{2s-5} = \frac{-k+\frac{5}{2}}{2} \cdot \frac{1}{s-\frac{1}{2}} + \frac{k-\frac{1}{2}}{2} \cdot \frac{1}{s-\frac{5}{2}} = (-\frac{k}{2} + \frac{5}{4}) \cdot \frac{1}{s-\frac{1}{2}} + (\frac{k}{2} - \frac{1}{4}) \cdot \frac{1}{s-\frac{5}{2}}$

and hence

$$x = \mathscr{L}^{-1}[X] = \left(-\frac{k}{2} + \frac{5}{4}\right) \cdot e^{\frac{1}{2}t} + \left(\frac{k}{2} - \frac{1}{4}\right) \cdot e^{\frac{5}{2}t}.$$

28. Use the Laplace method to solve $x'' - x = f(t), x(0) = 1, x'(0) = 0$, where $f \in \mathscr{C}$.
    Solution. $F = \mathscr{L}[f] \Leftrightarrow f = \mathscr{L}^{-1}[F]$, and $X = \mathscr{L}[x]$. One has $\mathscr{L}[x''] - \mathscr{L}[x] = \mathscr{L}[f] \Rightarrow s^2X - s - X = F$ whereby $X = \frac{F+s}{s^2-1}$. Thus

$$x(t) = \mathscr{L}^{-1}\left[\frac{F+s}{s^2-1}\right] = \mathscr{L}^{-1}\left[\frac{F}{s^2-1}\right] + \mathscr{L}^{-1}\left[\frac{s}{s^2-1}\right]$$

$$= (\sinh * f)(t) + \cosh t = \int_0^t \sinh(t - r)f(r)\, dr + \cosh t.$$

30. Find the generalized solution of $x'' + x = f(t), x(0) = 0, x'(0) = 1$, where

$$f(t) = \begin{cases} 1 & \text{if } 0 \leq t \leq a, \\ 0 & \text{if } t > a, \end{cases}$$

$a > 0$.
Solution. $\mathscr{L}[x''] + \mathscr{L}[x] = \mathscr{L}[f] \Rightarrow s^2X - 1 + X = \mathscr{L}[f] \Rightarrow,$

$$(s^2 + 1)X = 1 + \mathscr{L}[f] \Rightarrow X = \frac{1 + \mathscr{L}[f]}{s^2 + 1} = \frac{1}{s^2 + 1} + \frac{\mathscr{L}[f]}{s^2 + 1}.$$

Thus $x = \mathscr{L}^{-1}[X] = \sin t + (f * \sin)(t) = \sin t + \int_0^t \sin(t - r)f(r)\, dr$.
Since $f(r) = 0$ for $r \geq a$ and $f(r) = 1$ for $0 \leq r \leq a$, we infer
(a) $0 \leq t \leq a \Rightarrow \int_0^t \sin(t - r) \cdot a\, dr = a \cdot [\cos(t - r)]_0^t = a - a\cos t$,
(b) $t > a \Rightarrow \int_0^t \sin(t - r)f\, dr = a - a\cos a$.

Thus

$$x(t) = \begin{cases} \sin t + a - a\cos t, & \text{if } 0 \le t \le a, \\ \sin t + a - a\cos a, & \text{if } t > a. \end{cases}$$

Solutions to Chapter 13

2. Write the following in selfadjoint form:

$$t^2 x'' + t x' - x = 0, \quad t > 0.$$

Solution. Dividing by t, we have

$$t x'' + x' - \frac{1}{t} x = 0 \Rightarrow (t x')' - \frac{1}{t} x = 0.$$

4. Write the following in selfadjoint form:

$$t^2 x'' - t x' + x = 0.$$

Solution. Multiplying

$$x'' - \frac{1}{t} x' + \frac{1}{t^2} x = 0$$

by $\frac{1}{t}$, we obtain

$$\left(\frac{1}{t} x'\right)' + \frac{1}{t^3} x = 0.$$

6. Show that for any continuous function $p(t)$,

$$x'' - x' + p(t) x = 0$$

is nonoscillatory if $p(t) \le 0$.
Solution. Transforming it to selfadjoint form, it becomes

$$(e^{-t} x')' + e^{-t} p(t) x = 0$$

which is nonoscillatory, by Theorem 13.4.

8. Let $P(t) = a_p t^p + a_{p-1} t^{p-1} + \cdots + a_1 t + a_0$ and $Q(t) = b_p t^p + b_{p-1} t^{p-1} + \cdots + b_1 t + b_0$. Show that the equation $x'' + \frac{P(t)}{Q(t)} x = 0$ is oscillatory provided $\frac{a_p}{b_p} > 0$.
    Solution. Since $\lim_{t \to +\infty} \frac{P(t)}{Q(t)} = \frac{a_p}{b_p} > 0$, Corollary 13.2 applies.

10. Write the following in selfadjoint form and determine its oscillation status: $\sqrt{t} x'' + (t+1)x = 0$.

Solution. Write the equation as

$$(1 \cdot x')' + \frac{t+1}{\sqrt{t}} x = 0.$$

Then it follows from Theorem 13.5 that it is oscillatory.

12. Write it in selfadjoint form in order to determine a condition on $c(t)$ so that

$$x'' - \frac{1}{t} x' + c(t)x = 0$$

is nonoscillatory.

Solution. $(\frac{1}{t}x')' + \frac{1}{t}c(t)x = 0$. Hence $c(t) \leq 0$ will suffice.

14. Determine the oscillation status of

$$\left(\frac{2}{3t+2} x' \right)' + \frac{t}{t^2 + \sin t} x = 0, \quad t > 1. \tag{*}$$

Solution.

$$\frac{t}{t^2 + \sin t} \geq \frac{t}{t^2 + 1}.$$

Since

$$\left(\frac{2}{3t+2} x' \right)' + \frac{t}{t^2 + 1} x = 0$$

is oscillatory by Theorem 13.5, it follows from the Sturm comparison theorem that
(*) is oscillatory. It also follows from a direct application of Theorem 13.5.

16. Determine the values of the constant a for which

$$tx'' - x' + atx = 0, \quad t > 1,$$

is oscillatory. Are there values of a that will make it nonoscillatory?
Solution. We first write it in selfadjoint form

$$\left(\frac{1}{t} x' \right)' + \frac{a}{t} x = 0.$$

Then it follows from Theorem 13.5 that if $a > 0$, the equation is oscillatory. On the
other hand, if $a \leq 0$, then by Theorem 13.4, it is nonoscillatory.

18. Determine the oscillation status of

$$x'' - 2tx' + 2t^2 x = 0.$$

Solution. Eliminating the x'-term by letting $x = e^{\frac{t^2}{2}} y$ and simplifying, one gets

$$y'' + (1 + t^2)y = 0$$

which is oscillatory, by Corollary 13.2.

20. Prove (i) and (ii) of Corollary 13.4.

Solution. From Theorem 13.7 it follows that

$$\lambda_k[m_1] = \frac{k^2\pi^2}{m_1^2(b-a)^2}, \quad \lambda_k[m_2] = \frac{k^2\pi^2}{m_2^2(b-a)^2}.$$

Then, $m_1 \le m_2$ implies $\lambda_k[m_1] \ge \lambda_k[m_2]$, proving (ii).

Let $M = \max_{[a,b]} m > 0$. Since $m \le M$, the comparison property yields $\lambda_k[m] \ge \lambda_k[M]$. Since $\lambda_k[M] = \frac{k^2\pi^2}{M^2(b-a)^2}$, thus $\lambda_k[m] \ge \frac{k^2\pi^2}{M^2(b-a)^2} \Rightarrow \lim_{k\to+\infty} \lambda_k[m] = +\infty$., proving (i).

22. Show that the eigenvalues λ_k of $(1 + t^2)x'' + \lambda x = 0$, $x(0) = x(\pi) = 0$, satisfy $\lambda_k \ge k^2$.

Solution. Write the equation as $x'' + \lambda \cdot \frac{1}{1+t^2} \cdot x = 0$. Since $\frac{1}{1+t^2} \le 1$, then the comparison property of eigenvalues yields $\lambda_k[\frac{1}{1+t^2}] \ge \lambda_k[1] = k^2$.

24. Let $\lambda_k[r_i]$, $i = 1, 2$, denote the eigenvalues of $r_i x'' + \lambda p x = 0$, $x(a) = x(b) = 0$. If $r_1 \ge r_2 > 0$, show that $\lambda_k[r_1] \ge \lambda_k[r_2]$.

Solution. It suffices to rewrite the equation as $x'' + \lambda \frac{p}{r_i} x = 0$ and apply Theorem 13.8-(ii).

Solutions to Chapter 14

2. Let u be a solution of $u_x - u_y = 0$. Knowing that $u(0, 1) = 2$, find $u(1, 0)$.

Solution. The characteristics are $y + x = c$. The one passing through $(0, 1)$ is $y = 1 - x$. Since u is constant along the characteristics, then $u(x, 1 - x) = u(0, 1) = 2$ for all x. In particular, for $x = 1$ it follows $u(1, 0) = 2$.

4. Solve the ivp $u_x + 2u_y = 0$, $u(0, y) = y$.

Solution. $u(x, y) = \phi(y - 2x)$. Then $y = u(0, y) = \phi(y)$. In other words, ϕ is the map $r \mapsto r$ and hence $u(x, y) = y - 2x$.

6. Solve the ivp $2u_x - 3u_y = 0$, $u(x, 0) = x^2$.

Solution. One has $u(x, y) = \phi(y + \frac{3}{2}x)$. Then the initial condition yields $x^2 = u(x, 0) = \phi(\frac{3}{2}x) \Rightarrow \phi(r) = (\frac{2}{3}r)^2$. It follows that $u(x, y) = [\frac{2}{3}(y + \frac{3}{2}x)]^2 = (x + \frac{2}{3}y)^2$.

Alternatively, we can use the method discussed in Subsection 14.1.1.1. The given ivp has the form of (14.6) with $g(x) = 0$ and $h(x) = x^2$. Writing the equation as $u_x - \frac{3}{2}u_y = 0$ we find:

1. The characteristics through (x, y) are $\xi = s + x$, $\eta = -\frac{3}{2}s + y$.
2. Solving $\eta(s) = g(\xi(s)) = 0$ yields $-\frac{3}{2}s + y = 0$ and we find $s_0 = \frac{2}{3}y$.
3. One has $u(x, y) = h(\xi(s_0)) = h(s_0 + x) = (x + \frac{2}{3}y)^2$, as before.

8. Solve the ivp $2u_x - u_y = 0$, $u(x,x) = x - 1$.

Solution. One has $u(x,y) = \phi(2y + x) \Rightarrow x - 1 = u(x,x) = \phi(3x) \Rightarrow \phi(r) = \frac{1}{3}r - 1$. It follows that $u(x,y) = \frac{1}{3}(2y + x) - 1 = \frac{2}{3}y + \frac{1}{3}x - 1$.

Alternate solution:

1. The characteristics through (x,y) are $\xi = s + x$, $\eta = -\frac{1}{2}s + y$.
2. Solving $\eta(s) = g(\xi(s)) = 0 \Rightarrow \eta(s) = \xi(s) \Rightarrow s + x = -\frac{1}{2}s + y \Rightarrow s_0 = \frac{2}{3}(y - x)$.
3. One has $u(x,y) = \xi(s_0) - 1 = s_0 + x - 1 = \frac{2}{3}(y - x) + x - 1 = \frac{2}{3}y + \frac{1}{3}x - 1$.

10. Solve $u_x + u_y = 0$ such that $u = x$ on $y = \ln x + x$, $x > 0$.

Solution. One has $u(x,y) = \phi(y - x)$ whereby $\phi(\ln x + x - x) = x \Rightarrow \phi(\ln x) = x \Rightarrow \phi(r) = e^r \Rightarrow u(x,y) = e^{y-x}$.

Alternate solution:

1. The characteristics through (x,y) are $\xi = s + x$, $\eta = s + y$.
2. Solve $\eta(s) = g(\xi(s)) \Rightarrow \eta(s) = \ln \xi(s) + \xi(s) \Rightarrow s + y = \ln(s + x) + (s + x) \Rightarrow \ln(s + x) = y - x \Rightarrow s + x = e^{y-x} \Rightarrow s_0 = e^{y-x} - x$.
3. One has $u(x,y) = h(\xi(s_0)) = \xi(s_0) = s_0 + x = e^{y-x} - x + x = e^{y-x}$.

12. Solve the ivp $3u_x + 2u_y = 0$, $u(y,y) = \frac{1}{3}y$.

Solution. The general solution is $u = \phi(y - \frac{2}{3}x)$. The initial condition yields $\phi(y - \frac{2}{3}y) = \frac{1}{3}y$, namely $\phi(\frac{1}{3}y) = \frac{1}{3}y$. Thus $\phi(r) = r$ and hence $u(x,y) = y - \frac{2}{3}x$.

14. Solve the ivp $2yu_x - u_y = 0$, $u(x,0) = 2x$.

Solution. The ivp is like (14.6) with $a(x,y) = 2y$, $b(x,y) = -1$, $g(x) = 0$ and $h(x) = 2x$.

1. Find the characteristics by solving

$$\begin{cases} \xi'(s) = 2\eta(s), & \xi(0) = x, \\ \eta'(s) = -1, & \eta(0) = y. \end{cases}$$

The solution of the second equation is $\eta(s) = -s + y$. Substituting into the first equation we get $\xi'(s) = -2s + 2y$, $\xi(0) = x$, which yields $\xi(s) = x - s^2 + 2ys$.

2. Find s_0 such that $\eta(s_0) = g(\xi(s_0))$, namely $\eta(s_0) = 0$, which yields $-s_0 + y = 0 \Rightarrow s_0 = y$.
3. $u(x,y) = h(\xi(s_0)) = 2\xi(s_0) = 2(x - s_0^2 + 2ys_0) = 2(x - y^2 + 2y^2) = 2(x + y^2)$.

16. Solve the ivp $xu_x + 2u_y = 0$, $u(x,1) = -x$.

Solution. Here $a = x$, $b = 2$, $g(x) = 1$ and $h(x) = -x$.

1. Find the characteristics by solving

$$\begin{cases} \xi'(s) = \xi(s), & \xi(0) = x, \\ \eta'(s) = 2, & \eta(0) = y, \end{cases}$$

which yields $\xi(s) = xe^s$, $\eta(s) = 2s + y$.

2. Find the solution of $\eta(s) = g(\xi(s))$, namely $2s + y = 1$, which yields $s_0 = \frac{1}{2}(1 - y)$.
3. $u(x,y) = h(\xi(s_0)) = -\xi(s_0) = -xe^{s_0} = -xe^{\frac{1}{2}(1-y)}$.

18. Solve the ivp $u_x - yu_y = 0$, $u(0,y) = y^2$.

Solution. Solving

$$\begin{cases} \xi'(s) = 1, & \xi(0) = x, \\ \eta'(s) = -\eta(s), & \eta(0) = y, \end{cases}$$

we find $\xi(s) = s + x$, $\eta(s) = ye^{-s}$. For $s_0 = -x$ we have $\xi(s_0) = 0$. Then $\eta(s_0) = ye^{-s_0} = ye^x$. The initial condition $u(0,y) = y^2$ yields $u(\xi(s_0), \eta(s_0)) = u(0, \eta(s_0)) = \eta^2(s_0) = y^2 e^{2x}$ and thus $u(x,y) = y^2 e^{2x}$.

20. Solve $u_x - u_y = y$.

   Solution. We can find a particular solution u^* depending on y, only. So $-u_y^* = y \Rightarrow u^* = -\frac{1}{2}y^2$. Thus $u = \phi(y + x) - \frac{1}{2}y^2$.

22. Solve the ivp $u_x - 2u_y = y^2$, $u(0,y) = y^3$.

   Solution. A particular solution is $u^*(y) = -\frac{1}{6}y^3$. Thus $u = \phi(y - 2x) - \frac{1}{6}y^3$. The initial condition yields $\phi(y) - \frac{1}{6}y^3 = y^3 \Rightarrow \phi(r) = r^3 + \frac{1}{6}r^3 = \frac{7}{6}r^3$. Thus

$$u = \frac{7}{6}(y - 2x)^3 - \frac{1}{6}y^3.$$

24. Find the solution of Burger's equation $u_x - 3uu_y = 0$ such that $u(0,y) = 1 - \frac{1}{3}y$.

   Solution. Here $b = -3$ and $h(y) = 1 - \frac{1}{3}y$ and thus $z + bxh(z) = y$ (see page 251) becomes $z - 3x(1 - \frac{1}{3}z) = y$. Solving for z we find

$$z = z(x,y) = \frac{y + 3x}{1 + x}, \quad (x \neq -1),$$

and thus

$$u(x,y) = h(z(x,y)) = 1 - \frac{1}{3}z(x,y) = 1 - \frac{1}{3} \cdot \frac{y + 3x}{1 + x} = \frac{3 - y}{3(1 + x)}, \quad (x \neq -1).$$

Solutions to Chapter 15

Solutions to Section 15.4.1

2. Let u be such that $\Delta u = 0$ in $B = \{(x,y) \in \mathbb{R}^2 : x^2 + y^2 < 1\}$ and $u = x^4 + 2y^2$ on ∂B. Show that $1 \le u(x,y) \le 2$ for all $(x,y) \in \bar{B}$.

   Solution. Let $F(x,y) = x^4 + 2y^2$. Using, e.g., the Lagrange multipliers method we see that $\max_{\partial B} F = F(0, \pm 1) = 2$ and $\min_{\partial B} F = F(\pm 1, 0) = 1$. Using the Maximum principle (Theorem 15.4) it follows that $1 \le u(x,y) \le 2$ for all $(x,y) \in \bar{B}$.

4. Find the radial harmonic functions in $\mathbb{R}^2 \setminus \{(0,0)\}$.

   Solution. If u is radial harmonic, then $\Delta u = u_{rr} + \frac{1}{r}u_r = 0$, namely, $ru_{rr} + u_r = 0$. Since $ru_{rr} + u_r = (ru_r)'$, then $u = c_1 + c_2 \ln r$, $r > 0$.

6. Knowing that the Fourier series of

$$f(\theta) = \begin{cases} -\pi - \theta & \text{for } \theta \in [-\pi, -\frac{1}{2}\pi], \\ \theta & \text{for } \theta \in [-\frac{1}{2}\pi, \frac{1}{2}\pi], \\ \pi - \theta & \text{for } \theta \in [\frac{1}{2}\pi, \pi], \end{cases} \tag{f1}$$

is $f(\theta) = \frac{8}{\pi^2}(\sin\theta - \frac{\sin 3\theta}{3^2} + \frac{\sin 5\theta}{5^2} - \cdots)$, solve $\Delta u = 0$ for $r < 1$, $u(1, \theta) = f(\theta)$.
Solution. Since $u(r, \theta) = \sum_1^\infty (a_n \sin n\theta + b_n \cos n\theta) r^n$, where a_n, b_n are the Fourier coefficients of f, then

$$u(r, \theta) = \frac{8}{\pi^2}\left(r \sin\theta - r^3 \frac{\sin 3\theta}{3^2} + r^5 \frac{\sin 5\theta}{5^2} - \cdots\right).$$

8. Let u be harmonic in $r < 1$ and such that $u(1, \theta) = f(\theta)$. If f is odd, show that $u(r, 0) = u(r, \pi) = 0$.
Solution. If f is odd, then $f(\theta) = \sum_1^\infty a_n \sin n\theta$ and hence $u(r, \theta) = \sum_1^\infty a_n r^n \sin n\theta$. Then $u(r, 0) = u(r, \pi) = 0$.

10. Find $u(r)$ solving $\Delta u = 0$ for $1 < r < R$, $u = a$ for $r = 1$, $u = b$ for $r = R$.
Solution. For $r > 0$ radial harmonic functions are given by $u = c_1 + c_2 \ln r$. The boundary conditions $u(1) = a$, resp. $u(R) = b$, yield $c_1 = a$, resp. $b = a + c_2 \ln R$, namely $c_2 = \frac{b-a}{\ln R}$. Thus $u(r) = a + \frac{b-a}{\ln R} \ln r$.

12. Using separation of variables, find the solutions of $\Delta u = 0$ in the square $\Omega = \{(x, y) \in \mathbb{R}^2 : 0 < x < \pi, 0 < y < \pi\}$, such that $u_x(0, y) = u_x(\pi, y) = 0$ for all y and $u(x, 0) = u(x, \pi) = 1 + \cos 2x$.
Solution. Setting $u(x, y) = X(x)Y(y)$ we find (i) $X'' + \lambda X = 0$, $X'(0) = X'(\pi)$ and (ii) $Y'' - \lambda Y = 0$. As for (i) we find $\lambda = \lambda_n = n^2$, $n = 0, 1, 2, \ldots, X_n(x) = a_n \cos nx$. Then (ii) with $\lambda = n^2$ yields $Y_n(t) = B_n e^{ny} + C_n e^{-ny}$. Thus, relabeling $b_n = a_n B_n$, $c_n = a_n C_n$,

$$u(x, y) \sim \sum_0^\infty \cos nx \cdot (b_n e^{ny} + c_n e^{-ny}).$$

The condition $u(x, 0) = u(x, \pi) = \cos 2x$ implies

$$\underbrace{\sum_0^\infty \cos nx \cdot (b_n + c_n)}_{=u(x,0)} = \underbrace{\sum_0^\infty \cos nx \cdot (b_n e^{n\pi} + c_n e^{-n\pi})}_{=u(x,\pi)} = \cos 2x.$$

Then:
(i) $n = 0 \Rightarrow b_0 + c_0 = 1$
(ii) $n = 2 \Rightarrow b_2 + c_2 = 1$, and $b_2 e^{2\pi} + c_2 e^{-2\pi} = 1$,
(iii) $n \in \mathbb{N}, n \neq 0, 2 \Rightarrow b_n + c_n = 0$.
Solving (ii) we obtain $c_2 = 1 - b_2$ and $b_2 e^{2\pi} + (1 - b_2)e^{-2\pi} = 1$, namely $b_2 e^{4\pi} + (1 - b_2) = e^{2\pi}$, whereby

$$\begin{cases} b_2 = \frac{e^{2\pi}-1}{e^{4\pi}-1} = \frac{1}{e^{2\pi}+1}, \\ c_2 = 1 - \frac{1}{e^{2\pi}+1} = \frac{e^{2\pi}}{e^{2\pi}+1}. \end{cases}$$

In conclusion,

$$u(x,y) = (b_0 + c_0) + (b_2 e^{2y} + c_2 e^{-2y}) \cdot \cos 2x$$

$$= 1 + \left[\frac{1}{e^{2\pi}+1}e^{2y} + \frac{e^{2\pi}}{e^{2\pi}+1}e^{-2y}\right] \cdot \cos 2x$$

$$= 1 + \frac{1}{e^{2\pi}+1} \cdot \left[e^{2y} + e^{2\pi-2y}\right] \cdot \cos 2x.$$

14. Let u_a be harmonic in $r < 1$ and such that $u(1,\theta) = a(5 - 8\sin\theta)$. Using the Poisson integral formula, find a such that $u_a(\frac{1}{2}, \frac{\pi}{2}) = 1$.
Solution. The Poisson integral formula with $R = 1$, $r = \frac{1}{2}$, $\theta = \frac{\pi}{2}$ and $f(\vartheta) = 5a - 8a\sin\vartheta$ yields

$$u_a\left(\frac{1}{2}, \frac{\pi}{2}\right) = \frac{1}{2\pi}\int_{-\pi}^{\pi} \frac{(1-\frac{1}{4})(5a - 8a\sin\vartheta)}{1 - 2\frac{1}{2}\cos(\frac{\pi}{2} - \vartheta) + \frac{1}{4}}\,d\vartheta$$

$$= \frac{1}{2\pi}\int_{-\pi}^{\pi} \frac{\frac{3}{4}(5a - 8a\sin\vartheta)}{\frac{5}{4} - 2\sin\vartheta}\,d\vartheta = \frac{3}{2\pi}\int_{-\pi}^{\pi} \frac{5a - 8a\sin\vartheta}{5 - 8\sin\vartheta}\,d\vartheta = 3a.$$

Then $u_a(\frac{1}{2}, \frac{\pi}{2}) = 1$ implies $a = \frac{1}{3}$.

16. Let u be a harmonic function on $\mathbb{R}^2$. Show that for any $P_0 = (x_0,y_0) \in \mathbb{R}^2$ and $\epsilon > 0$ one has that $u(P_0) = \oint_{\partial B_\epsilon} f$ where B_ϵ is the ball centered in P_0 with radius ϵ and f is the restriction of u on the circle ∂B_ϵ.
Solution. Changing the variables $\tilde x = x - x_0, \tilde y = y - y_0$ and setting $\tilde u(x,y) = u(\tilde x, \tilde y)$ we find $\Delta\tilde u = \Delta u = 0$. Applying the mean value theorem to $\tilde u$ in $\tilde B_\epsilon = \{(x,y) : x^2 + y^2 < \epsilon^2\}$ and letting $\tilde f$ denote the restriction of $\tilde u$ on $\partial\tilde B_\epsilon$, we obtain

$$u(x_0,y_0) = \tilde u(0,0) = \frac{1}{2\pi}\int_{-\pi}^{\pi}\tilde f(\vartheta)\,d\vartheta = \oint_{\partial B_\epsilon} f.$$

Solutions to Section 15.4.2

2. Solve the heat equation (H) with

$$f(x) = \begin{cases} x & \text{for } x \in [0, \frac{1}{2}\pi], \\ \pi - x & \text{for } x \in [\frac{1}{2}\pi, \pi]. \end{cases}$$

Solution. Recall that $f(x) = \frac{8}{\pi^2}(\sin x - \frac{\sin 3x}{3^2} + \frac{\sin 5x}{5^2} - \cdots)$; see Exercise 15.4.1-7.

In this case the Fourier coefficients of f are $a_2 h = 0$ and $a_1 = \frac{8}{\pi^2}$, $a_3 = -\frac{8}{\pi^2} \cdot \frac{1}{3^2}$, $a_5 = \frac{8}{\pi^2} \cdot \frac{1}{5^2}$, etc. Thus (15.16) yields

$$u(x,t) = \frac{8}{\pi^2}\left(e^{-t}\sin x - \frac{1}{3^2}e^{-3^2 t}\sin 3x + \frac{1}{5^2}e^{-5^2 t}\sin 5x - \cdots\right).$$

4. Solve $u_t - 2u_{xx} = 0$, in $(0,1) \times (0,+\infty)$, $u(x,0) = 4\sin(3\pi x)$ for $x \in [0,1]$ and $u(0,t) = u(1,t) = 0$ for $t \geq 0$.

Solution. Using Remark 15.3-(iv) with $c = 2, L = 1$ we find

$$u(x,t) = 4e^{-18\pi^2 t}\sin(3\pi x).$$

6. Letting $c(t) > 0$, solve the following by separation of variables

$$\begin{cases} u_t - c(t)u_{xx} = 0, & \forall x \in [0,\pi], \ \forall t \geq 0, \\ u(x,0) = b\sin x, & \forall x \in [0,\pi], \\ u(0,t) = u(\pi,t) = 0, & \forall t \geq 0. \end{cases}$$

Solution. Letting $u(x,t) = X(x)T(t)$ we find $XT' = c(t)X''T$, that is

$$\frac{X''}{X} = \frac{T'}{c(t)T} = -\lambda.$$

Then we find

$$(i)\ X'' + \lambda X = 0,\ X(0) = X(\pi) = 0, \quad (ii)\ T' + \lambda c(t)T = 0.$$

Solving (i) we get $\lambda = n^2$ and $X_n(x) = c_n\sin nx$. Then (ii) becomes $T' + n^2 c(t)T = 0$. Separating the variables and integrating we find $T(t) = b_n e^{-n^2 C(t)}$, where $C(t) = \int_0^t c(s)\,ds$. Thus, setting $a_n = b_n c_n$,

$$u(x,t) = \sum a_n e^{-n^2 C(t)}\sin nx.$$

Since $C(0) = 0$, we obtain $u(x,0) = \sum a_n e^{-n^2 C(0)}\sin nx = \sum a_n \sin nx$. Thus the initial condition $u(x,0) = b\sin x$ yields $a_1 = b$ and $a_n = 0$ for $n = 2,3,\ldots$, and hence $u(x,t) = be^{-C(t)}\sin x$.

8. Solve the following Neumann problem for the heat equation

$$\begin{cases} u_t - u_{xx} = 0, & \forall x \in [0,\pi], \ \forall t \geq 0, \\ u(x,0) = a_1\cos x + a_2\cos 2x, & \forall x \in [0,\pi], \\ u_x(0,t) = u_x(\pi,t), & \forall t \geq 0. \end{cases}$$

Solution. Since the Fourier coefficients b_n of f are null for $n = 0,3,4\ldots$, whereas $b_1 = a_1, b_2 = a_2$, we find $u(x,t) = a_1 e^{-t}\cos x + a_2 e^{-4t}\cos 2x$.

10. Solve

$$\begin{cases} u_t - u_{xx} = 0, & \forall x \in [0, \pi], \ \forall t \geq 0, \\ u(x, 0) = |\cos x|, & \forall x \in [0, \pi], \\ u_x(0, t) = u_x(\pi, t), & \forall t \geq 0. \end{cases}$$

Solution. The Fourier coefficients of $f(x) = |\cos x|$ are

$$a_{2n+1} = 0, \quad a_{2n} = -\frac{4}{\pi} \frac{(-1)^n}{(2n-1)(2n+1)}$$

and thus

$$u(x, t) = -\frac{4}{\pi} \sum_0^\infty \frac{(-1)^n}{(2n-1)(2n+1)} e^{-(2n)^2 t} \cos 2nx.$$

12. Let $u(x, t)$ be a solution of $u_t - u_{xx} = h(x, t)$ in $A = [0, \pi] \times [0, T]$ for some $T > 0$. Show that if $h(x, t) < 0$ in the interior of A, then u achieves its maximum on the boundary of A.

Solution. Being continuous, u attains its maximum at some $(\bar{x}, \bar{t}) \in A$. If $(\bar{x}, \bar{t})$ belongs to the interior of A, then $\nabla u(\bar{x}, \bar{t}) = 0$, in particular $u_t(\bar{x}, \bar{t}) = 0$. From the equation it follows that $u_{xx}(\bar{x}, \bar{t}) = u_t(\bar{x}, \bar{t}) - h(\bar{x}, \bar{t}) = -h(\bar{x}, \bar{t}) > 0$. This is a contradiction because at any maximum point one has $u_{xx}(\bar{x}, \bar{t}) \leq 0$.

Solutions to Section 15.4.3

2. Solve $u_{tt} - u_{xx} = 0$, $u(x, 0) = x$, $u_t(x, 0) = 2x$.
   Solution. Using D'Alambert's formula with $c = 1$, one finds

$$u(x, t) = \frac{1}{2}((x + t) + (x - t)) + \frac{1}{2} \int_{x-t}^{x+t} 2s \, ds$$

$$= x + \frac{1}{2}[s^2]_{x-t}^{x+t} = x + \frac{1}{2}((x + t)^2 - (x - t)^2) = x + 2xt.$$

4. Show that the solution of $u_{tt} - c^2 u_{xx} = 0$, $u(x, 0) = f(x)$, $u_t(x, 0) = 0$ is even in t.
   Solution. One has $u(x, -t) = \frac{1}{2}(f(x - ct) + f(x + ct)) = u(x, t)$.

6. Let u be the solution of $u_{tt} - u_{xx} = 0$ in $[0, L] \times [0, +\infty)$, $u(x, 0) = f(x)$, $u_t(x, 0) = 0$.
   Knowing that $f(\frac{1}{8}L) = f(\frac{3}{8}L) = 2$, find $u(\frac{1}{4}L, \frac{1}{8}L)$.
   Solution. From $u(x, t) = \frac{1}{2}(f(x + t) + f(x - t))$ we infer

$$u\left(\frac{1}{4}L, \frac{1}{8}L\right) = \frac{1}{2}\left[f\left(\frac{1}{4}L + \frac{1}{8}L\right) + f\left(\frac{1}{4}L - \frac{1}{8}L\right)\right] = \frac{1}{2}\left[f\left(\frac{3}{8}L\right) + f\left(\frac{1}{8}L\right)\right] = 2.$$

8. Solve by separation of variables:

$$\begin{cases} u_{tt} - u_{xx} = 0, & \forall x \in (0,\pi),\ \forall t > 0, \\ u(x,0) = \sin x, & \forall x \in [0,\pi], \\ u_t(x,0) = \sin x, & \forall x \in [0,\pi], \\ u(0,t) = u(\pi,t) = 0, & \forall t \geq 0. \end{cases}$$

Solution. Using $u(x,t) = \sum r_n \sin nx \cdot [A_n \sin nt + B_n \cos nt]$, the condition $u(x,0) = \sin x$ yields $r_1 A_1 = 1$, $r_n A_n = 0$ for $n \in \mathbb{N}$, $n \neq 1$ and $r_n B_n = 0$ for $n \in \mathbb{N}$. Then $u(x,t) = \sin x \sin t$ which also satisfies $u_t(x,0) = \sin x$.

10. Solve by separation of variables:

$$\begin{cases} u_{tt} - u_{xx} = 0, & \forall x \in (0,\pi),\ \forall t > 0, \\ u(x,0) = 3 + 4\cos 2x, & \forall x \in [0,\pi], \\ u_t(x,0) = 0, & \forall x \in [0,\pi], \\ u_x(0,t) = u_x(\pi,t), & \forall t \geq 0. \end{cases}$$

Solution. Setting $u = X(x)T(t)$, we find the usual system

$$(i) \quad X'' + \lambda X = 0, \qquad (ii) \quad T'' + \lambda T = 0,$$

but now (i) is coupled with the boundary condition $X'(0) = X'(\pi)$. Solving, we find that the eigenvalues and the eigenfunctions are, respectively, $\lambda_n = n^2$, $n = 0,1,2,\ldots$ and $X_n = r_n \cos nx$, whence $u = \sum_0^\infty r_n \cos nx(A_n \sin nt + B_n \cos nt)$. From $u_t(x,0) = \sum_0^\infty nr_n A_n \cos nx = 0$ we infer that $r_n A_n = 0$. Then $u(x,t) = \sum_0^\infty r_n B_n \cos nx \cos nt$ and hence $u(x,0) = 3 + 4\cos 2x$ implies that $r_a B_0 = 3$, $r_2 B_2 = 4$ and $r_n B_n = 0$ for all $n \in \mathbb{N}$, $n \neq 0, 2$. Thus $u(x,t) = 3 + 4\cos 2x \cos 2t$.

Solutions to Chapter 16

2. Let $I[y] = \int_0^\pi [\frac{1}{2}y'^2 - \frac{1}{4}y^4]\,dx$ and $Y = \{y \in C^2([0,\pi]) : y(0) = y(\pi) = 0\}$. Show that $\inf_Y I[y] = -\infty$.
    Solution. Let $y = \sin x$ and evaluate $I[ry] = \frac{1}{2}r^2 \int_0^\pi \cos^2 x\,dx - \frac{1}{4}r^4 \int_0^\pi \sin^4 x\,dx$. Then one finds $\lim_{r \to +\infty} I[ry] = -\infty$.

4. Given $I[y] = \int_0^\pi [\frac{1}{2}y'^2 - \frac{\lambda}{2}y^2 + \frac{1}{4}y^4]\,dx$, $y \in Y = \{y \in C^2([0,\pi]) : y(0) = y(\pi) = 0\}$, show:
    (a) $I[y]$ is bounded below on Y, namely $\exists C \in \mathbb{R}$ such that $I[y] \geq C, \forall y \in Y$,
    (b) $\inf_Y I[y] < 0$ provided $\lambda > 1$,
    (c) knowing that this infimum is achieved at $\bar{y}$, what is the boundary value problem solved by $\bar{y}$?

    Solution.
    (a) One has $I[y] \geq \int_0^\pi [\frac{1}{4}y^4 - \frac{\lambda}{2}y^2]\,dx$. Since $\frac{1}{4}y^4 - \frac{\lambda}{2}y^2 \geq c$ we infer $I[y] \geq c\pi := C$.

(b) For $r \in \mathbb{R}$ let us evaluate $\phi(r) \overset{\text{def}}{=} I[r \sin x]$, yielding

$$\phi(r) = \frac{1}{2}r^2 \int_0^\pi \cos^2 x \, dx - \frac{\lambda}{2}r^2 \int_0^\pi \sin^2 x \, dx + \frac{1}{4}r^4 \int \sin^4 x \, dx$$

$$= \frac{1}{2}r^2 - \frac{\lambda}{2}r^2 + \frac{1}{4}r^4 \int \sin^4 x \, dx = \frac{1}{2}(1-\lambda)r^2 + \frac{1}{4}r^4 \int \sin^4 x \, dx.$$

It follows that $\phi(r) < 0$ for $r > 0$ small enough. As a consequence we have that $\inf_Y I[y] < 0$ for $\lambda > 1$.

(c) $\delta I = 0 \Rightarrow y'' + \lambda y - y^3 = 0$ and thus the minimum $\bar{y}$ is a solution of the boundary value problem $y'' + \lambda y - y^3 = 0$, $y(0) = y(\pi) = 0$.

6. Show that the time T taken by a body to descend from $A = (0,0)$ to $B = (1,1)$ along the Brachistochrone is less than or equal to $2/\sqrt{g}$, where g is the gravitational acceleration.

Solution. We know that $T = \min_Y I[y]$, with $I[y] = \int_0^1 \sqrt{\frac{1+y'^2}{2gy}} \, dx$ and $Y = \{y \in C^2([0,1]) : y(0) = 0, y(1) = 1\}$. Take $\bar{u}(x) = x$. Evaluating I on this specific $\bar{u} \in Y$ we find

$$I[\bar{u}] = \int_0^1 \sqrt{\frac{1+1}{2gx}} \, dx = \frac{1}{\sqrt{g}} \int_0^1 \frac{1}{\sqrt{x}} \, dx = \frac{2}{\sqrt{g}}$$

and therefore $T = \min_Y I[y] \leq I[\bar{u}] = \frac{2}{\sqrt{g}}$.

8. Consider a medium with refractive index $n(y) = 1/\sqrt{2gy}$ and check if the light rays are Brachistochrone curves.

Solution. If $n(y) = 1/\sqrt{2gy}$ the Fermat functional $I[y] = \int_a^b n(y)\sqrt{1+y'^2} \, dx$ equals the Brachistochrone functional $I[y] = \int_a^b \frac{\sqrt{1+y'^2}}{\sqrt{2gy}} \, dx$. Hence the two problems have the same stationary solutions.

10. As first noted by Newton, the problem of finding the shape of the solid of revolution with least resistance when it moves in a fluid with constant velocity parallel to the axis of revolution leads to looking for the minimum of $I[y] = \int_a^b y \cdot \frac{y'^3}{1+y'^2} \, dx$ on $Y = \{y \in C^2([a,b]) : y(a) = y(b) = 0\}$. Find the Euler–Lagrange equation of $I[y]$.

Solution. The Lagrangian $L(y,p) = \frac{yp^3}{1+p^2}$ is independent of x. So, using (EL) we find $y'L_p - L = k$. Since $L_p = y \cdot \frac{3p^2(1+p^2)-2p^4}{(1+p^2)^2} = y \cdot \frac{p^4+3p^2}{(1+p^2)^2}$ then $y'L_p - L = k$, namely $pL_p - L = k$, becomes

$$yp \cdot \frac{p^4 + 3p^2}{(1+p^2)^2} - \frac{y \cdot p^3}{1+p^2} = k.$$

Rearranging we find

$$\frac{yp^3}{(1+p^2)^2} \cdot [p^2 + 3 - (1+p^2)] = \frac{2yp^3}{(1+p^2)^2}$$

and hence the Euler–Lagrange equation is

$$\frac{2yy'^3}{(1+y'^2)^2} = k.$$

Bibliography

[1] M. Braun, *Differential Equations and Their Applications*, Springer-Verlag, 1975.

[2] S. L. Campbell, *An Introduction to Differential Equations and Their Applications*, Wadsworth, 1990.

[3] E. A. Codington and N. Levinson, *Theory of Ordinary Differential Equations*, McGraw-Hill, 1955.

[4] E. I. Ince, *Ordinary Differential Equations*, Dover Publication Inc., 1956.

[5] W. Leighton, *An Introduction to the Theory of Ordinary Differential Equations*, Wadsworth, 1976.

https://doi.org/10.1515/9783111185675-018

Index

https://doi.org/10.1515/9783111185675-019

www.ingramcontent.com/pod-product-compliance
Lightning Source LLC
Chambersburg PA
CBHW080712220326

41598CB00033B/5393